PERGAMON INTERNATIONAL LIBRARY
of Science, Technology, Engineering and Social Studies
The 1000-volume original paperback library in aid of education, industrial training and the enjoyment of leisure
Publisher: Robert Maxwell, M.C.

Introduction to Industrial Drying Operations

Other titles of interest

BOOTH: Industrial Gas Cleaning

COULSON & RICHARDSON: Chemical Engineering
Volume 1, 3rd edition: Fluid Flow, Heat Transfer and Mass Transfer
Volume 2, 2nd edition: Unit Operations
Volume 3
Volume 4: Solutions to the Problems in the Third Edition of Volume 1

*KEEY: Drying: Principles and Practice

STRAUSS: Industrial Gas Cleaning, 2nd edition

*Not available on inspection

Introduction to Industrial Drying Operations

by

R.B. KEEY
Reader in Chemical Engineering
University of Canterbury, New Zealand

PERGAMON PRESS

OXFORD · NEW YORK · TORONTO · SYDNEY · PARIS · FRANKFURT

U.K.	Pergamon Press Ltd., Headington Hill Hall, Oxford OX3 0BW, England
U.S.A.	Pergamon Press Inc., Maxwell House, Fairview Park, Elmsford, New York 10523, U.S.A.
CANADA	Pergamon of Canada Ltd., 75 The East Mall, Toronto, Ontario, Canada
AUSTRALIA	Pergamon Press (Aust.) Pty. Ltd., 19a Boundary Street, Rushcutters Bay, N.S.W. 2011, Australia
FRANCE	Pergamon Press SARL, 24 rue des Ecoles, 75240 Paris, Cedex 05, France
FEDERAL REPUBLIC OF GERMANY	Pergamon Press GmbH, 6242 Kronberg-Taunus, Pferdstrasse 1, Federal Republic of Germany

First edition 1978

British Library Cataloguing in Publication Data
Keey, Roger Brian
Introduction to industrial drying operations.
1. Drying
I. Title
660.2'8426 TP363 77-30467
ISBN 0-08-020594-1 (Hardcover)
ISBN 0-08-020593-3 (Flexicover)

In order to make this volume available as economically and as rapidly as possible the typescript has been reproduced in its original form. This method unfortunately has its typographical limitations but it is hoped that they in no way distract the reader.

Printed in Great Britain by William Clowes & Sons Limited
London, Beccles and Colchester

What would the world be,
once bereft of wet and of wilderness?

INVERSNAID, Gerard Manley Hopkins 1844-1889.

CONTENTS

Chapter 4 Humidification

Chapter 5 Process of Drying

Chapter 6 Performance of Dryers

Chapter 7 Continuous Drying

PREFACE

Drying is a common everyday activity which has its origins in antiquity. In mediaeval physiology, dryness was one of the fundamental qualities of elements, humours and planets, and strove for mastery with other qualities such as heat, cold and moistness. Today, shorn of all mystic purpose, considerable quantities of heat are used to dry moist materials of various kinds in numerous industries. Because of this enormous diversity of applications, a theoretical understanding of the drying behaviour of industrial equipment has been slow to mature. Only recently has the literature provided means whereby relatively simple, but effective, analyses can be undertaken as a basis for the design and operation of drying plant.

This book attempts to set out an introduction to those methods which seem most useful in determining the process conditions within solids-drying equipment. It began as an outgrowth of a series of lectures given to engineers in the pharmaceutical industry while the author was at the Swiss Federal Institute of Technology, Zürich in 1972/3. Other aspects of the book form the basis of an undergraduate course currently offered at the University of Canterbury, New Zealand.

It is hoped that this book will find a useful place, not only as a students' text on the principles of drying, but also as an engineer's guide to the better design and operation of drying plant. To this end, nearly seventy worked examples of various length are included in the book to illustrate the application of these principles to practical drying situations. Many involve only simple numerical analyses, while all examples are within the capability of portable programmable calculators currently available.

I wish to thank Mrs. P.C. Levicky for her patient skill in typing the first draft of a far from easy manuscript. My appreciation also goes to Pergamon Press for converting this typescript into camera-ready copy in an endevour to reduce costs of publishing. In writing such a work, one acquires certain monkish habits : for the forbearance of friends and family, I am grateful.

ACKNOWLEDGEMENTS

The Author and Publishers acknowledge with thanks the help given by the following individuals and organizations in giving permission for the reproduction of copyright materials

American Institute of Chemical Engineers for Fig. 5.6 and 5.7.

A/S Anhydro for Plate 3.

Babcock - BSH A. G. for Fig. 1.4.

William Boulton Ltd., Calmic Division for Plate 5.

British Steel (Chemicals) Corporation for Plate 2.

Buell Ltd. for Plate 1.

Butterworths and Co. (Publishers) Ltd. for Fig. 6.1, 6.14 and 8.1.

Casburt Ltd. for Plate 7.

Fleissner GmbH for Fig. 7.26.

Luwa - SMS GmbH for Plate 4.

McGraw-Hill Book Co. for Fig. 1.2 and 1.3.

Dr. Erich Mosberger for Fig. 6.8.

New Zealand Institution of Engineers, Chemical Engineering Group for Charts A1, A2 and A3.

Dr. P. Nordon for Fig. 8.4.

Pergamon Press Ltd. for Fig. 1.5.

Pergamon Press Journals Department for Fig. 5.13 and 6.6.

Dr. W. Poersch for Fig. 1.6.

Sandvik (UK) Ltd. for Plate 6.

Separation Processes Service, AERE Harwell for Charts A4 and A5.

Springer Verlag for Fig. 2.13, 5.23, 6.4, 6.7, 6.10 and 8.16.

The Textile Institute for Fig. 6.13.

Verlag Chemie GmbH for Fig. 5.9.

Verlag Sauerländer GmbH for Fig. 2.17.

VDI - Verlag GmbH for Fig. 6.12.

NOTATION

The following tabulation lists all symbols used in the text, except those used in a single context and specifically defined therein.

Moisture-concentrations on a dry basis are given the symbols X and Y and the flow of moisture-free streams L and G to emphasise mnemonically the analogies between drying and other mass-transfer operations, for which these symbols are commonly adopted in the chemical-engineering literature. The European convention of using λ for the thermal conductivity is employed to avoid confusion with the mass-transfer coefficient K and the drying coefficient k, otherwise the principal remaining symbols follow common practice.

Symbol	Meaning	Units	first used page
<u>Roman</u>			
a	ash content	1	80
a	exposed surface per unit volume of dryer	m^{-1}	62
a_o	exposed surface per unit volume of airspace	m^{-1}	259
b	thickness of layer	m	161
b_B	wall thickness	m	222
b_S	solids thickness	m	211
b_T	tray thickness	m	211
B_i	ith virial coefficient	1	16
c	total concentration	mol m^{-3}	108
c_W	moisture concentration	mol m^{-3}	108
C	capacitance	F	230
C_{LW}	heat capacity of liquid moisture	J kg^{-1} K^{-1}	27
C_P	heat capacity at constant pressure	J kg^{-1} K^{-1}	2
C_{PG}	heat capacity of dry gas	J kg^{-1} K^{-1}	26
C_{PW}	heat capacity of moisture vapour	J kg^{-1} K^{-1}	26
C_{PY}	humid heat	J kg^{-1} K^{-1}	26
C_S	heat capacity of dry solids	J kg^{-1} K^{-1}	42
d_o	orifice diameter	m	201
d_P	particle diameter	m	206
D	molar mass ratio	1	110
D	diameter	m	223
$\mathcal{D}_a$	apparent moisture diffusivity	m^2 s^{-1}	176
$\mathcal{D}_{WG}$	moisture-vapour diffusion coefficient	m^2 s^{-1}	108

Symbol	Meaning	Units	first used page
e	fractional approach to saturation	1	317
E	voltage gradient	$V\ m^{-1}$	230
E	external-age distribution	1	243
f	relative drying rate	1	152
f	frequency	Hz	230
f_s	vapour-pressure coefficient	1	19
f_W	wetted-area fraction	1	134
F	bone-dry solids charged	kg	61
F	mass-transfer coefficient	$mol\ m^{-2}\ s^{-1}$	109
g	mass-flow ratio	1	57
g_A	combustion-air ratio	1	81
g_A^*	minimum air ratio	1	80
g_A^+	excess air ratio	1	80
g_G	gas make	1	80
g_V	vapour make	1	80
G	specific dry gas rate	$kg\ m^{-2}\ s^{-1}$	65
G_i	dry gas flow of ith stream	$kg\ s^{-1}$	56
G_o	specific gas rate based on air-space	$kg\ m^{-2}\ s^{-1}$	259
G_T	total gas flow rate	$kg\ s^{-1}$	117
G_V	evaporation rate	$kg\ s^{-1}$	73
$\mathcal{G}$	conductance	$kg\ m^{-2}\ s^{-1}$	168
h	heat-transfer coefficient	$W\ m^{-2}\ K^{-1}$	117
h_C	convective heat-transfer coefficient	$W\ m^{-2}\ K^{-1}$	189
h_R	radiative heat-transfer coefficient	$W\ m^{-2}\ K^{-1}$	210
h_T	conductive heat-transfer coefficient	$W\ m^{-2}\ K^{-1}$	210
H	enthalpy	$J\ kg^{-1}$	23
H_{GG}	dry-gas enthalpy	$J\ kg^{-1}$	25
H_{GW}	moisture-vapour enthalpy	$J\ kg^{-1}$	25
H_S	solids enthalpy	$J\ kg^{-1}$	25
i_C	capacitive current	A	230
i_R	resistive current	A	230
I	humid enthalpy	$J\ kg^{-1}$	25
I_C	convective enthalpy	$J\ kg^{-1}$	132
I_G	humid enthalpy of damp gas	$J\ kg^{-1}$	25
I_L	enthalpy of added moisture	$J\ kg^{-1}$	43
I_S	humid enthalpy of moist solids	$J\ kg^{-1}$	43
I_T	total humid enthalpy	$J\ kg^{-1}$	132

Symbol	Meaning	Units	first used page
I_V	humid enthalpy of moisture vapour	$J\ kg^{-1}$	43
I_W	humid enthalpy of liquid moisture	$J\ kg^{-1}$	61
I_W	humid enthalpy for moist surface (eq.4.62)	$J\ kg^{-1}$	132
j_H	heat-transfer j-factor	1	118
j_M	mass-transfer j-factor	1	118
J_i	moisture flux	$mol\ m^{-2}\ s^{-1}$	162
J_W	diffusion flux	$mol\ m^{-2}\ s^{-1}$	108
k	drying coefficient	s^{-1}	304
K	mass-transfer coefficient	$kg\ m^{-2}\ s^{-1}$	111
K_o	limiting mass-transfer coefficient	$kg\ m^{-2}\ s^{-1}$	3
K_S	mass-transfer coefficient for solids	$kg\ m^{-2}\ s^{-1}$	169
ℓ	streamed length	m	204
L	characteristic length	m	198
L	specific dry-solids flow	$kg\ m^{-2}\ s^{-1}$	65
L_o	specific dry-solids flow based on solids space	$kg\ m^{-2}\ s^{-1}$	259
m_G	mass of dry air	kg	17
m_W	mass of moisture vapour	kg	17
m_{ij}	hygrothermal gradient between i and j	K^{-1}	169
M_G	molar mass of dry gas	$kg\ mol^{-1}$	18
M_W	molar mass of moisture	$kg\ mol^{-1}$	18
n	number of unit surfaces	1	232
$\tilde{N}_G$	molar gas flow per unit area	$mol\ m^{-2}\ s^{-1}$	108
N_t	number of transfer units	1	234
N_{te}	number of external transfer units	1	237
N_{ti}	number of internal transfer units	1	237
N_V	drying rate per unit exposed surface	$kg\ m^{-2}\ s^{-1}$	3
N_W	evaporation rate per unit exposed surface	$kg\ m^{-2}\ s^{-1}$	110
$\tilde{N}_W$	molar moisture-vapour flux	$mol\ m^{-2}\ s^{-1}$	108
$\mathcal{N}$	intensity of drying	1	182
p_G	partial pressure of dry gas	Pa	17
p_W	partial pressure of moisture vapour	Pa	17
p_W^o	vapour pressure	Pa	18
P	total pressure	Pa	16
P_S	suction pressure	Pa	44

Symbol	Meaning	Units	first used page
q	heat-transfer flux	$W\ m^{-2}$	118
$\tilde{q}$	specific heat release	$W\ m^{-3}$	230
q_C	convective heat-transfer flux	$W\ m^{-2}$	132
q_T	total heat-transfer flux	$W\ m^{-2}$	118
q_W	wetted-area heat-transfer flux	$W\ m^{-2}$	132
Q	heat quantity	J	24
$\tilde{Q}_E$	specific offgas heat	$J\ kg^{-1}$	87
Q_H	heat-input rate	W	68
Q_L	heat-loss rate	W	68
Q_V	evaporation-heat rate	W	87
$\tilde{Q}_V$	specific heat demand	$J\ kg^{-1}$	87
r	radius of curvature	m	36
r	recycle ratio	1	65
R	gas constant	$J\ mol^{-1}\ K^{-1}$	16
R_S	resistance	Ω	230
s	spacing	m	201
s	unit surface	m^2	232
s	shell/airspace ratio	-	260
S	entropy	$J\ K^{-1}$	90
S	cross-sectional area	m^2	259
S_T	total dryer cross-section	m^2	282
t	elapsed time	s	149
T	temperature	K	2
T_C	critical temperature	K	23
T_D	dewpoint temperature	K	55
T_E	evaporative temperature	K	169
T_F	fog temperature	K	55
T_G	gas temperature (dry-bulb)	K	55
T_{GS}	adiabatic-saturation temperature	K	108
T_S	saturation temperature (or surface)	K	28
T_T	tray temperature	K	210
T_W	wet-bulb temperature	K	23
u	velocity	$m\ s^{-1}$	206
$\hat{u}$	cup-mixing velocity	$m\ s^{-1}$	198
U	internal energy	$J\ kg^{-1}$	24
U	overall heat-transfer coefficient	$W\ m^{-2}\ K^{-1}$	110

Symbol	Meaning	Units	first used page
V	specific molar volume	$m^3\ mol^{-1}$	17
V	vapour flow	$kg\ s^{-1}$	60
V	voltage	V	228
V_L	specific molar volume of liquid moisture	$m^3\ mol^{-1}$	36
w	water content	1	81
w	width	m	201
W	moisture content (wet basis)	1	34
$\tilde{W}$	specific work	$J\ kg^{-1}$	34
W_F	fan-work	W	68
W_S	conveying (solids) work	W	72
X	moisture content (dry basis)	1	33
X^*	equilibrium moisture content	1	39
X^*_{max}	maximum hygroscopic moisture content	1	41
X_{cr}	critical moisture content	1	154
X_o	initial moisture content	1	178
y	fractional approach to saturation	1	232
y_G	mole fraction of dry gas	1	17
y_W	mole fraction of moisture vapour	1	17
Y	humidity (mass ratio vapour/dry gas)	1	2
Y_A	ambient humidity	1	81
Y_G	bulk-gas humidity	1	3
Y_S	saturation (or surface) humidity)	1	19
Y_W	wet-bulb humidity	1	3
z	distance from (solids) inlet	m	135
Z	compressibility factor	1	16
Z	dryer length (or height) in airflow direction	m	135
Z_t	extent of a transfer unit	m	234
Greek			
α	flow-ratio coefficient	1	277
γ	evaporative-resistance coefficient	1	172
δ	boundary-layer thickness	m	109
δ	relative depth of drying zone	1	175
ΔG	free energy change	$J\ kg^{-1}$	34
ΔH_C	calorific value	$J\ kg^{-1}$	82
ΔH_{GM}	residual gas-mixing enthalpy	$J\ kg^{-1}$	25

Symbol	Meaning	Units	first used page
ΔH_{Vx}	latent heat of vaporization at T_x	J kg^{-1}	26
ΔH_W	heat of wetting	J kg^{-1}	43
ΔS_L	specific entropy increase of moist solid	J K^{-1} kg^{-1}	92
ΔS_V	specific entropy increase of vapour	J K^{-1} kg^{-1}	92
ε	loading ratio	1	246
ε	voidage of bed	1	207
ε_G	gas emissivity	1	218
ε_S	surface emissivity	1	220
ε_T	tray emissivity	1	220
ζ	relative distance	1	330
ζ_{cr}	critical relative distance	1	338
η	relative depth	1	176
η	effectiveness ratio	1	241
η_H	heat-input ratio (eq. 7.10)	1	260
η_{LG}	heat-loss ratio (eq. 7.11)	1	260
θ	temperature difference	K	25
θ	relative distance along band	1	245
θ	relative time	1	330
λ	number of transfer units	1	344
λ	thermal conductivity	W m^{-1} K^{-1}	117
λ_B	thermal conductivity of wall	W m^{-1} K^{-1}	223
λ_S	thermal conductivity of solids	W m^{-1} K^{-1}	211
λ_T	thermal conductivity of tray	W m^{-1} K^{-1}	211
Λ	moisture-sorption factor	1	239
κ	thermal diffusivity	m^2 s^{-1}	164
μ_D	diffusion-resistance coefficient	1	167
ν	kinematic viscosity	m^2 s^{-1}	198
ξ	relative depth	1	152
ξ	tortuosity	1	167
Π	humidity potential	1	295
Π_A	ambient-air potential	1	325
Π_o	inlet humidity potential	1	296
ρ_G	gas density	kg m^{-3}	202
ρ_L	liquid-moisture density	kg m^{-3}	199
ρ_S	density of bone-dry solids	kg m^{-3}	62

Symbol	Meaning	Units	first used page
σ	surface tension	$N\ m^{-1}$	41
σ	psychrometric coefficient	1	122
σ	Stefan-Boltzmann coefficient	$W\ m^{-2}\ K^{-4}$	210
τ	time	s	61
τ_R	residence time	s	243
ϕ	humidity-potential coefficient	1	110
ϕ_E	Ackermann correction	1	117
Φ	characteristic moisture content	1	156
Φ_o	initial characteristic moisture content	1	174
ψ	relative humidity	1	22
Ψ	porosity	1	167
ω	relative distance into bed	1	295

Overlines

^	cup-mixing average
–	distance average

Superscripts

*	equilibrium value

Subscripts

A	ambient	o	initial or inlet conditions
C	convective	R	radiative
cr	critical	S	surface or saturation
D	dewpoint	V	vaporization
E	evaporative	W	wetted surface or wet-bulb
G	bulb gas or dry-bulb	W	moisture vapour
L	moist solids	Z	at distance Z from solids inlet
L	based on length L		

Commonly used dimensionless groups

Bi	Biot number	eq. 5.4
Le	Lewis number	Table 4.2
Lu	Luikov number	Table 4.2
Nu	Nusselt number	eq. 6.14
Pr	Prandtl number	Table 4.2
Re	Reynolds number	eq. 6.14
Sc	Schmidt number	Table 4.2

PLATE 1. Rotary dryers under construction. Particulate solids are fed to the upper end of a large, slowing revolving, inclined barrel which contains longitudinal particle-lifting baffles or "flights" to cascade the particles through the hot gas stream. [Buell Ltd., with permission.]

Chapter 1

SCOPE

1.1 Introduction

This book is concerned with the determination of process conditions in equipment for drying solids by heat on an industrial scale. Up to the turn of this century such conditions were set by rule of thumb rather than by scientific principles. Formal methods of designing dryers became a commercial necessity as installations had to be adapted to meet more widely different requirements. Further, as improved manufacturing processes speeded production, so the slow traditional methods of drying could no longer be tolerated.

For instance, Williams-Gardner[(13)] cites the following example from the ceramic industry. The old method of drying tableware was to use juvenile labour to carry the ware on its mould into a heated room where the pottery was allowed to dry slowly for a period lasting up to a day. Capital was tied up, in the form of clay bodies and moulds, for an excessive length of time; young persons had to enter unhealthy atmospheres, hot and humid, often dust-laden through people treading on clay scrap. Such intolerable conditions have been circumvented by using progressive dryers under controlled conditions, for which drying times have been reduced to a few minutes for cups and a few hours for large dinner plates.

Drying has been defined as the removal of volatile substances (we call moisture) by heat from a mixture that yields a solid product.[(3)] This definition excludes concentration of a solution or slurry by evaporation and mechanical dewatering with filter-presses or centrifuges. Excluded also are thermal methods akin to distillation such as the azeotropic dehydration of some organic liquids. The definition does, however, take in a wide variety of kinds of equipment, ranging from simple ovens to large chambers in which solutions of solids are sprayed into a hot gas to yield a dried product.

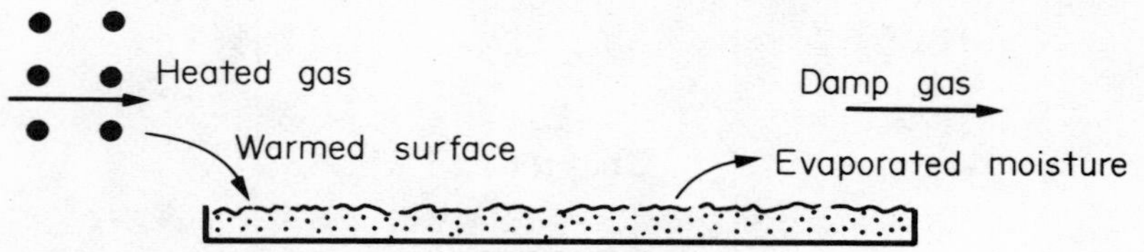

Fig. 1.1 Drying process

For drying, as defined above, to take place, the moist material must obtain heat from its surroundings, the moisture in the body evaporate and be received by a carrier gas. The drying process is sketched in Fig.1.1. A gas, heated for instance by being passed through hot coils, warms up the surface of the material to be dried. Moisture evaporates from the body into the gas which becomes progressively damper as it streams over the surface. Clearly the temperature and concentration of moisture vapour above the drying surface influence the rate of drying, and the correct manipulation of these hygrothermal conditions is crucial in any successful drying process.

Often the carrier gas is air and the moisture water, so the moisture vapour can be discharged to the atmosphere directly; but should the moisture be a valuable solvent, then the solvent vapour is recovered in a suitable condenser. Sometimes, particularly if the material is thermally sensitive, the operation takes place under vacuum when the carrier gas is almost absent, and the evaporated moisture is also removed by condensation.

Drying is an energy-intensive operation, and ways of reducing the use of heat have constantly been sought. The incentive to increase the moisture capacity and steam economy of existing dryers led Grosvenor[2] in 1907 to devise the first effective process-design method for dryers. He recommended that calculations should be done on an dry-gas basis for ease of following changes in the moisture-vapour content of the air. He suggested that the extent to which the air humidifies on contact with a solid containing water is given by the relationship

$$\frac{dY}{dT} = \frac{aC_p}{b-T} \tag{1.1}$$

in which a,b are coefficients,

C_p is the specific heat of the moist gas,

T is the temperature at which the evaporation is stopped,

Y is the moisture fraction in the gas or humidity.

A succession of straight lines representing eq.1.1 on a graph of temperature (T) against humidity (Y) enabled one to do hygrometric calculations more easily than by the tedious methods then available of interpolating tabulated values of the moisture content of damp air under selected conditions.

Fifteen years later Mollier[7] introduced a new chart, which surprisingly has yet to become universally adopted. Mollier's diagram for moist air consisted of a plot of enthalpy or heat content against humidity for mixtures of water vapour in air. Since these parameters are linearly related, the chart became a powerful tool for calculating the properties of mixed humid airstreams.

The search for simple, yet effective, methods of characterising the drying solid has been more arduous, and only recently have methods with some promise been proposed. Most descriptions of drying behaviour are based on the experimental observation of Sherwood and Comings[10] in 1933 that a number of materials, when very wet and placed in a pan exposed to warm cross-flowing air, dried at a rate that was almost equal to the rate of evaporation of water from the same pan under identical external conditions. These tests led to the idea, now firmly entrenched in the literature, of a critical point dividing unhindered from hindered drying when the solid offers resistance to the loss of moisture. By using this critical point as a datum, Van Meel[12] twenty-five years afterwards (1958) suggested that under certain conditions a drying curve characteristic of the moist material might be drawn up from a single experimental rate-of-drying curve. Whence the rate of moisture loss from a unit exposed surface could be described by an expression of the type

$$N_V = fK_O(Y_W - Y_G) \qquad (1.2)$$

in which f is a relative drying rate found from the characteristic curve,

K_O is a coefficient which depends upon the geometric arrangements and the airflow conditions,

Y_W is the humidity of air adjacent to a fully wetted surface,

Y_G is the humidity of the bulk air.

Drying-rate profiles thus follow from known humidity levels. This concept has been applied and extended in a recent monograph[4], in which process-design methods were advanced for both batch and continuously worked dryers of various kinds.

The charts proposed by Grosvenor and Mollier together with the postulate of the characteristic drying curve provide us with the essential conceptual tools to shape the calculational methods outlined in this book.

1.2 Need for Drying

The reasons for drying are as diverse as the materials which may be dried. A product must be suitable for either subsequent processing or sale. Materials need to have a particular moisture content for pressing, moulding or pelleting. Powders must be dried to suitably low moisture contents for satisfactory packaging. Whenever the product must be heated to a high temperature, as in many ceramic and metallurgical processes, predrying at lower temperatures ahead of firing kilns of high capital and operating costs can result in overall economy. Sometimes a two-stage drying process is adopted for the same reason. Costs of transport often depend upon the moisture content of the product, and a balance must be struck between the cost of conveying moisture on the one hand and the cost of drying on the other. Excessive drying, however, is wasteful: not only is more heat (and thus expense) involved than is necessary, but often overdrying results in a degraded product.

To illustrate some of these points, it is worthwhile to look in some greater detail at three common kinds of material: timber, clay and milk.

Timber A live tree contains copious quantities of moisture, but as soon as it is felled it loses moisture at a slow and dimishing rate until a time is reached when the moisture in the wood is in equilibrium with the moisture in the surrounding air. The newly logged timber, if it is to be satisfactorily used in joinery or cabinetmaking, must be dried to a suitable moisture content to accelerate this natural seasoning under controlled conditions which minimize degrade of the material. An article of furniture made from a hardwood, for instance, with a moisture content of 16 per cent would probably remain free from splits and distortion in an unheated bedroom, but if placed in a centrally heated room would very quickly warp and show other signs of degrade. Complex drying schedules have been evolved, mainly by batch experiments with small lots, to produce timbers that will not unduly distort in use. As energy costs rise, so the optimization of these somewhat lengthy drying processes is becoming a matter of concern and research.

Clay A large number of drying operations may be seen in the ceramic industries. The clays, felspar and other raw materials used are dried before delivery to the works. Glazes and pigments are often made wet and are also received dry. Domestic tableware moulded from a plastic clay must be dried to release the ware from the mould, while the ware that is cast must also be dried for the same reason. With cast ware, the clay is added as a creamy slurry to the porous mould which adsorbs considerable amounts of water to consolidate the casting before subsequent processing and drying can begin. The moulds, in turn, must be dried to remove the adsorbed moisture. Glazes applied to the fired biscuit ware by dipping or spraying have to be dried before final kiln-firing.

Milk Milk powders have become an important commodity of international trade as a means of providing a liquid milk supply by reconstitution in water in places far distant from dairying centres. Moreover, powders of varying fat content are used as dry ingredients in numerous products: meat smallgoods, dietary preparations, babyfoods, ice-cream mixes, beverage whiteners and animal foodstuffs. Milk is never dried in its original form owing to its high water content, but is evaporated to a solids concentration of 45 to 52 per cent before drying. Almost all milk products are dried by passing the milk concentrates through atomizing discs or nozzles from which a spray of fine droplets is thrown into a hot airstream to yield a granular powder. Sometimes, when the gritty taste of a spray-dried product is unacceptable, the milk concentrate is passed over slowly rotating, heated drums and the dried material is peeled off as flakes. Very fine powders on dispersing in water reconstitute poorly, but by agglomerating the particles or adding edible wetting agents such as lecithin a so-called "instant" product can be made with much better properties of reconstitution. Thus many modern spray-drying installations have secondary units involving drying and cooling to enable the specified properties of each powder to be met.

1.3 Types of dryers

The varied nature and history of drying operations have begotten a wealth of dryer types. Some operations are long-standing, such as the drying of fellmongered and scoured wool: other techniques are recent innovations, as, for example, in the preparation of ore concentrates from mineral slurries by spray-drying. There have been a number of attempts to classify dryers, the most comprehensive of which is that devised by Kröll[5] who presents a decimal coding system. The principal factors that define the nature of the dryers are:

1. The manner whereby the moist material receives heat.
2. The pressure and temperature of operation.
3. The way in which the material is supported in the dryer.

These points will be considered in turn.

Heating Methods

<u>Convection</u> Perhaps the commonest way in which a moist material is heated in a dryer is by convection. In convective heating, the carrier gas for the evaporated moisture is preheated before passing over or through the material, and the drying conditions can be readily controlled by the temperature and humidity of the warm gas. Fig.1.2 illustrates a common form of continuously worked, convective dryer. The moist material is loaded onto shallow trays which are stacked on trolleys being slowly moved through the drying tunnel. Heaters, usually steam coils if the air temperature is less than 150°C, are set in the air-space above the train of trolleys, and internal fans circulate the air through the hot coils and over the material spread out on the trays. If the material to be dried is open or granular, it is usual to pass the air upwards through perforated trays for better contact, however the air

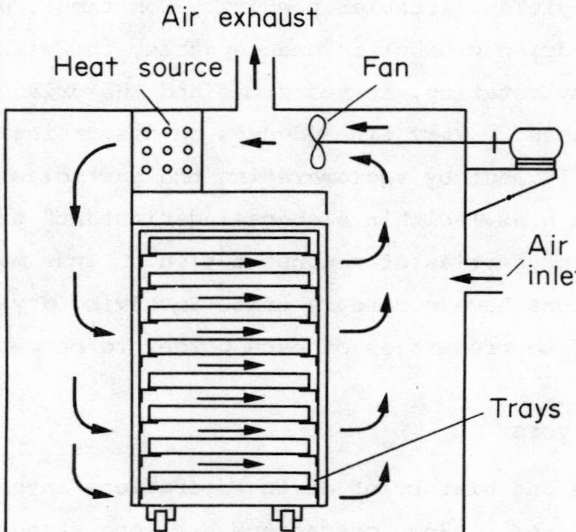

Fig.1.2 A drying tunnel for trucked trays. After Sloan.[11]

is still recirculated to improve the thermal efficiency of the system.

The installation lends itself to be subdivided into a series of zones at different temperatures for fine control of drying rates, but the labour of loading and unloading the trays is costly. Thus, particularly if the particles are free-flowing, other systems involving less labour are employed.

Conduction If the material to be dried is very thin or very wet, conductive heating may be employed. All the heat for evaporating moisture passes through the material from hot surfaces supporting or confining the material, so the material temperatures are higher than in convective drying. For this reason, thermally sensitive materials must be dried under vacuum. Fig. 1.3 shows a twin-drum dryer which accepts a thin slurry free of lumps. The rotating drums drag the slurry around their slowly revolving peripheries and knives peel off the dried product which falls into conveying ducts. With single-drum units, the wet paste may be splashed onto the surface of the drum, or the drum may dip into a pool of the feed material which is the commoner practice. Thin materials in the form of sheets, such as paper webs, can be drawn over a train of hot cylinders to provide a discontinuous thermal contact. Papermaking machines are built with upwards of 60 cylinders in series over which the sheet of pulp is drawn at speeds exceeding 7.5 m s^{-1}

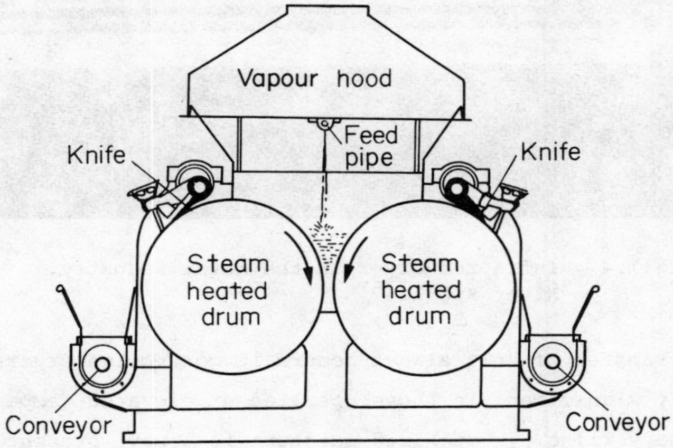

Fig. 1.3 A twin-drum dryer for thin slurries. After Sloan.[11]

Radiation Energy may be supplied from various sources of electromagnetic radiation, whose wavelengths range from those of solar radiation to those of microwaves (0.2 m to 0.2 μm). Radiation within this waveband barely penetrates beyond the surface of the exposed material and the host material only accepts most of the incident energy for radiation of certain wavelengths. Infra-red radiant heating is often used in drying thin coatings, films and webs since absorption in the 4 to 8- μm waveband is good. Fig.1.4 shows an internal view of a drying tunnel fitted with infra-red heaters for drying and baking the enamelled finish of motor vehicles.

Fig.1.4 Infra-red dryer in the motor industry. After Kröll.[5]

Radiative transfer of heat always occurs in convective dryers and is particularly significant in those operated at elevated temperatures such as rotary kilns. Vaporized moisture is always present in the air-space above drying solids, and carbon dioxide in the atmosphere of direct-fired units. These gases absorb and emit radiation, mainly in the infra-red region. The presence of airborne dust will also enhance radiation.

<u>Dielectric heating</u> Although most moist materials, especially when nearly dry, are poor conductors of 50 Hz current, at radio-frequencies the impedance of such materials falls to make electrical heating a feasible technique. The current lags the applied field, and there is a large energy loss which appears as heat. The energy loss diminishes as the material dries out, so that dielectric heating is a potentially useful way of smoothing variations in moisture content of the drying goods without causing undue drying stresses. The technique is very expensive, and only occasional industrial applications have been reported. One of these is depicted in Fig. 1.5 which shows a high-frequency unit used for drying beech shoe-lasts. The timber pieces pass on a conveying band through a drying chamber which contains the high-potential electrodes and is ventilated to release moisture. Another reported application[6] is the preparation of dicalcium phosphate which must be completely free of contamination.

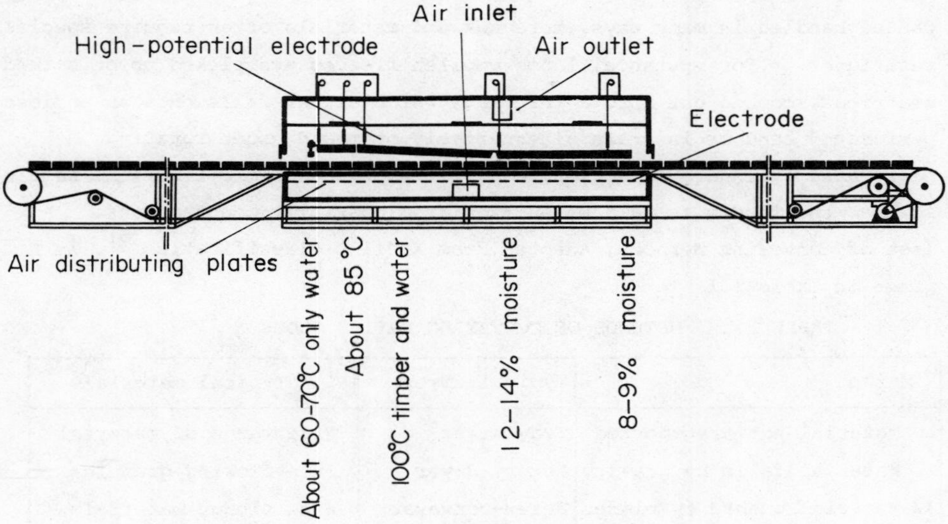

Fig. 1.5 High-frequency dryer for wood pieces. After Brown[1]

Process Conditions

One of the major concerns of this book will be the consideration of the influence of process conditions on the course of drying. In practice, these conditions may be severely constrained by the nature of the material to be dried and sometimes by the heating systems available. The thermal sensitivity of the material fixes the maximum temperature to which the stuff may be heated during drying. This temperature varies with the time the goods are held in the dryer. For many materials, the rate of thermal degradation follows the Arrhenius relationship and the maximum permissible temperature falls exponentially with increase in holding time. Polymer chips of PVC, for example, may be safely processed in a fluid-bed dryer, where the residence time might be of order 20 min. In a static bed, the drying conditions would have to be very mild, and the associated dryer large and costly. Should slow batch operation be unavoidable, the use of vacuum working to reduce operating temperature may be necessary. Vacuum shelf dryers were once common, for example, in the drying of organic dyes at low temperatures with the exclusion of as much air as possible.

Conveying Methods

The outward appearance of a dryer depends largely upon the way the drying goods move through the equipment. While free-flowing granules can be handled in many ways, more awkward materials often require special techniques. For instance, loose woollen fleeces are picked up on spiked brattices, combed out into a thin mat, which either falls onto an endless perforated band or is drawn alternatively over and under rotating perforated cylinders in series, each fitted with a hemicylindrical baffle so that the fleece is held by suction from coaxially mounted fans. A list of conveying methods, adapted from Kröll's classification[5], is given in Table 1.1

TABLE 1.1 METHODS OF CONVEYING DRYING GOODS

Method	Typical dryer	Typical materials
1. Material not transported	Tray dryer	Wide range of materials
2. Material falls by gravity	Rotary dryer	Free-flowing granules
3. Material pushed by blades	Screw-conveyer dryer	Wet, sludgy materials
4. Material conveyed on trucks	Tunnel dryer	Wide range of materials
5. Material pulled over rolls	Hot-cylinder dryer	Thin webs, sheets and boards
6. Material conveyed on band	Band dryer	Wide range of rigid materials
7. Material vibrated on band	Vibrating-band dryer	Free-flowing granules
8. Material suspended in air	Fluid-bed dryer	Granules
9. Material thrown into air	Spray dryer	Solutions, slurries and thin pastes

Most modern dryers are operated continuously, or semi-continuously over the working day, as a continuous dryer will require less labour, fuel and floor-space than a batch dryer of the same capacity. However, batch drying would be chosen whenever the production rate is small (under 200 kg h^{-1}), or a large number of products have to be handled in the same unit, or whenever large bulky items have to be dried under extensive and complex schedules for which the drying of porcelain sanitary ware and sawn timber boards provide examples.

1.4 Design of Dryers

It follows from the discussion in the previous section that there may be more than one kind of dryer that will be suitable for a particular job. The engineer concerned with the process design has to choose for a given dryer those conditions which enable the specified properties of the product to be obtained. In this way, the performance characteristics of alternative systems can be assessed as a basis for the ultimate choice of a specific plant. Almost always some small-scale tests are needed to determine the material's drying characteristics that are required to predict the way in which the stuff will dry in a commercial plant. A flow chart illustrating the various steps needed to design a dryer is shown in Fig. 1.6.

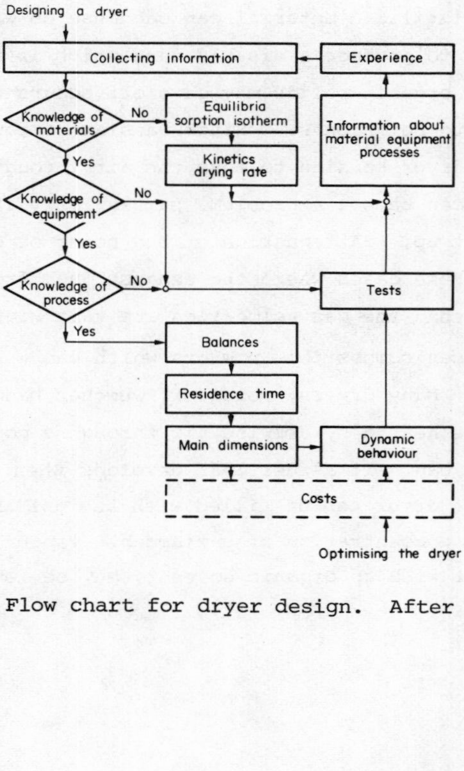

Fig. 1.6. Flow chart for dryer design. After Poersch (8)

This process design, whilst being fundamental in making a logical choice, is by itself not enough, since malfunctioning of the solids-conveying system and other equipment associated with the dryer can lead to the unsatisfactory operation of the whole process. The feed must be presented to the dryer in as uniform condition as possible and, if very wet, some mechanical dewatering is highly desirable with the use of squeeze rolls, filters or centrifuges as appropriate. Plastic and pasty materials can be preformed into ribbons and somewhat stiffer materials can be extruded into pellets. With conveyor dryers, the wet material should be distributed evenly over the whole width of the band by an oscillating spreader with no dwell time on reversal. Direct feeding from a suitably designed hopper may be adequate for free-flowing stuff. The heating equipment must be appropriately sized. If the heaters are too small, production will falter: should the equipment be too large, uneven heating difficult to control will result.

Even if the product meets its specification in all respects, the process will still be unsatisfactory should unacceptable discharges from the plant occur. There is always the risk of fine airborne dust from drying plant, especially if particles being dried are thrown or suspended in the gas-stream or if lint and fluff can break free. Dust emissions are now subject to increasing legislative scrutiny, even in rural areas. Volatilized material can condense on water particles in a "steamy" exhaust to produce a visible fume which is a source of public irritation. Some organic solids may give off odours which can be equally obnoxious to neighbours. Since fans are frequently installed integral with the dryer housing to move the air through the dryer, the emission of noise can become a problem, particularly if resonant frequencies are set up. Attenuation of the noise of discharging gases may be needed in these cases where the exhaust duct from the dryer is too restricted so that the gas velocities are very high.

Drying often produces combustible powders which cause substantial fire and explosion hazards. Many dryers, moreover, whether handling combustible powders or not, are heated by drawing air through a combustion chamber fired directly by oil or gas. If a fuel leak develops when the plant is not working, the entire dryer can be filled with an ignitable cloud of gas or vapour. While the concentration of a flammable vapour, produced from drying solids moist with an organic solvent, may be less than the lower explosive limit, the hybrid mixture of any dust with the vapour can be much more susceptible to ignition. Ways to reduce these hazards have recently been surveyed.[9]

REFERENCES

1. Brown, W.H., An Introduction to the Seasoning of Timber, p.155, Pergamon, Oxford (1965).
2. Grosvenor, W.M., Calculations for dryer design, Trans.AIChE., 1, 184-202 (1907).
3. Keey, R.B., Drying Principles and Practice, p.1, Pergamon, Oxford (1972).
4. ibid., pp.204-270.
5. Kröll, K., Trockner, einteilen, ordnen, benennen, benummern, Vol.6, Schilde, Bad Hersfeld (1965).
6. Macdonald, J.O.S., Progress in drying, Part 2, Proc.Technology, 18, 203-6 (1973).
7. Mollier, R., Ein neues Diagramm für Dampfluftgemische, Zeits.VDI, 67, 869-872 (1923).
8. Poersch, W., Present state of drying technology, IChE solids drying course, Birmingham (1977).
9. Reay, D., Fire and Explosion Hazards in Dryers, Proc.4 Int.Powder Technol. Bulk Solids Conf., Powtech 77, 62-8 Harrogate (1977).
10. Sherwood, T.K. and E.W. Comings, The drying of solids, V. Mechanism of drying of clays, Ind.Eng.Chem., 25, 311-6 (1933).
11. Sloan, C.P., Drying systems and equipment, Chem.Eng. 74, (14), 169-200 (1967).
12. Van Meel, D.A., Adiabatic convection batch drying with recirculation, Chem. Eng.Sci., 9, 36-44 (1958).
13. Williams-Gardner, A., Industrial Drying, p. 1, Leonard Hill, London (1971).

PLATE 2. Night view of dryer to remove moisture and preheat coking coal. [British Steel (Chemicals) Corporation, with permission.]

Chapter 2

MOISTURE IN GASES AND SOLIDS

2.1 Humidity of Moist Gases

We consider a moist gas to be composed of two constituents, the utterly dry gas and the moisture vapour, even if the dry gas be a gaseous mixture (as air) and the moisture a mixed vapour, say of co-evaporating solvents. Since the specific volume of a moist gas varies with the concentration of moisture as well as temperature and pressure, it is usual to describe the fraction of moisture present in terms of mass ratios. A dry-gas basis is chosen for ease of bookkeeping. The humidity of a gas then is defined as the ratio of the mass of moisture vapour to the mass of completely dry gas. This ratio is dimensionless, but at low moisture levels it is often convenient to record humidities in g kg^{-1}.

Since mixtures of air and water vapour commonly occur in drying, it is appropriate to look at such humid mixtures in some detail. Clean, dry air is obtained when atmospheric air, uncontaminated with minor constituents from man-made activity or natural sources such as forest fires and volcanic action, has had all of the moisture abstracted from it. The nitrogen content is substantially uniform over the earth's surface to a height of at least 100 km, but the small amount of carbon dioxide is variable. At night photosynthesis ceases, and a carbon dioxide content of up to 0.06 per cent by volume has been found over woodland in early morning. By contrast, the equilibrium partial pressure of the gas is low over cold seas, and a value of 0.015 per cent has been recorded at Spitzbergen (latitude 77°N).[2] Goff[3], in his final report to an international committee for psychrometric data, has defined dry air as shown in Table 2.1. Under this definition, the mean molar mass of dry air is 0.028 966 kg mol^{-1}. The molar mass for water is 0.018 016 kg mol^{-1}, so the water/air molar-mass ratio is 0.622 to three significant figures.

TABLE 2.1 COMPOSITION OF DRY AIR

Substance	Molar Mass $kg\ mol^{-1}$	Composition mole fraction
Nitrogen	0.028 016	0.7809
Oxygen	0.032 000	0.2095
Argon	0.039 944	0.0093
Carbon dioxide	0.044 01	0.0003

The pressure-temperature behaviour of air can be expressed by the virial equation of state

$$Z = \frac{PV}{RT} = 1 + \frac{B_2}{V} + \frac{B_3}{V^2} + \frac{B_4}{V^3} \qquad (2.1)$$

where V is the specific molar volume having units of $m^3\ mol^{-1}$ and R is the universal gas constant which takes a value of 8.314 $J\ mol^{-1}\ K^{-1}$. The coefficients B_i are the so-called virial coefficients and are sole functions of temperature. These coefficients account for molecular interactions not considered by simple kinetic theory and must be determined experimentally. Corresponding values of the compressibility factor Z are listed in Table 2.2 from values listed in "Tables of Thermal Properties of Gases" prepared by the American Bureau of Standards.[6]

TABLE 2.2 VALUES OF $(Z-1) \times 10^5$ FOR CLEAN DRY AIR WHERE $Z = PV/RT$

Temperature/K	Pressure / kPa		
	10	40	100
300	-3	-12	-30
400	+2	+ 8	+19
500	3	14	34
700	4	15	38
1000	3	13	33

Clearly the compressibility factor for clean dry air is essentially unity under conditions likely in drying.

Equation 2.1 can also be used to represent the behaviour of water vapour, but the accurate determination of virial coefficients is not easy and reported values are of uncertain accuracy. Some values of the compressibility factor for steam, which are presented by Hilsenrath[6], are listed in Table 2.3.

TABLE 2.3 COMPRESSIBILITY FACTOR (Z) FOR STEAM AT ATMOSPHERIC PRESSURE (101.325 kPa)

Temperature/K	380	430	480	530	580	630
Z	0.98591	0.99219	0.99509	0.99667	0.99763	0.99824

Steam too may be regarded as an ideal gas with a compressibility factor of unity for drying calculations.

In moist air, various interactions between like and unlike molecules are possible, so that a possible equation of state would be

$$\begin{aligned} PV &= RT \\ &- [y_G^2 B_{GG} + 2\, y_G y_W B_{GW} + y_W^2 B_{WW}]\, P \\ &- [y_G^3 B_{GGG} + 3\, y_G^2 y_W B_{GGW} + 3\, y_G y_W^2 B_{GWW} + y_W^3 B_{WWW}]\, P^2 \qquad (2.2) \\ &- \ldots\ldots\ldots\ldots\ldots \end{aligned}$$

where y_G is the mole fraction of dry air and y_W the mole fraction of water vapour. The virial coefficient for a single molecule of air reacting with another is B_{GG}, B_{GW} represents the coefficient for the interaction of one molecule of dry air with one of water vapour, B_{GGG} is the coefficient for a triple interaction of air molecules, and so on. Values of these coefficients are listed by Goff[2] in his report to which reference has already been made.

Should a mixture of m_G kg of air and m_W kg of water vapour behave as an ideal gas, one has

$$p_W V = (m_W/M_W)\, RT \qquad (2.3)$$

and $$p_G V = (m_G/M_G)\, RT \qquad (2.4)$$

in which p_G and p_W are the partial pressures of air and water respectively.

These pressures sum to the total pressure P, i.e.

$$P = p_G + p_W \tag{2.5}$$

and from back-substitution into equation 2.4 the expression

$$(P - p_W)V = (m_G/M_G)\ RT \tag{2.6}$$

is obtained. From the definition of the humidity Y, we find from equations 2.3 and 2.6 that

$$Y = \frac{m_W}{m_G} = \left[\frac{M_W}{M_G}\right] \frac{p_W}{(P-p_W)} \tag{2.7}$$

The molar-mass ratio (M_W/M_G) has a value of 0.622.

The maximum value that p_W can normally reach at any given temperature is the saturated-vapour pressure p_W^o, often known simply as the vapour pressure. A pressure-temperature diagram for the substance water is sketched in Fig.2.1. The curve AC, dividing the liquid and vapour regions, represents the vapour-pressure curve for water and ends abruptly at the critical state. This curve follows the unique pressure at which the liquid is in equilibrium with its vapour over a plane interface. There is a similar curve AB for the vapour pressures of subliming ice.

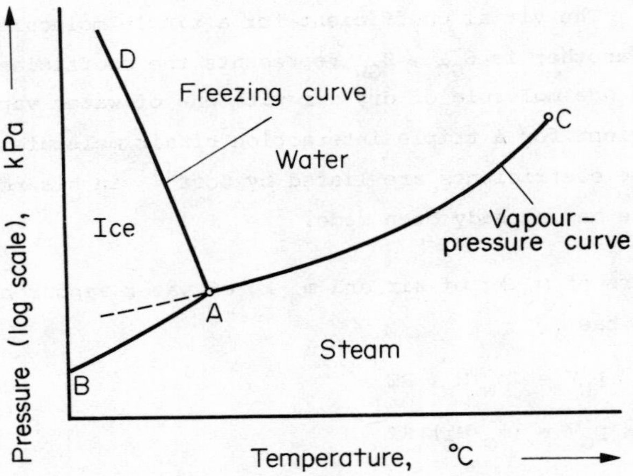

Fig. 2.1 Pressure-temperature diagram for the substance water.

The ideal saturation humidity Y_S is thus

$$Y_S = \left(\frac{M_W}{M_G}\right) \frac{p_W^o}{(P-p_W^o)} \tag{2.8}$$

and follows from substituting p_W^o for P_W in equation 2.7. The true saturation humidity will presumably take a different value and correspond to an apparent vapour pressure $f_S p_W^o$, for which the coefficient f_S is likely to be a function of pressure P and temperature. Values of this coefficient for a total pressure of 100 kPa are reproduced in Table 2.4 from Goff's calculations.[3]

TABLE 2.4 VALUES OF GOFF'S VAPOUR-PRESSURE COEFFICIENT f_S AT 100 kPa TOTAL PRESSURE

Temperature /°C	f_S	temperature / °C	f_S
0	1.0043	50	1.0056
10	1.0043	60	1.0060
20	1.0045	70	1.0062
30	1.0047	80	1.0058
40	1.0051	90	1.0040

The error in ignoring this correction is between 0.4 and 0.62 per cent and is thus negligible in drying calculations.

Most published tables of data for the vapour pressure of water are based on those of the International Critical Tables, and more recent compilations such as those of Mayhew and Rogers[17] do not differ significantly. Selected values from this latter source are listed in the Appendix together with values of the corresponding saturation humidity Y_S for a total pressure of 100 kPa.

Equation 2.8 is, of course, not limited to mixtures of air and water vapour. The expression holds for any ideal vapour-gas pair, and may be invoked whenever vapour-pressure data are available. Comprehensive tables of data may be found in many works, such as the handbook of Chemistry and Physics[24]. Some selected data for common solvents are plotted in Fig.2.2. Since the change of vapour pressure with temperature dp^o/dT is inversely proportional to the absolute temperature, these data are plotted with log.-linear co-ordinates to yield straight lines for each substance (see the following Example 2.2).

Example 2.1 Estimate the saturation humidity of acetone in nitrogen at 40°C when the total pressure is 110 kPa.

From Fig.2.2 (line 7), the vapour pressure p^o is 55.5 kPa. The molar mass for acetone is 0.058 08 kg mol^{-1} and that for nitrogen is 0.028 02 kg mol^{-1}. We substitute in equation 2.8, retaining three significant figures, to get

$$Y_S = \left(\frac{M_W}{M_G}\right) \frac{p_W^o}{(P-p_W^o)}$$

$$= \left[\frac{0.0581}{0.0280}\right] \frac{55.5}{(110 - 55.5)}$$

$$= \underline{2.11 \text{ kg kg}^{-1}}$$

Example 2.2 Weast[24] states that the vapour pressure of cyclohexane is 53.3 kPa at 60.8°C and 101.325 kPa at 80.7°C. Estimate the vapour pressure at 70°C.

Given $\frac{dp^o}{dT} = \frac{a}{T}$ where a is a constant

$$\therefore \frac{d \ln p^o}{dT} = \frac{a}{T^2}$$

$$\therefore \frac{d \ln p^o}{d(1/T)} = -a$$

We may use this relationship to interpolate the given data noting that 0°C ≡ 273.15 K.

temperature/ °C	1/T /K^{-1}	p^o/kPa	ln p^o
60.8	0.002 994	53.3	3.98
80.7	0.002 826	101.325	4.62

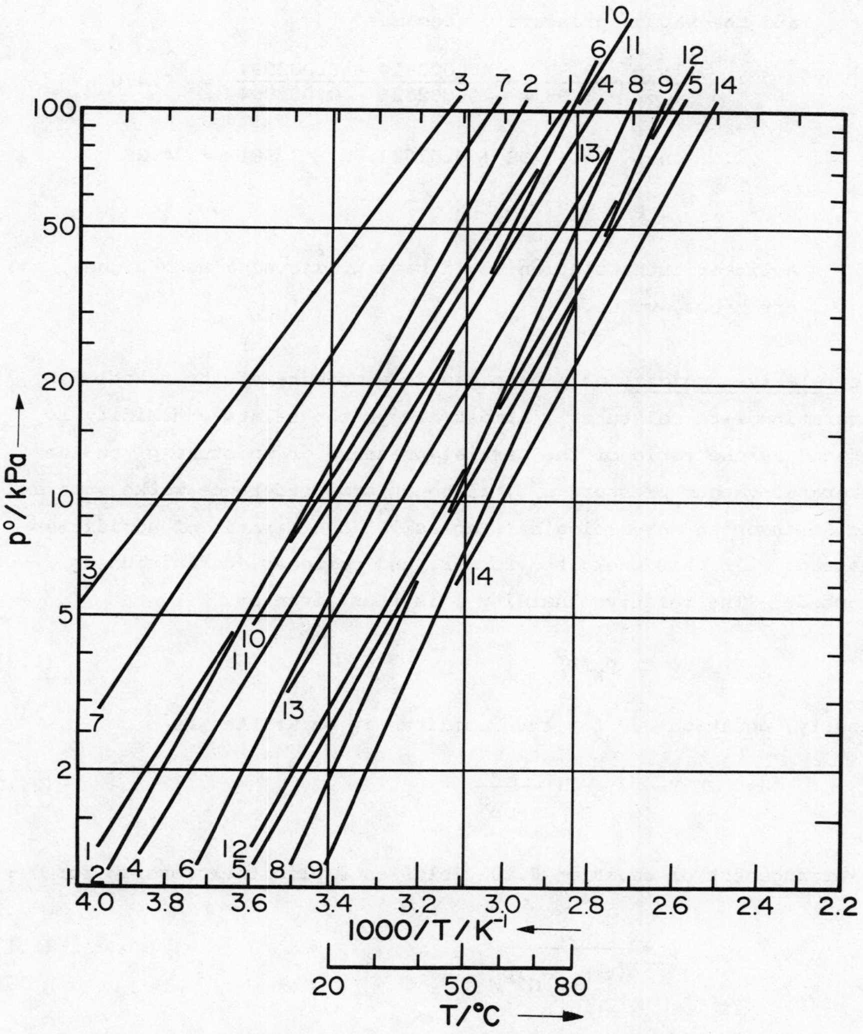

Fig.2.2 Vapour-pressure chart for common solvents

<u>Key</u>

1. carbon tetrachloride CCl_4
2. methanol CH_4O
3. carbon disulphide CS_2
4. trichlorethylene C_2HCl_3
5. 1,1,2,2-tetrachlorethane $C_2H_2Cl_4$
6. ethanol C_2H_6O
7. acetone C_3H_6O
8. n-propanol C_3H_8O
9. isobutanol $C_4H_{10}O$
10. benzene C_6H_6
11. cyclohexane C_6H_{12}
12. toluene C_7H_8
13. n-heptane C_7H_{16}
14. n-octane C_8H_{18}

A temperature of 70°C corresponds to a value of 1/T of 0.002 914 and the vapour pressure p^o becomes

$$\frac{\ln p^o - 3.98}{4.62 - 3.98} = \frac{0.002914 - 0.002994}{0.002826 - 0.002994} = 0.476$$

$$\therefore \quad \ln p^o = 3.98 + 0.476(4.62 - 3.98) = 4.28$$

$$\therefore \quad p^o = 72.6 \text{ kPa}$$

A linear interpolation would have predicted a much higher, and erroneous, value.

The relative humidity of a damp gas is a measure of its fractional saturation with moisture. Almost always the relative humidity is defined as the ratio of the partial pressure of moisture p_W to the saturated-vapour pressure p_W^o, although some reference works such as the Smithsonian Meteorological Tables[13] use a ratio of humidities instead. In this book, the former, and commoner, definition is adopted. The relative humidity ψ is thus given by

$$\psi = p_W/p_W^o \qquad (2.9)$$

Clearly, equation 2.7 for the humidity may be written as

$$Y = \left(\frac{M_W}{M_G}\right) \frac{\psi p_W^o}{P - \psi p_W^o} \qquad (2.10)$$

Re-arrangement of equation 2.10 yields an explicit expression for ψ :

$$\psi = \frac{YP}{(Y + M_W/M_G)p_W^o} \qquad (2.11)$$

Example 2.3 What is the relative humidity of acetone vapour in nitrogen at 40°C when the humidity is 1 kg kg^{-1} and the total pressure is 110 kPa ?

We have $P = 110$ kPa, $Y = 1.0$ kg kg^{-1}, $M_W = 0.0581$ kg mol^{-1}, $M_G = 0.0280$ kg mol^{-1}, $p_W^o = 55.5$ kPa.

On substitution in equation 2.11, we find

$$\psi = \frac{110 \times 1.0}{[1.0 + 0.0581/0.0280] \times 55.5}$$

$$= 0.645$$

Note that the humidity ratio Y/Y_S is $1/2.11 = 0.474$.

2.2 Enthalpy of Moist Gases

All substances will normally have internal energy which can be identified with the energies of molecular motion. Absolute values of the internal energy U are unknown, but we can assign values of U relative to an arbitrary zero at a particular temperature. In any steady-flow process, there is additional energy in forcing streams into the system against pressure and forcing streams out. This flow-work per unit mass is PV, where P is the pressure and V is the specific volume. The internal and flow energies are conveniently lumped together to give a composite energy known as the enthalpy H . The enthalpy is thus defined by the expression

$$H = U + PV \tag{2.12}$$

and has units of energy per unit mass (N.m kg^{-1} or J kg^{-1}). Absolute values of the enthalpy, like those of internal energy, cannot be fixed. One convenient reference state for zero enthalpy is liquid water under its own vapour pressure of 611.2 Pa at the triple-point temperature of 273.16 K (0.01^{o}C).

The isobaric variation of enthalpy with temperature is shown in Fig.2.3 for a pure fluid. At low pressures in the gaseous state, where gas behaviour is ideal, the enthalpy is almost independent of pressure, so the isobars nearly superpose upon each other. The envelope VCL represents the enthalpies of saturated vapour over the segment VC and the enthalpies of saturated liquid over the section CL, with the two curve segments meeting at the critical point C. The enthalpy difference V'L' corresponds to the latent heat of vaporization at a temperature T_S, and this difference wanes with increasing temperature to become zero at critical temperature T_C .

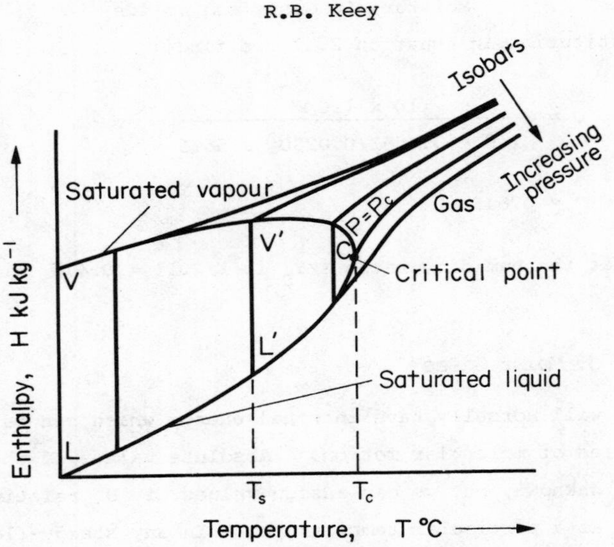

Fig. 2.3 Isobaric enthalpy-temperature diagram for a pure fluid.

The heat capacity of a body is defined by the heat required to raise the temperature of a unit mass. For a constant-pressure process, the heat capacity C_P is given by

$$C_P = \left(\frac{\partial Q}{\partial T} \right)_P \tag{2.13}$$

where the heat flow dQ is the sum of the internal energy change dU and the work done against pressure PdV. Thus equation 2.13 may be expanded as follows

$$C_P = \left(\frac{\partial U}{\partial T}\right)_P + P\left(\frac{\partial V}{\partial T}\right)_P = \left(\frac{\partial H}{\partial T}\right)_P \tag{2.14}$$

whence the isobars of Fig. 2.3 yield the heat capacity.

In drying calculations, it is more convenient to use mean values of the heat capacity over a finite temperature step, i.e.

$$\bar{C}_P = \left(\frac{\Delta Q}{\Delta T} \right)_P = \frac{1}{(T_2 - T_1)} \int_{T_1}^{T_2} C_P dT \tag{2.15}$$

While second-order polynomials in temperature can be fitted to C_P to describe very well heat-capacity data for gases over the range 300 to 1500 K[(7)], the quadratic term may be neglected for temperature ranges normally occurring in drying.

If

$$C_P = a + bT \tag{2.16}$$

it follows from equation 2.15 that

$$\bar{C}_P = a + \tfrac{1}{2}b(T_2 + T_1) = C_P(T_{av}) \tag{2.17}$$

The mean heat capacity is the value at the arithmetic-mean temperature T_{av}.

From equations 2.14 and 2.15 we can estimate the enthalpy of a pure substance from its heat capacity, since

$$H = \bar{C}_P\theta \tag{2.18}$$

where θ is the temperature excess above the zero-enthalpy datum. This temperature is, for practical purposes, equal to 0^oC, so $\bar{C}_P$ is evaluated at a temperature $\theta/2\ ^oC$. The heat capacity of dry air is 1.0049 kJ $kg^{-1}K^{-1}$ at 300 K and 1.0135 kJ $kg^{-1}K^{-1}$ at 400 K, so that at low temperatures the enthalpy in units of kJ kg^{-1} is numerically almost equal to the temperature in degrees Celsius for this substance. Water vapour has a higher heat capacity, having values of 1.864 and 1.901 kJ $kg^{-1}K^{-1}$ respectively at these temperatures. Heat capacities of vapours of certain organic solvents are listed in Table 2.5 following.

Since enthalpy is an extensive property, we would expect that the enthalpy of a moist gas is the sum of the partial enthalpies of the constituents and a small term to account for heats of mixing and similar effects. For convenience we define the so-called <u>humid enthalpy</u> I, which is the enthalpy of unit mass of dry gas and its associated moisture, since this dry-gas basis is similar to that adopted for expressing moisture concentrations in terms of humidities. Under this definition of enthalpy,

$$I_G = H_{GG} + H_{GW}Y + \Delta H_{GM} \tag{2.19}$$

where H_{GG} is the enthalpy of the dry gas, H_{GW} is the enthalpy of the moisture vapour and ΔH_{GM} is the residual enthalpy due to mixing and other effects. In air saturated with water vapour, this residual is only - 0.63 kJ kg^{-1} at 60^oC or about 1% of H_{GG},[13] and thus it is customary to ignore the influence of the residual enthalpy.

It is often useful to consider the specific heat of unit mass of dry gas with its associated moisture vapour. By analogy with equation 2.18, we may write

$$I = \bar{C}_{PY}\theta \tag{2.20}$$

wherein $\bar{C}_{PY}$ may be called the <u>humid heat capacity</u>, but usually it is known simply as the <u>humid heat</u>. Combination of equations 2.19 and 2.20 yields the expression

$$\bar{C}_{PY} = \bar{C}_{PG} + \bar{C}_{PW}Y \tag{2.21}$$

in which $\bar{C}_{PG}$ and $\bar{C}_{PW}$ are the mean heat capacities for the dry gas and moisture vapour respectively. Inspection of equation 2.21 shows that the humid heat of wet, saturated air is greater by less than 10 per cent of the heat capacity for completely dry air at temperatures below $40^{\circ}C$. The humid heat of moist combustion gases differs only slightly from that of damp air. For a low-humidity gas containing 20 mol per cent carbon dioxide, the humid heat is 3 per cent less than that of air at the same humidity. There are numerous possible paths whereby the moisture vapour could have arisen from its liquid. For instance, the liquid may be vaporized directly at $0^{\circ}C$, where the enthalpy is zero by definition, and superheated directly to a given temperature T, as shown in Fig.2.4. Alternatively, the liquid could have been heated to T_S, the temperature at which the cooled vapour would condense to form a dew, vaporized at T_S and superheated to the temperature T. This path yields the true humid enthalpy. However, the final enthalpy is almost independent of the vaporization path since the isothermal pressure coefficient $(\partial H/\partial P)_T$ is very small. Under these conditions, we can conveniently evaluate the humid enthalpy at T from the first-mentioned route involving vaporization at $0^{\circ}C$.

From equation 2.19 for $\Delta H_{GM} = 0$, we get

$$I_G = \bar{C}_{PG}T + [\bar{C}_{PW}T + \Delta H_{VO}]Y \tag{2.22}$$

which, on introducing the humid heat, simplifies to

$$I_G = \bar{C}_{PY}T + \Delta H_{VO}Y \tag{2.23}$$

The humid heat $\bar{C}_{PY}$ is evaluated at the mean temperature T/2, and the latent heat of vaporization ΔH_{VO} at $0^{\circ}C$. For air saturated with water vapour, I_G estimated through equation 2.23 differs by only 0.3% from the

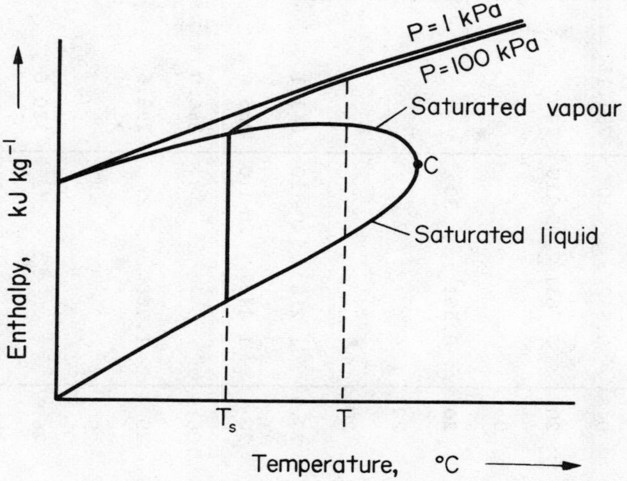

Fig. 2.4 Alternative vaporization paths

true humid enthalpy at 20°C, but at 50°C the difference is 5%. Despite the handiness of equation 2.23, its use is not recommended should the humidity be greater than 0.05 kg kg^{-1}.

For more accurate work, the unique vaporization path that yields the correct state must be followed, and the vapour enthalpy thus becomes:

$$H_{GW} = \bar{C}_{LW} T_S + \Delta H_{VS} + \bar{C}_{PW}(T - T_S) \qquad (2.24)$$

where $\bar{C}_{LW}$ is the mean heat capacity of the liquid moisture from 0 to T_S, $\bar{C}_{PW}$ is the mean heat capacity of the moisture vapour from T_S to T and ΔH_{VS}, the latent heat of vaporization, is estimated at T_S. Thermophysical data, needed to use equation 2.24, are given in Table 2.5 for some common organic solvents. The value of ΔH_{VS} can be estimated roughly from the known value of the latent heat at the specified temperature T_O from the cube-root formula[11]

$$\frac{\Delta H_{VS}}{\Delta H_{VO}} = \left[\frac{T_S - T_c}{T_O - T_c}\right]^{1/3} \qquad (2.25)$$

where T_c is the critical temperature. Whenever data are not available, estimates often sufficiently accurate for many engineering calculations can be made from predictive methods such as those outlined by Perry[21].

TABLE 2.5 THERMOPHYSICAL PROPERTIES OF SOME ORGANIC SOLVENTS

(Principal data source: International Critical Tables, Vol.5)

Substance	Molar mass/ $kg\ mol^{-1}$	$\bar{C}_{LW}$/ $kJ\ kg^{-1}\ K^{-1}$	at T/ ^oC	ΔH_{VO}/ $kJ\ kg^{-1}$	at T/ ^oC	$\bar{C}_{PW}$/ $kJ\ kg^{-1}\ K^{-1}$	at T/ ^oC	T_C / ^oC
1. carbon tetrachloride CCl_4	0.153 84	0.845	20	217.8	0	0.552	30	283.15
2. methanol CH_4O	0.032 04	2.50	18.8	1200	20	0.7611	40-110	240
3. carbon disulphide CS_2	0.076 14	0.96	0	373.8	0			
4. trichlorethylene C_2HCl_3	0.131 39	0.96	20	236	20	0.561	30	
5. tetrachlorethane $C_2H_2Cl_4$	0.167 86	1.113	16	230.3	20			
6. ethanol C_2H_6O	0.046 07	2.52	30	938.1	20			
7. acetone C_3H_6O	0.058 08	2.15	17-20	555.6	27.5	1.216	40-110	243.1
8. n-propanol C_3H_8O	0.060 09	2.28	0	753.5	20	1.452	26-110	235
9. isobutanol $C_4H_{10}O$	0.074 12	2.52	30	577.7	106.9			256.0
10. benzene C_6H_6	0.078 11	1.73	26.8	432.6	20	1.060	27	288.6
11. cyclohexane C_6H_{12}	0.084 16			392.5	25			
12. toluene C_7H_8	0.092 13	1.62	0	427.2	26.87			320.6
13. n-heptane C_7H_{16}	0.100 19	2.223	20	391.6	0			266.9
14. n-octane C_8H_{18}	0.114 22	2.20	25.1	358.2	0	1.656	27	296.2

Example 2.4 Calculate the humid enthalpy of acetone vapour in nitrogen at 80°C and 110 kPa when the humidity is 0.75 kg kg^{-1}.

(i) Firstly we estimate the partial pressure of acetone vapour to find the temperature T_S at which dew would form on cooling the humid gas. We use equation 2.7,

$$Y = \left(\frac{M_W}{M_G}\right) \frac{p_W}{(P-p_W)}$$

which can be re-arranged to give

$$p_W = \frac{YP}{(Y+M_W/M_G)} \quad .$$

On substituting the values, $Y = 0.75$ kg kg^{-1}, $P = 110$ kPa, $M_W/M_G = 0.0581/0.0280 = 2.075$, we find

$$p_W = \frac{0.75\text{x}110}{(0.75+2.075)} = 29.2 \text{ kPa} \quad .$$

Reference to the vapour pressure chart (Fig.2.2, line 7) suggests that the equilibrium temperature (T_S) corresponding to a saturated-vapour pressure of 29.2 kPa is 25°C.

(ii) Next we calculate the value of the latent heat of vaporization from the given value in Table 2.5, using the cube-root expression (equation 2.25). The tabular value gives $\Delta H_{VO} = 555.6$ kJ kg^{-1} for $T_O = 27.5$°C, thus

$$\Delta H_{VS} = \Delta H_{VO} \left[\frac{T_S - T_c}{T_O - T_c}\right]^{1/3}$$

$$= 555.6 \left[\frac{25 - 235}{27.5 - 235}\right]^{1/3}$$

$$= 557.8 \text{ kJ kg}^{-1}$$

In this case, the correction is scarcely necessary, but the method is illustrated.

(iii) Equation 2.24 can now be used to estimate H_{GW}, the vapour enthalpy, from the thermophysical data in Table 2.5

$$H_{GW} = \bar{C}_{LW} T_S + \Delta H_{VS} + \bar{C}_{PW}(T - T_S)$$

$$= (2.15 \times 25) + 557.8 + 1.452 \times (80 - 25)$$

$$= 691.4 \text{ kJ kg}^{-1}$$

(iv) The thermal properties of nitrogen are similar to those of air, and at the temperature, averaged between the datum and process values (40°C), the heat capacity $\bar{C}_{PG}$ is 1.0056 kJ $kg^{-1}K^{-1}$. The dry-gas enthalpy is thus

$$H_{GG} = \bar{C}_{PG} T$$

$$= 1.0056 \times 80 = 80.5 \text{ kJ kg}^{-1}$$

(v) The humid enthalpy now follows from its definition (equation 2.19) for $\Delta H_{GM} = 0$

$$I_G = H_{GG} + H_{GW} Y$$

$$= 80.5 + (691.4 \times 0.75)$$

$$= \underline{599.1 \text{ kJ kg}^{-1}}$$

Clearly at high humidities the enthalpy of the evaporated moisture is dominant.

2.3 Enthalpy-Humidity Charts

To follow the progress of a drying operation, we need to have access to a field of enthalpy-humidity values. The advent of widespread availability of minicomputers has freed the engineer from repetitive routine calculations of the sort illustrated in Example 2.4, but there is still place for the graphical storage of data; indeed, there is no cheaper way! Further, complex hygrothermal changes can be readily charted so that the design engineer sometimes prefers to use computing facilities for calculating data and assembling these into chart form.[18]

Mollier's original enthalpy-humidity chart[17] was drawn up with standard rectangular co-ordinates, but nowadays oblique co-ordinates are used to expand the area of the chart which corresponds to the unsaturated-vapour region. The basic geometry is shown in Fig.2.5.

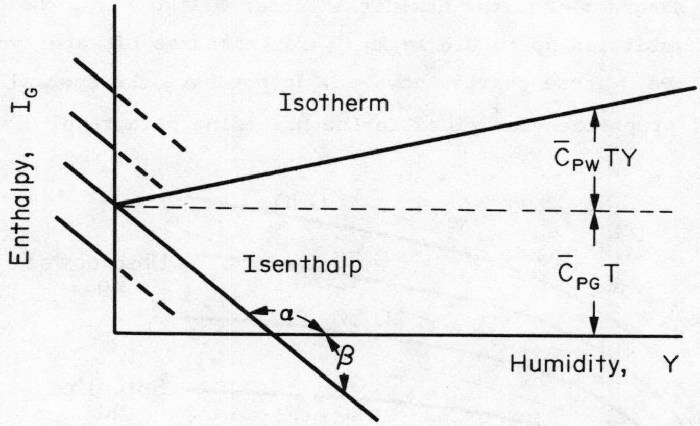

Fig.2.5 Basic geometry of enthalpy-humidity chart.

The isenthalpic lines are inclined to the horizontal axis at an angle α, usually about 135°, and there is thus an angle of approximately 45° between the isenthalpic and isohumid lines in this case. Threlkeld[22] discusses the selection of the angle α and scale factors for the co-ordinates.

One particular choice of co-ordinate system results in a simple pattern for the isotherms. In the unsaturated-air region, there is approximately a linear relationship between the enthalpy I_G, the humidity Y and the temperature T, whenever zero temperature is taken to be the datum for zero enthalpy, since from equation 2.24

$$I_G = \bar{C}_{PG}T + \Delta H_{VO}Y + \bar{C}_{PW}TY \qquad (2.24a)$$

If the isenthalpic lines are so inclined that they fall with a slope $\tan \beta = -\Delta H_{VO}$, then the isotherm cuts the ordinate axis at a value $\bar{C}_{PG}T$ (the dry-gas enthalpy) and the isotherm rises with a positive gradient of $\bar{C}_{PW}T$.

The elements of an enthalpy-humidity chart for a moist gas are shown in Fig.2.6 Contours of relative humidity ψ are plotted, the region above the curve ψ = 1 corresponding to an unsaturated moist gas, the region below the curve to fogging conditions. Isotherms, which cross over the relative-humidity curves, are shown dashed. Placed in the Appendix are more accurate diagrams, one for humidities under 0.1 kg kg^{-1}, the other encompassing humidities up to 0.6 kg kg^{-1}, for mixtures of water vapour in air at 100 kPa. These charts, however, do not have the special thermophysical properties described in the preceding paragraph.

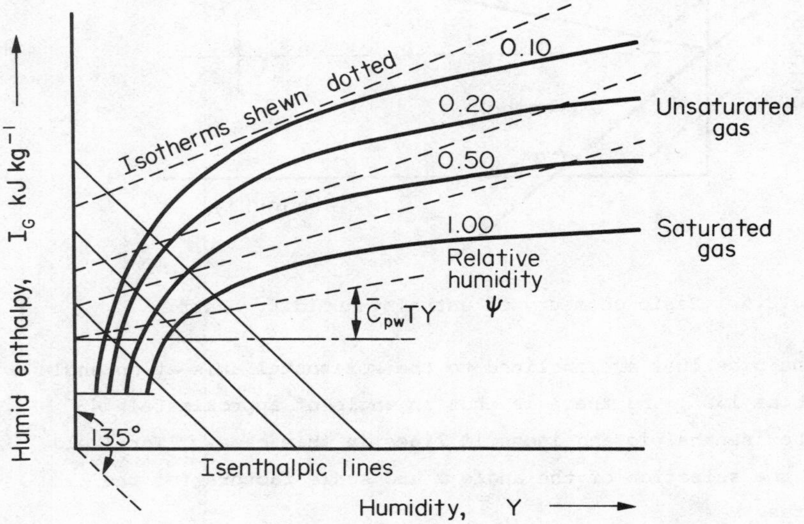

Fig.2.6 An enthalpy-humidity diagram for a moist gas

A humidity chart, such as that depicted in Fig.2.6, is not only limited to a specific system of gas and vapour, but is also limited to a particular total pressure, as inspection of equation 2.10 shows.

Re-arrangement of equation 2.10 yields the expression

$$Y = \left(\frac{M_W}{M_G} \right) \frac{p_W^o}{(P/\psi - p_W^o)} \tag{2.26}$$

At the same vapour pressure p_W^o (and thus temperature T), the humidity at a different total pressure will thus be the same provided the same ratio P/ψ prevails. Therefore the relative-humidity contours at one pressure P can be transposed to those at another P' through the expression

$$\psi' = (P'/P)\,\psi \tag{2.27}$$

So, for example, the saturated-humidity curve at 200 kPa corresponds to the 50% relative-humidity contour at 100 kPa.

The thermophysical properties of air may be used with reasonable accuracy for diatomic gases generally, so that charts developed for mixtures in air can be used to describe the properties of the same moisture vapour in a gas such as nitrogen.

Charts other than those for moist air are often required in the drying of fine chemicals and pharmaceutical products. Kienzle[10] presents data for a few systems, including air-benzene and carbon dioxide - methanol mixtures, while a small-scale diagram for nitrogen-isopropanol mixtures is given by Moller and Hansen[18]. Other charts for nitrogen-methanol and nitrogen-toluene mixtures at 100kPa may be found in the Appendix.

Sometimes charts are prepared with the humidity scaled in terms of the mass (or mole) fraction of moisture. Such charts cover mixtures over the whole range in composition from a dry gas to a pure vapour, and are particularly useful in following operations in which the vapour content is high, such as condensation in the presence of an uncondensable gas. In drying, the amount of moisture vapour is normally much less than the associated gas, so these charts offer no advantages.

2.4 Moisture Content of Solids

A wet solid is usually swollen compared with its condition when free of moisture. For this reason, it is not convenient to express the moisture content of a wet solid in terms of volumetric concentrations because of the uncertainty of what volume basis is being employed. A similar convention to that adopted for defining moisture concentrations in gases is used instead. The moisture content X of a solid is therefore defined as the mass ratio of moisture to bone-dry material in the solid. The usefulness of employing this dimensionless ratio, which like the humidity has a dry-substance basis, is illustrated in the following worked example.

Example 2.5 Wholemilk with a solids content of 10 per cent by weight is concentrated in a multiple-effect evaporator to 48 per cent solids and spray-dried to yield a powder having a moisture content of 4 per cent (dry basis). What fraction of the total moisture to be removed is evaporated before the drying stage ?

Moisture content of feed wholemilk = 90/10 = 9.000 kg kg^{-1}

Moisture content of evaporated milk = 52/48 = 1.083 kg kg^{-1}

Moisture content of milk powder = 0.040 kg kg^{-1}

$$\therefore \frac{\text{moisture removed in evaporation}}{\text{total moisture removed}} = \frac{9.000 - 1.083}{9.000 - 0.040} = 0.884$$

i.e. 88.4 % of the moisture is taken off in the evaporator. However, the spray-drying plant is visually the dominant item in a milk-powder factory, which perhaps illustrates the relative difficulty in recovering the last amounts of moisture.

Sometimes one sees a so-called wet-basis moisture content W, which is the moisture/solid ratio based on the total mass of wet material. For instance, a moisture content (wet basis) of 20 per cent is equivalent to a dry-basis moisture content, used above, of 20/0.8 = 25 per cent. The two moisture contents are related by the expression

$$X = W/(1-W) \tag{2.28}$$

The ratio W varies between 0 and 1, but X can become infinitely large as the solids content vanishes.

Moisture associated with a wet material may be freely attached, or may be bound to it so that no longer does the moisture exert its full vapour pressure. The work per unit quantity of substance (in J mol^{-1}) in driving off moisture can be expressed in terms of the relative vapour pressure p_W/p_W^o for an isothermal, reversible process without change of composition:

$$\tilde{W} = -\Delta G = -RT\ln(p_W/p_W^o) = -RT\ln\psi \tag{2.29}$$

Therefore, as the free energy of the moisture-solid bond rises, so the corresponding relative humidity ψ falls. There is thus an equilibrium moisture content for a given relative humidity. Very wet material, on exposure to an atmosphere of fixed relative humidity, will lose moisture until remaining is that with a minimum bond energy given by equation 2.29. Further moisture can only be removed by reducing the relative humidity, or by doing work as in mechanical expression, and a completely dry material can only exist in a moisture-free environment.

Bond energies can be less than 100 J mol^{-1} for moisture held in "large" pores and spaces (greater than 0.1 μm diameter) within the material. On the other hand, chemically attached moisture, such as water of crystallization, can have a bond energy of order 5000 J mol^{-1}.[8] Clearly, the removal of lightly bound moisture will not change the material's form and properties significantly, but the removal of strongly bound moisture will induce changes in the character of the solid. Broadly, we can classify materials as being colloidal or capillary-porous, however materials, particularly those of a biological origin, often have complex structures that defy simple classification.

1. Colloidal materials. Clays provide examples of materials which are colloidal and infinitely swelling. They change in properties from a fluid paste at high moisture contents to a rock-hard solid when very dry. The partial pressure of moisture p_W is always less than the vapour pressure p_W^o over the whole moisture-content range, although the relative humidity is very close to unity at moisture contents above 0.3 kg kg^{-1}.[9] A material such as gelatine has a limited swelling range. Therefore above a certain moisture content the excess moisture is unbound and so exerts its full vapour pressure. This limiting condition of moistness is called the maximum hygroscopic moisture content. All moisture is bound at lesser amounts and the corresponding relative humidity is less than unity.

2. Capillary-porous materials. A capillary-porous material is a non-swelling substance in which moisture moves by capillarity within the pores and voids between the solid particles of the body. As long as these spaces between the solid particles are large enough, no significant lowering of the vapour pressure of moisture takes place. Such a material is said to be non-hygroscopic, and examples include crushed stone and plastic pellets. The lowering of the vapour pressure, due to the energy associated with the curvature of the moisture interfaces within the

curvature of the moisture interfaces within the substance, is given by the Kelvin equation

$$- \Delta G = - RT \ln \psi = (2 \sigma V_L \cos \theta)/r \qquad (2.30)$$

where V_L is the specific molar volume ($m^3\ mol^{-1}$) of moisture liquid, σ is the surface tension, r is the radius of curvature and θ is the angle of wetting. For water at 50^oC, equation 2.30 predicts that ψ is only 0.1 % less than 1 when the radius is 1 μm.

A capillary-porous material thus becomes hygroscopic when only micropores smaller than 1 μm in radius remain filled with moisture which has formed weak links with the confining capillary walls. Such hygroscopic behaviour is seen with porous materials such as charcoal and coke. Materials like red brick, having a wide range of capillary sizes, can thus soak up much more moisture than that which is bound to the walls, as Table 2.6 demonstrates.

TABLE 2.6 MOISTURE-UPTAKE CHARACTERISTICS OF A FEW POROUS MATERIALS (From Luikov[15])

Material	Porosity	Maximum sorption		Full saturation		Saturation / sorption ratio
		Filling of pores %	$100\ W^*_{max}$ %	Filling of pores %	$100 W_{sat}$ %	W_{sat}/W^*_{max}
Silicate brick	0.31	6.4	2.0	80	24.8	12.4
Red brick	0.36	2.9	1.09	89	34.0	31.2
Slag concrete	0.65	5.2	3.4	-		-
Foamed glass	0.85	2.2	1.9	53	53.5	28.2

3. Complex cellular materials. A biological material such as timber has a complex cellular-capillary structure which is not dimensionally stable on the removal of moisture. Even with softwoods such as pines, which have a comparatively regular anatomy, the moisture-solid behaviour is complex. When a green plank of a softwood timber is being dried out, for instance, the large interconnecting vessels in the material empty from the exposed surfaces inwards through cell cavities and other openings until the wood fibres alone are saturated. Thenceforward in the drying

process, the cellular moisture in the fibres is driven off, and the whole material shrinks. The maximum hygroscopic moisture content appears to be almost co-incident with this fibre-saturation point.[5]

The foregoing considerations lead to curves for the isothermal variation of the relative humidity with the equilibrium moisture content as sketched in Fig.2.7. The maximum hygroscopic moisture content is ill-defined as each curve asymptotically approaches the line $\psi = 1$. For precise work, we must define this moisture content in terms of a specific relative humidity close to, but not equal to, 1.

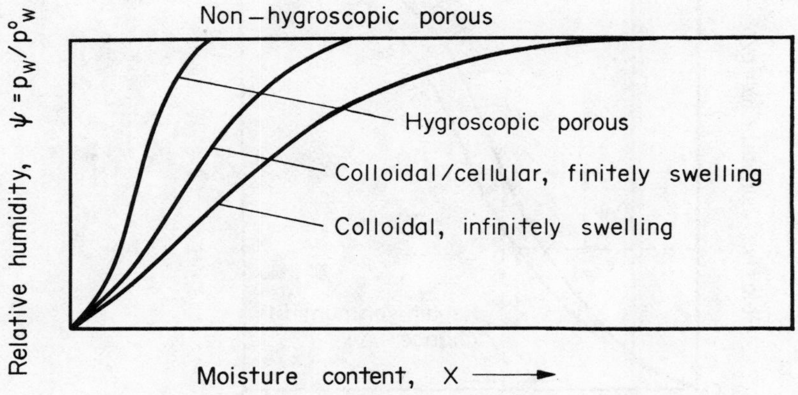

Fig. 2.7 Isothermal variation of relative humidity of moisture vapour for various solid materials under drying conditions

An exception to this kind of behaviour is that of inorganic crystalline solids which have multiple hydrates. The relative humidity falls stepwise with loss of moisture as each hydrate in turn disappears.

2.5 Moisture Isotherm

Curves, such as those drawn up in Fig.2.7, are normally called <u>moisture isotherms</u>, being derived under isothermal conditions. The amount of moisture retained by a body at a given temperature and relative humidity depends upon whether the hygrothermal equilibrium has been approached by wetting (adsorption) or by drying (desorption). There is a closed-loop hysteresis and the desorption isotherm always shows the larger equilibrium moisture content at a given relative humidity. In drying operations, we are interested in desorptive conditions. However, sometimes we are concerned with cases in which moisture is taken up by dried solids on storage in damp air when the adsorption isotherm would be appropriate. Illustrated in Fig. 2.8 is the change of moisture content for a sudden increase of $\Delta\psi$ in the relative humidity of the surroundings.

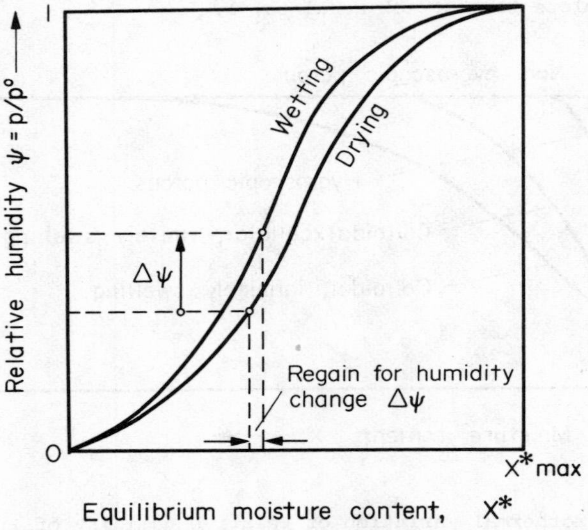

Fig. 2.8 Sorption behaviour of a hygroscopic solid

Some isotherms for a few substances are drawn in Fig.2.9 from data obtained mainly at room temperature. These curves may be compared with those sketched in Fig. 2.7. Despite differences in material type, each of the isotherms of Fig. 2.9 is sigmoid with a stretched exponential tail at the higher moisture contents.

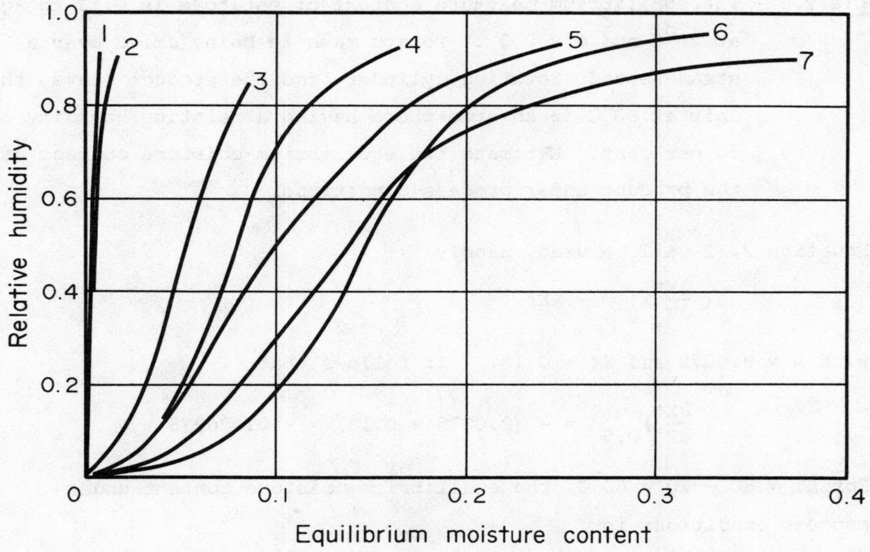

Fig. 2.9 Moisture isotherms for a few substances mainly at room temperature. After Keey[8].

1. Asbestos fibre. 2. Polyvinyl chloride powder ($50^{\circ}C$).
3. Wood charcoal. 4. Kraft paper. 5. Jute.
6. Wheat. 7. Potatoes.

Temperature variation. Although the moisture isotherms pertain to a particular temperature, the variation of equilibrium-moisture content for small changes in temperature (of order $10^{\circ}C$ and less) is usually ignored. To a first approximation, the temperature coefficient of the equilibrium-moisture content is proportional to the moisture content at a given relative humidity:[12]

$$\left(\frac{\partial X^*}{\partial T} \right)_{\psi} = - AX^* \tag{2.31}$$

The coefficient A lies between 0.005 and 0.01 K^{-1} for relative humidities between 0.1 and 0.9 for materials such as natural and synthetic fibres, wood and potatoes. The value of A to be taken increases with relative humidity, so that at $\psi = 0.5$ there is about a 0.75 per cent fall in moisture content for each degree rise in temperature. The extent of sorption hysteresis also becomes smaller with increasing temperature.

<u>Example 2.6</u> The equilibrium-moisture content of potatoes is 0.13 kg kg^{-1} at $20^{o}C$ and $\psi = 0.5$. Potato mash is being dried over a steam-heated, rotating cylinder, and the product leaves the unit at $80^{o}C$ in an atmosphere having a relative humidity of 50 per cent. Estimate the equilibrium-moisture content of the product under process conditions.

Equation 2.31 will be used, namely

$$\left(\frac{\partial X^*}{\partial T}\right)_{\psi} = - AX^*$$

with A = 0.0075 and X* = 0.13. It follows that

$$\left(\frac{\partial X^*}{\partial T}\right)_{0.5} = - (0.0075 \times 0.13) = - 0.000975$$

For $\Delta T = 80 - 20 = 60^{o}C$, the equilibrium-moisture content under process conditions is

$$X^*_{80} = 0.13 - (0.000975 \times 60) = 0.0715 \text{ kg kg}^{-1} .$$

Görling[4] finds values of 0.065 ~ 0.07 kg kg^{-1} at this temperature. This calculation illustrates that under normal drying conditions the equilibrium-moisture content of a solid will be considerably less than the value found at room temperature at the same relative humidity.

<u>Pressure variation</u>. For a given amount of moisture, at equilibrium, the relative humidity of the moisture vapour is not only temperature-dependent, but also hinges upon the external pressure. The effect is small over the range of pressures encountered in drying practice, and may be neglected.

2.6 Correlation of Moisture-equilibrium Data

Most laboratory data for moisture-desorption equilibria are recorded at room temperature, but other data at other temperatures do exist. Tabular data for selected substances are given by Keey[8], while Krischer[12] presents an extensive graphical compilation of moisture isotherms.

Since the moisture isotherms are predominantly exponential over most of the moisture-content range, it should be possible to correlate isotherms by expressions of the type

$$-a \ln \psi = 1/X^* + b \qquad (2.32)$$

where the coefficients a and b are specific to given substance. When $\psi = 1$, $\ln \psi = 0$, so $b = -1/X^*_{max}$, where X^*_{max} is the maximum hygroscopic moisture content (see Fig. 2.10). Equation 2.32 can thus be re-written as

$$\frac{1}{X^*} = \frac{1}{X^*_{max}} - a \ln \psi \qquad (2.33)$$

Comparison of equation 2.33 with 2.29 suggests that the coefficient a is directly proportional to RT.

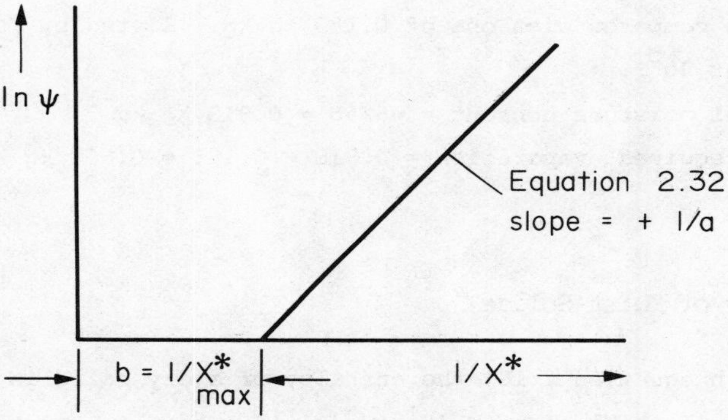

Fig. 2.10 Possible correlation of moisture-equilibrium data

Luikov[16] claims that equation 2.33 is good over the range in relative humidity from 0.1 to 1 and presents values for the coefficient a and the maximum hygroscopic moisture content X^*_{max} for 57 materials at temperatures ranging between -40°C and $+40^{\circ}$C, the temperature range reflecting perhaps the rigours of the Russian climate. Equilibrium-moisture contents at higher temperatures may be estimated roughly by the method outlined in Example 2.6. Luikov's values for the parameters a and X^*_{max} are listed in the Appendix. The following example illustrates the use of equation 2.33 to estimate equilibrium-moisture contents when raw data are not available.

Example 2.7 Scoured fleece wool, after passing through the hot squeeze rolls, is delivered to the dryer at 45 per cent water content (i.e. wet-basis moisture content). The dried wool should be in equilibrium with ambient air of 60 per cent relative humidity. Estimate the evaporation needed per unit mass of dry stuff.

The Luikov parameters† for wool at an unspecified temperature are: $a = 6.7$, $X^*_{max} = 0.312$. From equation 2.33

$$\frac{1}{X^*} = \frac{1}{X^*_{max}} - a \ln \psi$$

$$= \frac{1}{0.312} - 6.7 \ln 0.6 = 6.63$$

$$\therefore X^* = 0.151 \text{ kg kg}^{-1}$$

This value compares with one of 0.153 kg kg^{-1} listed by Keey[9] for wool at 30°C.

The initial moisture content = 45/55 = 0.818 kg kg^{-1}

Thus the required evaporation = 0.818 - 0.151 = 0.667 kg kg^{-1} .

2.7 Enthalpy of Moist Solids

By analogy with equation 2.18, the enthalpy of a dry solid is defined by the product of the heat capacity or specific heat C_S and the temperature excess θ, i.e.

$$H_S = C_S\theta \tag{2.34}$$

The specific heat C_S ranges from about 0.8 kJ $kg^{-1}K^{-1}$ for minerals up to about 2.7 kJ $kg^{-1}K^{-1}$ for timber. Values of C_S for some solids of interest in drying operations are listed below in Table 2.7 from data assembled by Perry.[20]

† Footnote: Luikov's original table[16] gives a = 67, but clearly, by comparison with his data for woollen thread, there is a typographical error.

TABLE 2.7 HEAT CAPACITIES OF SOME SOLIDS ($\bar{c}_S$)

Substance	$\bar{c}_S$ / kJ kg^{-1}K^{-1}	Temperature range / ^oC
brickwork	~ 0.8	
cellulose	1.33	
clay	0.937	
concrete	0.653	21 - 156
earthenware (green)	0.778	20 - 100
gypsum	1.084	16 - 46
limestone	0.908	
sand	0.799	
sugar (cane)	1.26	22 - 51
wood (oak)	2.38	
wool	1.36	

An expression similar to that of equation 2.34 is appropriate to define the enthalpy of moisture within a body provided it is non-hygroscopic. Should the moisture be bound to the host material, then the enthalpy of the attached moisture is smaller than that of the pure liquid by the energy of this binding. Let I_L be the moisture enthalpy for unit mass of dry stuff, then

$$I_L = \left[c_{LW}\theta - \Delta H_W\right] X \tag{2.35}$$

where ΔH_W is the so-called enthalpy of wetting which includes all forms of enthalpy change such as heats of sorption, hydration and solution. The enthalpy of the moist solid on a unit dry-matter basis is thus

$$I_S = H_S + I_L = c_S\theta + \left[c_{LW}\theta - \Delta H_W\right] X \tag{2.36}$$

The heat capacity of air-free water varies from 4.2177 kJ kg^{-1}K^{-1} at 0^oC to 4.2160 kJ kg^{-1}K^{-1} at 100^oC with a minimum value which is only slightly smaller (4.1782) at 35^oC. Thus the heat capacity of liquid water is virtually independent of temperature in this range and is also little affected by saturation with air at 100 kPa.[23] The heat capacities of organic solvents are considerably less, and a few have been listed in Table 2.5 as $\vec{C}_{LW}$.

Enthalpy of wetting. The lowering of the vapour pressure of moisture, when bound, can be related to a "suction pressure" P_S holding back the moisture in the material. By considering isothermal free-energy changes, we get the Gibbs equation

$$VdP = V_L dp_W \tag{2.37}$$

where V and V_L are the specific molar volumes ($m^3 mol^{-1}$) of the moist gas and the liquid moisture respectively. If the vapour gas mixture behaves as an ideal gas, then

$$V = RT/P \tag{2.38}$$

On inserting equation 2.38 into 2.37 and integrating, we find

$$- \ln \frac{p_{W1}}{p_{W2}} = \frac{V_L(P_1 - P_2)}{RT} \tag{2.39}$$

whenever the molar volume of the liquid moisture is independent of pressure. The moisture will exert its full vapour pressure when the suction pressure is zero, so we may put $p_{W2} = p_W^o$ for $P_2 = 0$. Equation 2.39 can then be re-arranged to an expression for the relative humidity $\psi = p_W/p_W^o$ in terms of the suction pressure P_S :

$$- \ln \psi = V_L P_S/RT \tag{2.40}$$

Since there is no change of internal energy at constant temperature, the enthalpy of wetting ΔH_W may be substituted for the product $V_L P_S$ from the definition of enthalpy (equation 2.12); thus

$$- \ln \psi = \Delta H_W/RT \tag{2.41}$$

It follows from equation 2.41 that

$$\left.\frac{d(\ln\psi)}{d(1/T)}\right|_X = - \frac{\Delta H_W}{R} \tag{2.42}$$

Therefore, if we plot the natural logarithm of the relative humidity of the reciprocal absolute temperature at a given moisture content, then the gradient yields the enthalpy of wetting or $P_S V_L$ for that moisture content. To do this, we need knowledge of the moisture isotherms at various temperatures. The method of estimation is illustrated in Fig. 2.11.

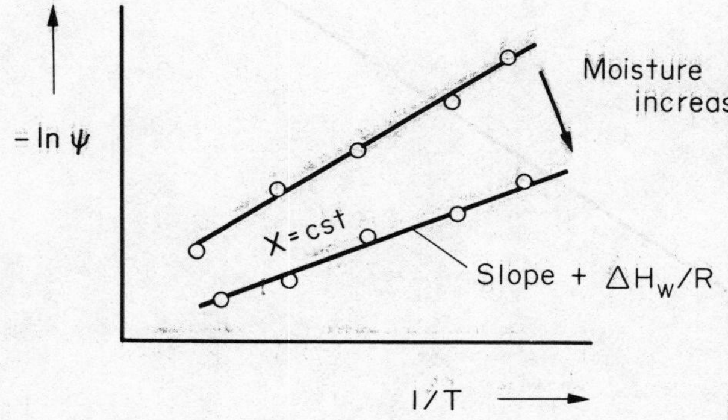

Fig. 2.11 Determination of enthalpy of wetting from moisture isotherms

It is more useful to know the mean enthalpy of wetting over a moisture-content range rather than the value so calculated which pertains to a specific moisture content. This mean, or integral, enthalpy is calculated from the point values by integration :

$$\overline{\Delta H_W} = \frac{1}{(X_2 - X_1)} \int_{X_1}^{X_2} \Delta H_W dX \tag{2.43}$$

The following example illustrates this calculation.

Example 2.7. Görling[4] presents moisture isotherms for potatoes over a range in temperature from 0 to 100°C. Smoothed values of the relative humidity ψ at an equilibrium moisture content of 0.1 kg kg^{-1} may be read as follows:

T/°C	0	20	40	60	80	100
ψ	0.23	0.31	0.43	0.56	0.68	0.77

From these data calculate the heat required to evaporate 1 kg of moisture at 50°C at a moisture content of 0.1 kg kg^{-1}.

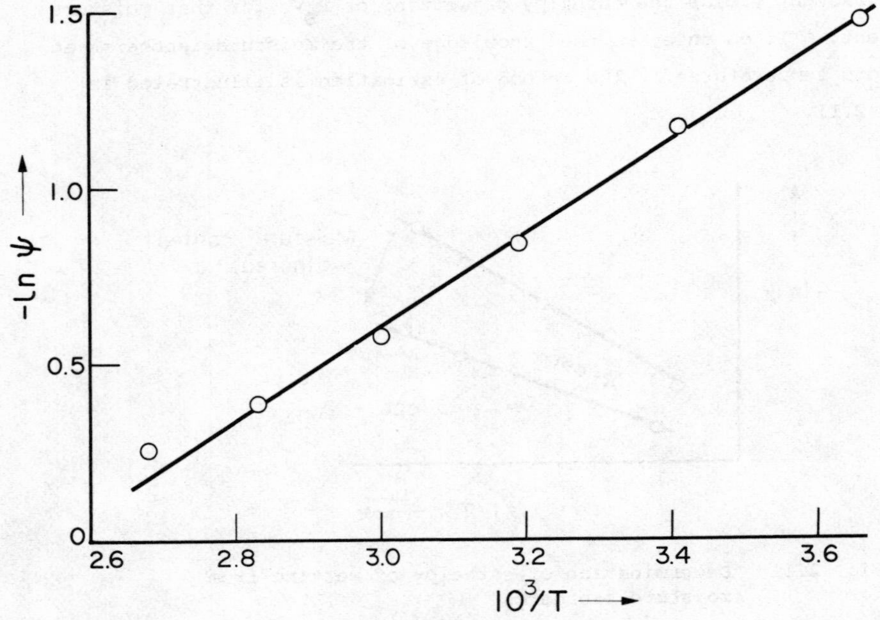

Fig. 2.12 Graph for calculating enthalpy of wetting

We evaluate $-\ln\psi$ and $10^3/T$ (in K^{-1}) from the above data:

$-\ln\psi$	1.47	1.17	0.84	0.58	0.39	0.26
$10^3/T$	3.66	3.41	3.19	3.00	2.83	2.68

The graph of $-\ln\psi$ against $10^3/T$ should yield a straight line of slope $\Delta H_W/10^3R$ (equation 2.42). All the data points do fall rectilinearly (Fig. 2.12), except for the point at 100°C which is probably subject to error. The slope has a value of 1.33 K. Thus

$$\Delta H_W = 1.33 \times 8.314 \times 10^3$$

$$= 11.1 \times 10^3 \text{ J mol}^{-1} \text{ or } 11.1 \text{ kJ mol}^{-1}$$

The molar mass of moisture is 0.018 kg mol^{-1}, so the enthalpy of wetting per unit mass is 11.1/0.018 = 616 kJ kg^{-1}.

The latent heat of vaporization of water at 50°C is 2406 kJ kg^{-1},[17] whence the total heat required for evaporation ΔH is given by

$$\Delta H = \Delta H_V + \Delta H_W$$

$$= 2406 + 616 = 3022 \text{ kJ kg}^{-1}$$

By repeated calculations in the above way, graphs of ΔH_W as a function of moisture content can be drawn up, as shown in Fig. 2.13.

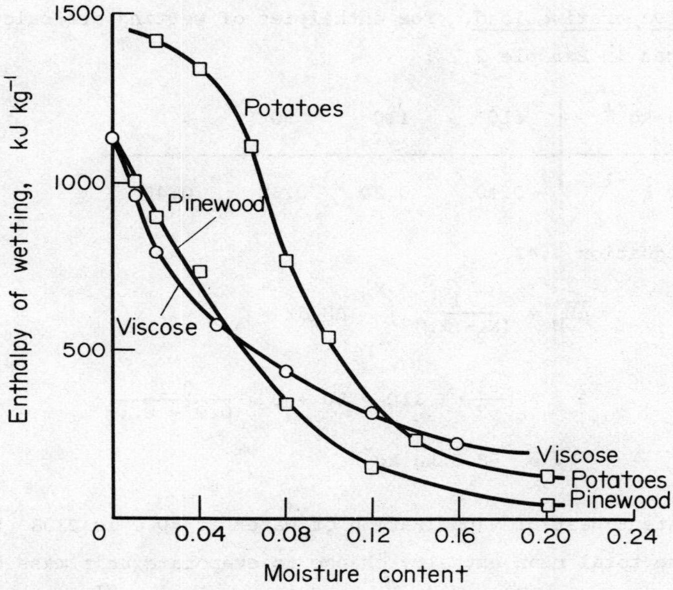

Fig. 2.13 Variation of enthalpy of wetting with moisture content.[8]

Since the enthalpy of wetting is directly related to the vapour-pressure lowering with bound moisture, materials, irrespective of composition, at the same relative humidity will have the same enthalpy of wetting ΔH_W. This observation may be used to estimate ΔH_W for those materials for which only a single isotherm is available.

In the foregoing worked example, the enthalpy of wetting accounts for 20 per cent of the total enthalpy change, but often it is less significant. When slightly hygroscopic materials are dried, the enthalpy needed to sunder the moisture-solid bonds is often negligible compared with that for vaporization, especially if the moisture is water. Moreover, when one considers the whole drying process, the enthalpy of wetting may often play only a minor role, even if the material is highly hygroscopic, as the following example shows.

Example 2.8 A potato mash of moisture content 0.9 kg kg^{-1} is dried down to 0.1 kg kg^{-1} at a mean temperature of 80°C. Estimate the heat required, neglecting losses, per 100 kg of dried product if the feed enters warm at 40°C.

(i) Evaporative load. The enthalpies of wetting are calculated as outlined in Example 2.7 :

ΔH_W/kJ kg^{-1}	616	110	50	~ 0
X / kg kg^{-1}	0.10	0.20	0.30	0.40

From equation 2.43

$$\overline{\Delta H}_W = \frac{1}{(X_2 - X_1)} \int_{X_1}^{X_2} \Delta H_W dX$$

$$= [\frac{616}{2} + 110 + 50 + 0] \frac{0.1}{(0.9 - 0.1)}$$

$$= 58.5 \text{ kJ kg}^{-1}$$

The latent heat of vaporization of water at 80°C is 2308 kJ kg^{-1}, and the total mean enthalpy change to evaporate unit mass of moisture is thus

$$\overline{\Delta H} = \Delta H_V + \overline{\Delta H}_W$$

$$= 2308 + 58.5 = 2366.5 \text{ kJ kg}^{-1}$$

(Note that $\overline{\Delta H}$ is larger than ΔH_V by only 2.5%)

The total evaporation is 0.90 - 0.10 = 0.80 kg per kg dry foodstuff. In 1 kg of product there is 1/(1 + 0.1) = 0.909 kg of dry matter. The total evaporative load is thus

$$2366.5 \times 0.80 \times 100/0.909 = 2.08 \times 10^5 \text{ kJ.}$$

(ii) Sensible-heating load. The specific heat of the dry solid matter is not known, but a value of 1.3 kJ $kg^{-1}K^{-1}$ will be guessed from inspecting the limited data of Table 2.7. From equation 2.36 the enthalpy change per unit mass of dry foodstuff is given by the following expression for $\Delta H_W=0$ (the entering potato mash is non-hygroscopic) :

$$\Delta I_S = \bar{C}_S \Delta T + \bar{C}_{LW} X \Delta T$$

$$= 1.3\ (80-40) + 4.2 \times 0.9 \times (80-40)$$

$$= 203 \text{ kJ kg}^{-1}$$

The total sensible load is thus $203 \times 100/0.909 = 0.22 \times 10^5$ kJ .

(iii) Total heating load. This is the sum of the loads calculated in sections (i) and (ii), that is:

$$\text{total load} = (2.08 + 0.22) \times 10^5 = 2.30 \times 10^5 \text{ kJ} .$$

To obtain this quantity, a figure for $\bar{C}_S$ had to be guessed. The error in this estimate is unlikely to be larger than 0.5 kJ $kg^{-1}K^{-1}$ and is probably much less. Therefore at worst, ΔI_S might be 10 per cent in error from this source, and the total heating load 1 per cent in error. Clearly, some physiochemical data can be very rough indeed without loss of overall accuracy.

2.8 Representation of Moisture Equilibria on Humidity Charts

If a number of sorption isotherms at various temperatures are available, then these equilibria can be cross-plotted on an enthalpy-humidity chart. Such a representation will be useful in following the state path of the moist gas adjacent to the solids being processed in a dryer, as will be shown later.

When the moisture content is equal to or is above the maximum hygroscopic value X^*_{max} , the equilibrium-air conditions at the exposed surface of the solids will be found on the moisture-saturation curve, $\psi = 1$. At a given moisture content in the hygroscopic region ($X < X^*_{max}$), one can find a set of values for the relative humidity at various values of the temperature excess θ and thus enthalpy I_S. The points (I_S, ψ) can then be located on an enthalpy-humidity chart, as illustrated in Fig.2.14.

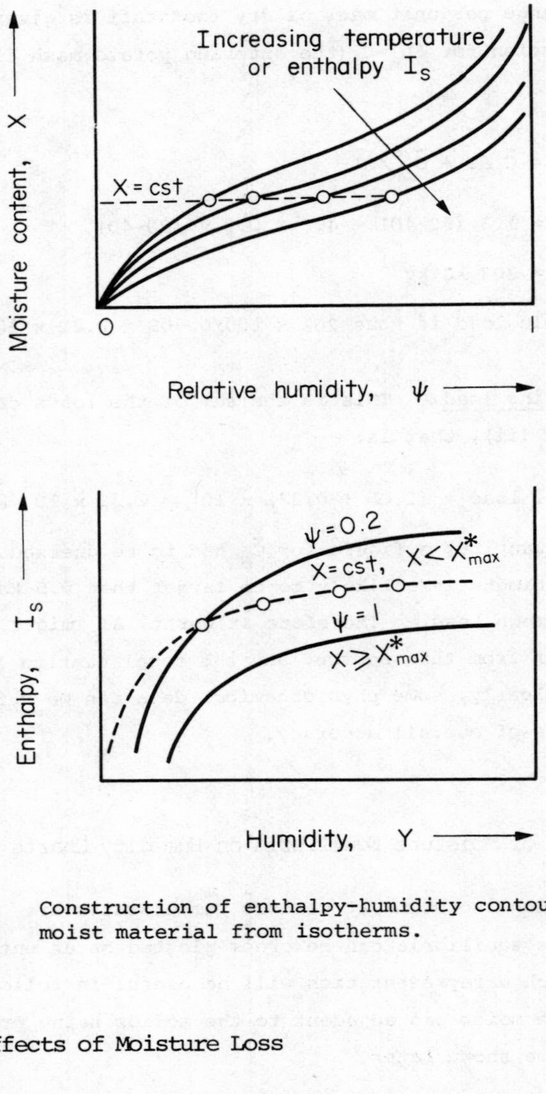

Fig.2.14. Construction of enthalpy-humidity contour for a moist material from isotherms.

2.9 Side-effects of Moisture Loss

The removal of moisture from a wet material may result in physical, chemical or biological changes which can downgrade the quality of the dried product. The need to avoid undesirable changes in properties often severely circumscribe the possible processing conditions, and thus the choice of drying technique.

<u>Drying stresses</u>. The dried product has almost always contracted in size compared to the original material, but the shrinkage may not have taken place throughout the whole drying process. Should this shrinkage be anisotropic or hindered, the material will develop stresses due to the local movement at any part of the body being different from the free shrinkage. A material, if stretched beyond its yield point, retains stresses at the end of drying. Since there can be no net load, the area-smoothed stress is zero, and parts of the body are in tension while others are compressed

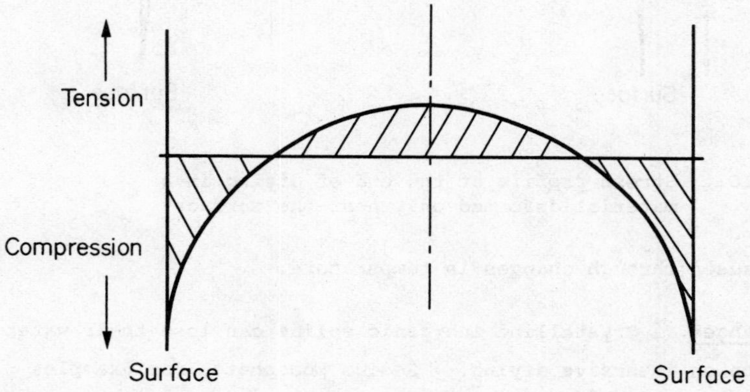

Fig. 2.15 Stress profile in a permanently deformed material at the end of drying.

Such stress patterns can lead to case-hardening of the surface or cracking along planes of weakness, particularly when only the surface layers are deformed permanently, as illustrated in Fig. 2.16. Timber, which shrinks only at low moisture contents, is especially prone to split and show surface "checks" unless great care is exercised in the later stages of drying. A stress profile, similar to that drawn in Fig. 2.16, will occur during the course of drying when an elastic material at the surface shrinks onto a plastic core.[1] Pasty materials such as macaroni dough can show this stress pattern on rapid drying.

In general, the faster the drying, the more likely are such degradative phenomena to occur. An introduction to estimating stresses induced by drying is given by Keey.[8] One might also expect to see certain similarities between the appearance of drying stresses and stresses in

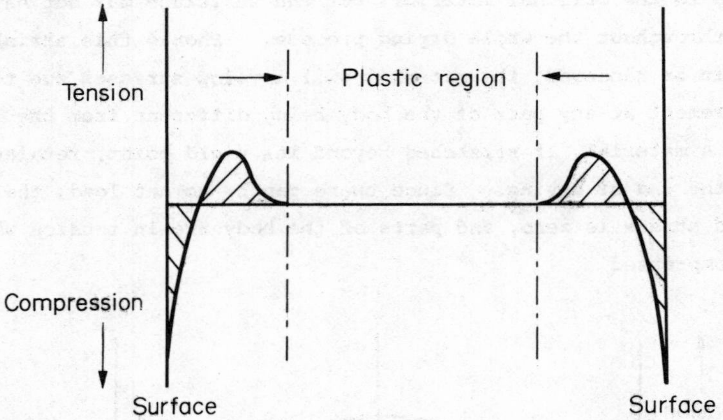

Fig. 2.16. Stress profile at the end of drying in a material deformed only near the surface.

materials caused through changes in temperature.

Chemical changes. Crystalline inorganic solids can lose their water of crystallation on intensive drying. Sodium phosphate, for example, besides existing in anhydrous form, can hydrate with two, seven and twelve molecules of water for each molecule of phosphate. Sometimes complete or partial dehydration is sought, but often this is unwanted. Crystals are often sold on appearance, so highly undesirable are visible flecks of anhydrous material or bits of unwanted crystal forms. In another instance, the formation of anhydrite in gypsum plasterboard by overdrying leads to a loss of cohesiveness which renders the board faulty in service.

Organic materials often oxidise in the presence of air, even at moderate temperatures. Thus inert gases such as nitrogen, despite their inconvenience and cost, are sometimes used as drying media for handling pharmaceutical and similar heat-sensitive materials of high cost. Foodstuffs, in particular, are susceptible to oxidative damage which often shows up as scorching of the product. Proteins can become denatured, so that the dried product is less attractive to the consumer as it is more difficult to rehydrate in water.

Biological changes. Thermal drying usually irreversibly affects the elasticity of cell walls in biological materials, and the dried tissues do not fully recover their original properties on rehydration. On the other hand, the immunity of foodstuffs to microbiological attack is much enhanced by the removal of moisture. The growth of bacteria is impossible below a relative humidity of 90 per cent, while most fungus growth is checked at relative humidities below 80 per cent. Nevertheless, the killing of micro-organisms by heat is less effective as the relative humidity is reduced (see Fig. 2.17). Bacteria, moreover, have an optimum temperature for growth and resistance to heat. Some psychrophilic bacteria grow at temperatures just above the freezing-point of water, while the so-called thermophilic organisms can survive temperatures above 65°C. Such thermoduric bacteria can be troublesome in making dried food products with a low bacterial count.

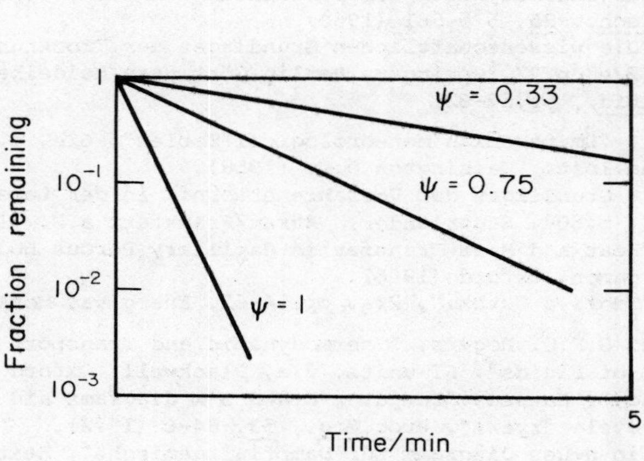

Fig. 2.17 Thermal annihilation of micro-organisms. After Loncin[14].

By contrast, the yellowing of wholemilk powder, which is caused by fat-splitting enzymes known as lipases, appears to be most susceptible at an intermediate relative humidity of about 70 per cent.[14]

REFERENCES

1. Beuscher, H. in Kröll, K., "Trockner und Trocknenverfahren", p.86, Springer, Berlin/Göttingen/Heidelberg (1959).
2. Callender, G.S., Variations of the amount of carbon dioxide in different air currents, Quart.J.Roy.Meteorol.Soc., 66, 395 (1940).
3. Goff, J.A., Standardization of the thermodynamic properties of moist air, Trans.ASHVE, 55, 459-484 (1949).
4. Görling, P., "Untersuchungen zur Aufklärung des Trocknungsverhaltens pflanzlichen Stoffe", VDI Forsch.Heft 458 (1956).
5. Hawley, L.F., "Wood-liquid relations", p.8, Tech.Bul.248, U.S. Dept. Agric., Washington D.C. (1931).
6. Hilsenrath, J. et al., "Tables of Thermal Properties of Gases", Natl. Bur.Std. Circ. 564, 74, U.S. Govt.Printing Office, Washington D.C. (1955).
7. Hougen, O.A., Watson, K.M. and Ragatz, R.A., "Chemical Process Principles", 2/e, 1, p.257, Wiley, New York (1954).
8. Keey, R.B., "Drying Principles and Practice", Chapter 2, Pergamon (1972).
9. Keey, R.B., ibid., pp.342-3.
10. Kienzle, K., "Zur Thermodynamik der Verdunstung von Flüssigkeiten", Chem.Ing.Tech., 25, 575-581 (1953).
11. Krischer, O., "Die wissenschaftlichen Grundlagen der Trocknungstechnik", 2/e, p.22, Springer, Berlin/Göttingen/Heidelberg (1962).
12. Krischer, O., ibid., pp.52-62.
13. List, R.J., ed., "Smithsonian Meteorological Tables", 6/e, p.340, Smithsonian Inst., Washington D.C. (1958).
14. Loncin, M., "Die Grundlagen der Verfahrenstechnik in der Lebensmittelindustrie", p.608, Sauerländer, Aarau/Frankfurt a.M. (1969).
15. Luikov, A.V., "Heat and Mass Transfer in Capillary Porous Bodies", p.199, Pergamon, Oxford (1966).
16. Luikov, A.V., "Teoriya Sushki", 2/e, pp.54-62, Energiya, Moskva (1968).
17. Mayhew, Y.R. and G.F.C. Rogers, "Thermodynamic and Transport Properties of Fluids", SI units, 2/e, Blackwell, Oxford (1971).
18. Moller, J.T. and O. Hansen, "Computer-drawn H-x diagrams aid design of closed-cycle dryers", Proc.Eng., 53, 84-6 (1972).
19. Mollier, R., "Ein neues Diagramm für Dampfluftgemische", Zeits.VDI, 67, 869-872 (1923).
20. Perry, R.H. and C.H. Chilton, ed., "Chemical Engineers' Handbook", 5/e, p 3-136, McGraw-Hill, New York (1973).
21. Perry, R.H. and C.H. Chilton, ed., ibid., pp 3-226 - 3-350.
22. Threlkeld, J.L., "Thermal Environmental Engineering", 2/e, p.180, Prentice-Hall, New York (1970).
23. Weast, R.C., ed., "Handbook of Chemistry and Physics", 48/e, pp.D94-5, Chemical Rubber Co., Cleveland, Ohio (1967-8).
24. Weast, R.C., ibid., pp.D121-140.

Chapter 3

HEAT AND MASS BALANCES

3.1 State Changes in Moist Air

The air within drying equipment undergoes substantial changes in temperature and humidity from place to place, and thus following changes in hygrothermal conditions is of prime importance in assessing the performance of dryers. The enthalpy-humidity chart, which graphically presents the laws of conservation of mass (humidity) and of energy (enthalpy) provides a useful means of demonstrating these changes.(1)

Fogging: If a moist gas at state A (T_G, Y_G) is cooled, the state path of the gas follows a vertical line in the direction of the humidity axis on the enthalpy-humidity chart, see Fig. 3.1, as there is neither gain nor loss of moisture. At state B (T_D, Y_G), the dewpoint, the air is saturated and moisture is about to precipitate. On further cooling to state C (T_F, Y_G), the air becomes foggy as moisture droplets condense out. The liquid-air mixture is composed of saturated gas at state D (T_F, Y_F) and liquid moisture. States C and D lie on the same isenthalpic line, and the amount of dew formed is ($Y_G - Y_F$) per unit mass of dry air.

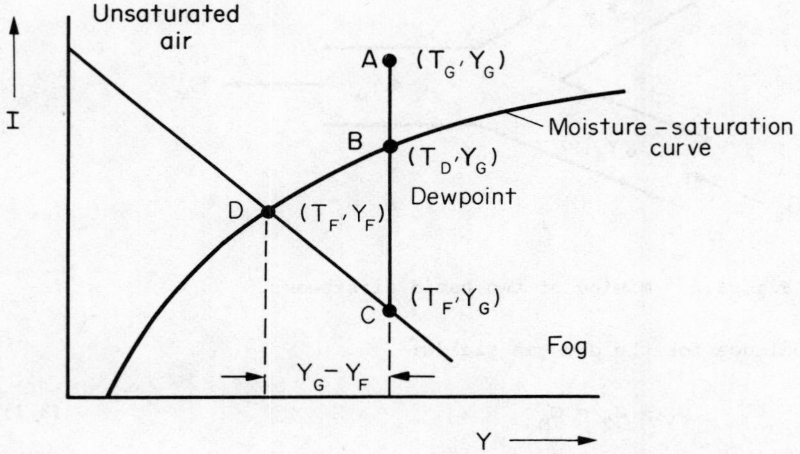

Fig. 3.1 Fog formation on cooling moist air.

The following example illustrates the numerical magnitude of the quantities involved.

<u>Example 3.1</u> Air at 100°C and a humidity 0.1 kg kg^{-1} is cooled. What is the dewpoint temperature, and how much dew would condense if the moist air were cooled to 30°C ?

From the enthalpy-humidity chart, the dewpoint temperature (T_D) is 52°C. At T_F = 30°C the saturation humidity (Y_F) is 0.0276 kg kg^{-1}.

$\therefore$ Dew formed = $(Y_G - Y_F) = (0.1 - 0.0276) = 0.0724$ kg kg^{-1} .

Thus the wetness fraction of the fog is 0.0724/1.0724 = 0.0675 .

The formation of fog in dryers is always undesirable because the wet condensate can corrode metal surfaces or dirty droplets can trickle back onto the drying goods. Thus particular attention has to be paid to conditions inside the exhaust ducting which can easily cool down below the dewpoint of the offgases.

<u>Airstream mixtures.</u> Suppose two airstreams, carrying respectively G_1 and G_2 kg s^{-1} of dry air with corresponding humidities Y_1 and Y_2 , are mixed together to form a common stream of dry-mass flow G_M and humidity Y_M.

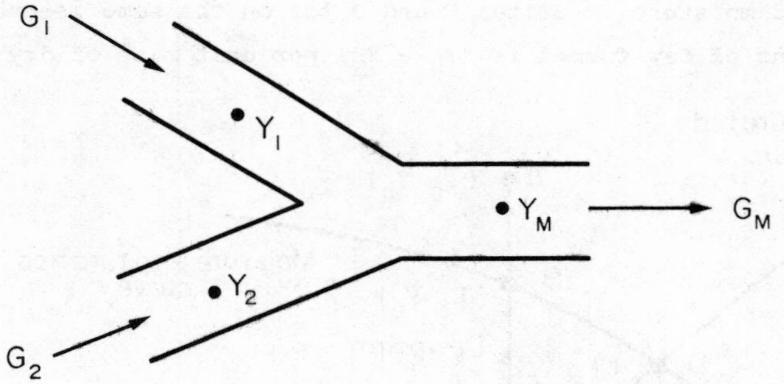

Fig. 3.2 Mixing of two humid airstreams.

A mass balance for the dry gas yields:

$$G_1 + G_2 = G_M \tag{3.1}$$

The corresponding mass balance for the moisture vapour gives:

$$Y_1G_1 + Y_2G_2 = Y_MG_M \tag{3.2}$$

By combining these equations, we get an expression for the humidity Y_M of the emergent mixture:

$$Y_M = \frac{G_1Y_1 + G_2Y_2}{G_1 + G_2} \tag{3.3}$$

If the mixing is adiabatic, enthalpy is also conserved and a balance, analogous to equation 3.2, may be struck for the enthalpies of the streams, i.e.

$$I_1G_1 + I_2G_2 = I_MG_M \tag{3.4}$$

It thus follows that the enthalpy of the mixture is given by

$$I_M = \frac{G_1I_1 + G_2I_2}{G_1 + G_2} \tag{3.5}$$

which is analogous in form to equation 3.3.

Equations 3.3 and 3.5, however, can be put into a more useful form. Let $g_1 = G_1/(G_1 + G_2)$ and $g_2 = G_2/(G_1 + G_2)$, so that $g_2 = 1 - g_1$, then equation 3.3 can be rewritten as

$$Y_M = g_1Y_1 + (1 - g_1)\,Y_2 \tag{3.6}$$

and

$$\frac{Y_2 - Y_M}{Y_M - Y_1} = \frac{g_1Y_2 - g_1Y_1}{(1-g_1)Y_2 - (1-g_1)Y_1} = \frac{g_1}{1-g_1} = \frac{g_1}{g_2} \tag{3.7}$$

In a similar way, equation 3.5 can be transformed into the expression

$$\frac{I_2 - I_M}{I_M - I_1} = \frac{g_1}{1-g_1} = \frac{g_1}{g_2} \tag{3.8}$$

Thus, if point P_1 on the enthalpy-humidity chart represents the state condition (Y_1, I_1), P_2 the condition (Y_2, I_2) and P_M the state of the mixture (Y_M, I_M), then the ratio of the distances P_2P_M/P_MP_1 gives us the mass-fraction ratio g_1/g_2. This linear mixing relationship is known as the <u>Lever Rule</u>. In other words, the state point P_M for the mixture divides the line connecting the state points P_2 and P_1 for the individual streams in the ratio $g_1:g_2$, see Fig. 3.3a. Interestingly, it is possible to get a fog by mixing two unsaturated airstreams, as Fig. 3.3b shows, a phenomenon observed in the formation of coastal fog with the drift of a cold onshore wind into warmer air over land.

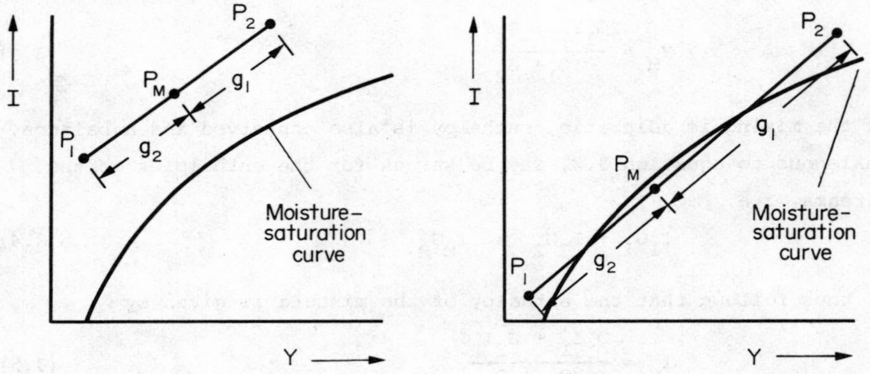

(a) Unsaturated air mixture (b) Supersaturated air mixture

Fig. 3.3 Application of Lever Rule to mixed moist airstreams.

<u>Example 3.2</u> Calculate the mixture properties when the following airstreams are mixed adiabatically:-

item	stream 1	stream 2
enthalpy I/kJ kg^{-1}	100	500
humidity Y/kg kg^{-1}	0.01	0.12
mass fraction g/kg kg^{-1}	0.2	0.8

From equation 3.6

$$Y_M = g_1Y_1 + g_2Y_2$$

$$= (0.2 \times 0.01) + (0.8 \times 0.12)$$

$$= 0.098 \text{ kg kg}^{-1}$$

Likewise, $$I_M = g_1I_1 + g_2I_2$$

$$= (0.2 \times 100) + (0.8 \times 500)$$

$$= 420 \text{ kJ kg}^{-1}$$

To find P_M on the enthalpy-humidity chart, we note that $g_1/g_2 = 0.2/0.8 = 1/4$. So P_M divides the line P_1P_2 in the ratio 4:1. Whence

$$\frac{Y_M - Y_1}{Y_2 - Y_1} = \frac{4}{5}$$

and

$$Y_M = 0.01 + \frac{4}{5}(0.12 - 0.01) = 0.098 \text{ kg kg}^{-1} .$$

The calculation is usually done by mensuration on the chart.

Airstream mixing with heat addition. If a certain quantity of heat Q is added as the airstreams mix, then the enthalpy of the mixture is increased by Q/G_M. The mixture point is now shifted from P_M a distance Q/G_M to a new point P'_M. The result is immaterial whether the heat is added to either inlet stream or partitioned between them (Fig. 3.4a). A heat loss Q_2 results in an enthalpy decrement of Q_L/G_M (Fig. 3.4b). The constructions shown in Fig. 3.4 follow from simple geometric considerations. One notes that the various line segments retain the same proportions, i.e.

$$\frac{P_2P_M}{P_MP_1} = \frac{P_2P_M}{P'_MP'_1} = \frac{P'_2P'_M}{P_MP_1} = \frac{P'_2P'_M}{P'_MP_1} = \frac{g_1}{g_2}$$

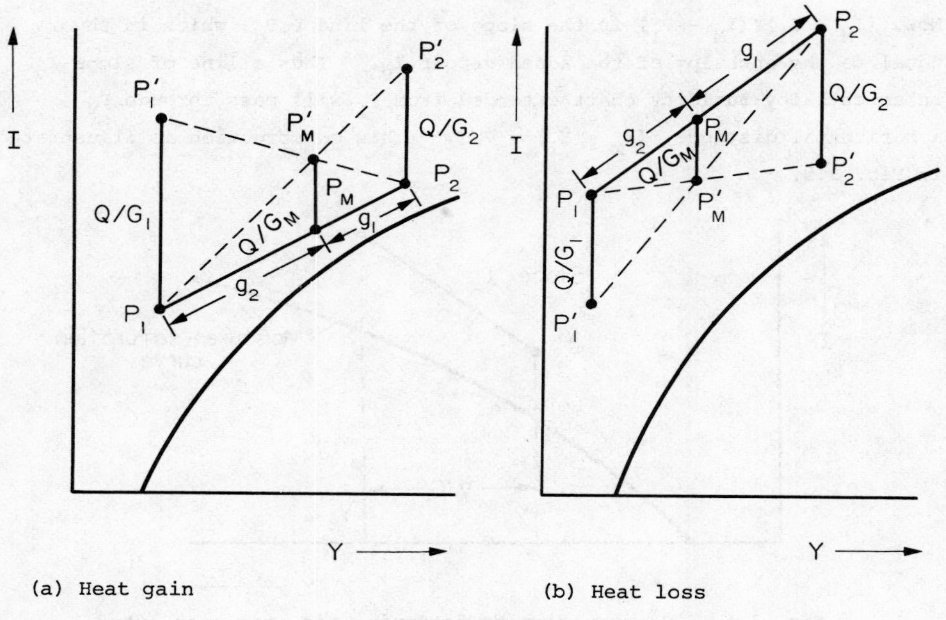

Fig. 3.4 Non-adiabatic mixing of moist airstreams.

Moisture addition to moist air. Moisture is sometimes added directly into the gas in a dryer, either deliberately as in a timber-drying schedule or inadvertently through a leak in a steam coil. The humidity of water vapour is infinite (there is no dry air!), so the corresponding state point cannot be put on the chart. Suppose V kg s^{-1} of moisture vapour of enthalpy I_V is added to G kg s^{-1} of dry gas. Then the moisture balance becomes

$$GY_1 + V = G_M Y_M \tag{3.9}$$

and the corresponding enthalpy balance is

$$GI_1 + VI_V = G_M I_M \tag{3.10}$$

However, $G_M = G$, so it follows from these equations that

$$\frac{G}{V} = \frac{1}{Y_M - Y_1} = \frac{I_V}{I_M - I_1} \tag{3.11}$$

Whence, by re-arranging equation 3.11, one finds

$$\frac{I_M - I_1}{Y_M - Y_1} = I_V \tag{3.12}$$

Now, $(I_M - I_1)/(Y_M - Y_1)$ is the slope of the line P_1P_M, which is thus equal to the enthalpy of the added vapour I_V. Thus a line of slope I_V on an enthalpy-humidity chart extended from P_1 will pass through P_M a horizontal distance, $(Y_M - Y_1) = V/G$. This construction is illustrated in Fig. 3.5.

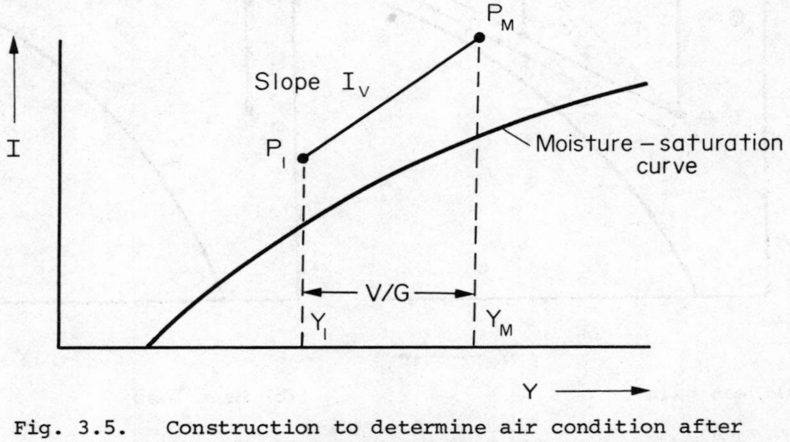

Fig. 3.5. Construction to determine air condition after admixing vapour of enthalpy I_V with moist air. Vapour/ dry gas ratio = V/G.

A similar expression to equation 3.12 holds should water be sprayed into the air rather than steam. If the enthalpy of the water is I_W, then

$$\frac{I_M - I_1}{Y_M - Y_1} = I_W \tag{3.13}$$

This gradient will appear negative on those moist-air charts for which the isenthalpic lines decline obliquely at such an angle that the gradient, $\Delta I/\Delta Y = \Delta H_{VO} = 2501\ \mathrm{kJ\ kg^{-1}}$, is horizontal. The state path for admixing water into damp air is shown in Fig. 3.6 for such a chart.

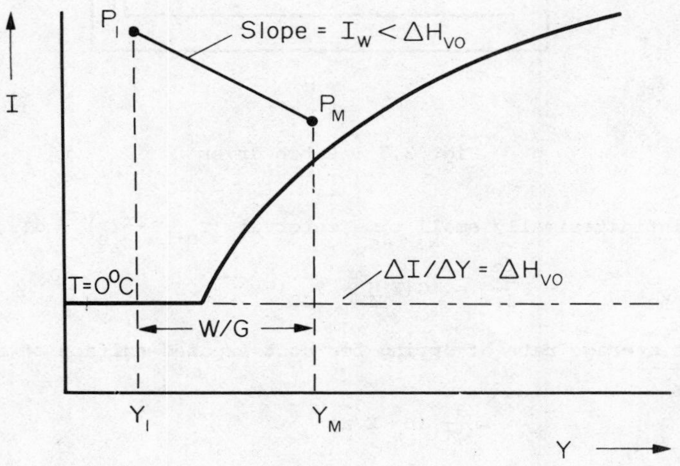

Fig. 3.6. Construction to determine air condition after admixing water of enthalpy I_W with moist air. Water/dry gas ratio = W/G.

3.2 Mass Balances

Batch drying. Let us consider a drying chamber, as shown in Fig. 3.7, in which F kg of bone-dry solids are charged and there is a through-flow of G kg s^{-1} of utterly dry air. The moisture content of the solids, averaged over the whole dryer, is $\bar{X}$ at any instant.

Over a time interval from τ_n to τ_{n+1}, the fall in moisture content of the solids charged is related to the increase in air humidity over the drying chamber from the inlet value Y_{GO} to that in the offtake duct Y_{GE}:

$$-F(\bar{X}_{n+1} - \bar{X}_n) = G\left[\int_{\tau_1}^{\tau_2} Y_{GE}\,d\tau - Y_{GO}(\tau_{n+1} - \tau_n)\right] \tag{3.14}$$

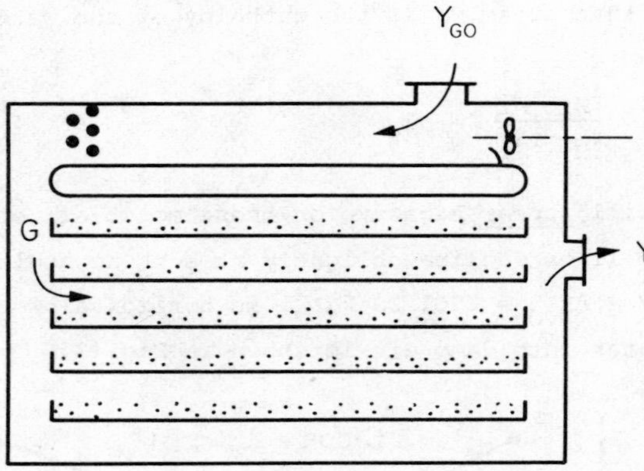

Fig. 3.7 Batch dryer

For an infinitesimally small time interval $(\tau_{n+1} - \tau_n) \to d\tau$,

$$- F\frac{d\bar{X}}{d\tau} = G(Y_{GE} - Y_{GO}) \tag{3.15}$$

Now, the average rate of drying for unit exposed surface of solids is given by

$$\bar{N}_V = - \frac{d}{d\tau} (\rho_s X/a) \tag{3.16}$$

where ρ_s is the density of the bone-dry solids and a is the exposed surface of solids per unit volume. Combination of equations 3.15 and 3.16 yields

$$Y_{GE} = Y_{GO} + aF\bar{N}_V/\rho_s G \tag{3.17}$$

so the exit-gas humidity Y_{GE} reflects the mean drying rate $\bar{N}_V$ as one would expect. Over a given batch, Y_{GE} rises rapidly from Y_{GO} as drying begins to reach a maximum value, then falling as the drying rate dwindles towards the end of the process. The variation of Y_{GE} with time is shown in Fig. 3.8. The extent of the induction period depends upon whether the chamber is warm or cold when the solids are charged.

Clearly the hygrothermal conditions must be chosen so that the exhausted humid gases are never saturated during the period of the fastest drying rates. There are, however, certain cases when saturated-air conditions

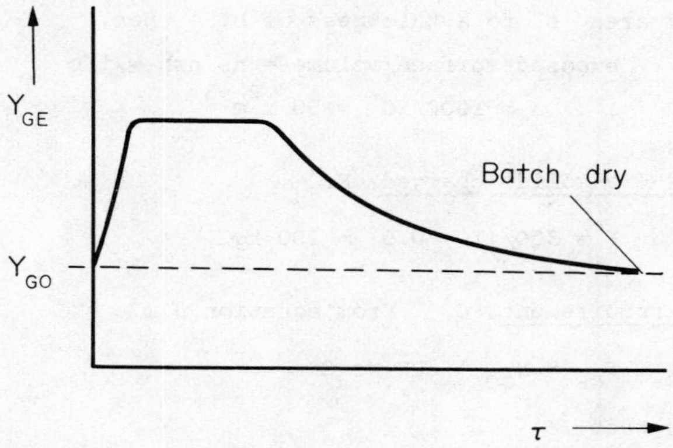

Fig. 3.8 Variation of exit-gas humidity Y_{GE} with time during batch drying.

are deliberately sought during some stage of drying when the material being dried is passing through a critical period of dimensional instability. Methods of estimating N_V will be given in Chapter 6. The following example illustrates the calculation of air requirements for batch drying when N_V is known.

Example 3.3. A batch of solids (ρ_s = 1800 kg m^{-3}) of mass 300 kg and moisture content 0.5 kg kg^{-1} is charged to a dryer in which the humid air is exhausted at 60°C and a relative humidity of 50% when the solids dry at maximum rate. This maximum solids-drying rate is 0.2 kg m^{-2}h^{-1} when the solids are dumped to a thickness of 20 mm on the trays. Estimate the air requirement when the dewpoint of the fresh air is 10°C and the total pressure is 100 kPa.

(i) Outlet-air humidity, Y_{GE}. The saturated-air vapour pressure at the outlet-air temperature of 60°C is 19.92 kPa. From equation 2.10

$$Y_{GE} = \left(\frac{M_W}{M_G}\right) \frac{\psi p_W^o}{(P-\psi p_W^o)}$$

$$= \frac{0.6228 \times 0.5 \times 19.92}{(100 - (0.5 \times 19.92))}$$

$$= 0.0689 \text{ kg kg}^{-1}$$

(ii) Specific surface area, a. Suppose the solids are dumped over n trays of area s to a thickness of b. Then,

$$\text{exposed surface/volume} = ns/nsb = 1/b$$

$$\therefore \quad a = 1000/20 = 50 \text{ m}^2\text{m}^{-3} .$$

(iii) Bone-dry solids charged, F.

$$F = 300/(1 + 0.5) = 200 \text{ kg}.$$

(iv) Air requirement, G. From equation 3.17

$$Y_{GE} = Y_{GO} + aF\bar{N}_V/\rho_s G$$

is follows that

$$G = \frac{aF\bar{N}_V}{\rho_s(Y_{GE}-Y_{GO})}$$

$$= \frac{50 \times 200 \times 0.2}{1800 \times (0.0689 - 0.0077)}$$

$$= 18.2 \text{ kg h}^{-1}$$

Since the inlet-air humidity is 0.0077 kg kg^{-1}, the air drawn is

$$18.2 \times 1.0077 = 18.3 \text{ kg h}^{-1} .$$

Continuous drying. The analysis of continuous drying is somewhat simpler as the process conditions remain steady apart from minor process upsets. Suppose a solid is being continuously dried from a moisture content X_0 to X_Z with corresponding air humidities of Y_{GO} and Y_{GZ} at the solids inlet and outlet positions respectively (Fig. 3.9). Leakages at places where the solids enter and leave will be assumed to be negligible.

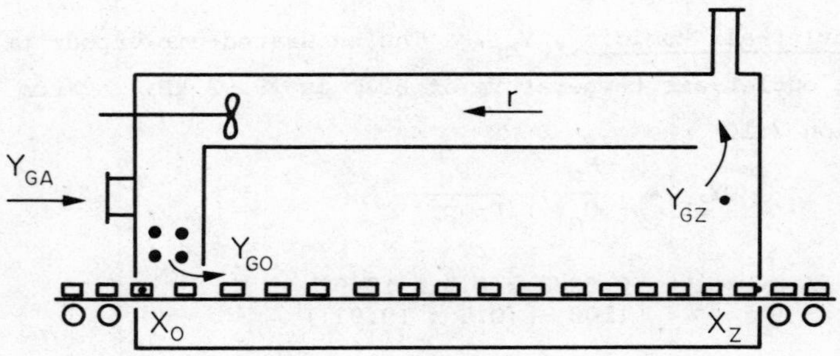

Fig. 3.9 A continuous dryer with recycle.

The overall mass balance is obtained by equating moisture lost from the solids to moisture gained by the air:

L = specific dry solids flow

$$\pm\, G(Y_{GO} - Y_{GZ}) = L(X_O - X_Z) \tag{3.18}$$

in which the negative sign is taken when the solids and air progress in the same direction and the positive sign when the solids and air go in opposite directions. Since often $\Delta Y/\Delta X \sim 0.01/1$, the airflow can be of order one hundredfold greater than the dry-solids flow.

Suppose some of the moist air to be discharged is sent back to air intake to improve the thermal economy of the dryer. Let r be the mass of dry gas so recycled per unit mass of gas flowing through the dryer. A mass balance over the mixture of the recycled and freshly admitted airstreams yields for the co-current case shown in Fig. 3.9

$$Y_{GO} = r\,Y_{GZ} + (1-r)Y_{GA} \tag{3.19}$$

Re-arranging this equation, one gets for the recycle ratio r

$$r = (Y_{GO} - Y_{GA}) \,/\, (Y_{GZ} - Y_{GA}) \tag{3.20}$$

$$= 1 - (Y_{GZ} - Y_{GO}) \,/\, (Y_{GZ} - Y_{GA}) \tag{3.21}$$

As the recycle ratio r increases, so the humidity change across the dryer $(Y_{GZ} - Y_{GO})$ becomes smaller with respect to the humidity rise above ambient $(Y_{GZ} - Y_{GA})$. A smaller quantity (1-r) of fresh air has to be heated, but the gas in the dryer has a lower potential for picking up moisture and a bigger dryer is needed. The greater capital cost of the dryer is offset by the smaller heating charges, but eventually the gain in thermal economy ceases to become worthwhile.

The state paths for the air, when drying with recirculating air, are shown in Fig. 3.10. Point A corresponds to the state of the fresh air, B the mixed air at the inlet, C the same air after heating, D the air leaving the dryer. Since AB/BD = r/(1-r) by the Lever Rule, AB/AD = r. This result also follows directly from equation 3.20.

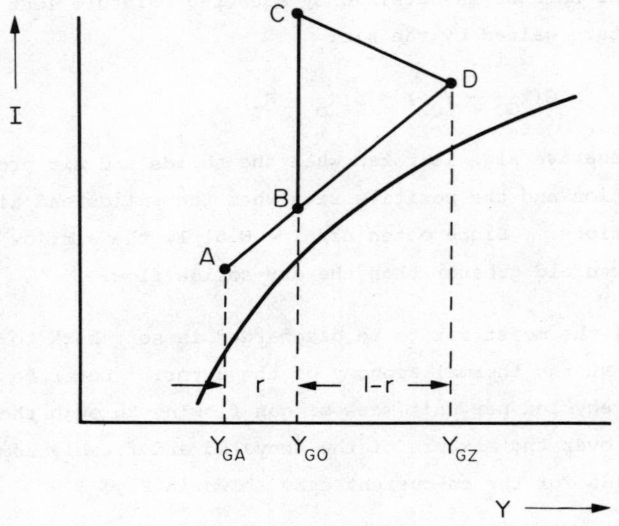

Fig. 3.10 State paths for air on drying with recirculation

Example 3.4. Moist PVC granules are injected into the throat of a duct through which a hot, high-velocity airstream conveys them until fairly dry. Since the contact time is short, some of the dried material is recycled. In this case, there is a recirculation of air and solids, rather than air alone (Fig. 3.11). The following conditions pertain:

Inlet-solids moisture content	=	0.40 kg kg^{-1}
Outlet-solids moisture content	=	0.015 kg kg^{-1}
Gas/solids flow ratio (dry basis)	=	15
Outlet-air temperature	=	50°C
Moisture isotherm for solids	$X^* =$	$0.03\ \psi$
Fresh-air humidity	=	0.0077 kg kg^{-1}

If the solids at the air outlet essentially reach equilibrium with the air, calculate the recycle ratio of solids.

(i) Outlet-air humidity, Y_{GZ}. The relative humidity of the outlet air is given by the moisture content of the solids at the end of the duct.

$$\psi = X_Z^*/0.03$$
$$= 0.015/0.03 = 0.5$$

From the enthalpy-humidity chart in the Appendix, air at 50°C and a relative humidity of 0.5 has a humidity of 0.042 kg kg^{-1}.

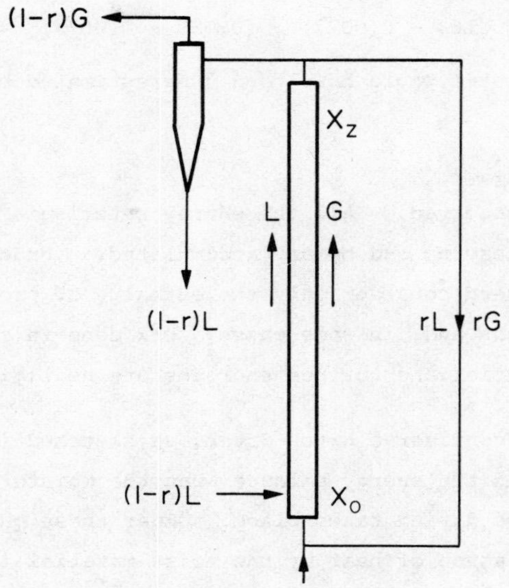

Fig. 3.11 Airlift dryer.

(ii) <u>Inlet-air humidity</u>, Y_{GO}. From an overall moisture balance, equation 3.18

$$- G(Y_{GO} - Y_{GZ}) = L(X_O - X_Z)$$

$$\therefore Y_{GO} = Y_{GZ} - \frac{L}{G}(X_O - X_Z)$$

$$= 0.042 - \frac{1}{15}(0.40 - 0.015)$$

$$= 0.0163 \text{ kg kg}^{-1}$$

(iii) <u>Air-recycle ratio</u>, r. Equation 3.20 applies, namely

$$r = \frac{Y_{GO} - Y_{GA}}{Y_{GZ} - Y_{GA}}$$

$$= \frac{(0.0163 - 0.0077)}{(0.042 - 0.0077)} = 0.251$$

Thus only one quarter of the outlet stream need be diverted to the intake. The assumption of equilibrium at the outlet is rather rough and thus Y_{GZ} is likely to be somewhat less than estimated. If, instead of 0.042, the true value should have been 10 per cent smaller, then

$$r = (0.0163 - 0.0077) / (0.038 - 0.0077) = 0.284$$

i.e., the recycle ratio would have been underestimated by 12 per cent.

3.3 Energy Balances

Energy, like mass, is conserved. All the energy entering a system must be accounted for by that leaving and by any accumulated. Under commercial drying conditions, we need consider only the enthalpy of process streams, the work done by the fans and, in some cases, work done in conveying the solids: kinetic, potential and surface energies are negligible.

Batch drying. Let us consider a batch dryer, as sketched in Fig.3.12. Of principal interest is the energy balance when the moisture is lost most speedily when unhindered drying takes place. Under these quasi-steady conditions, the accumulation of heat by the moist material is zero. The energy balance thus becomes for an intake of G kg s^{-1} of dry gas

$$I_{GA}G + W_F + Q_H - I_{GE}G - Q_L = 0 \tag{3.22}$$

or

$$I_{GA} + W_F/G + Q_H/G - I_{GE} - Q_L/G = 0 \tag{3.23}$$

where W_F is the fan-work, Q_H the heat input from the heater and Q_L the heat loss from the chamber walls to the outside. All the quantities in equation 3.23 can be conveniently displayed on an enthalpy-humidity diagram, as Fig. 3.13 shows. The flow of gas across the drying goods is G/(1-r), where r is the internal recycle ratio which can be found from the enthalpy-humidity chart in the way already shown.

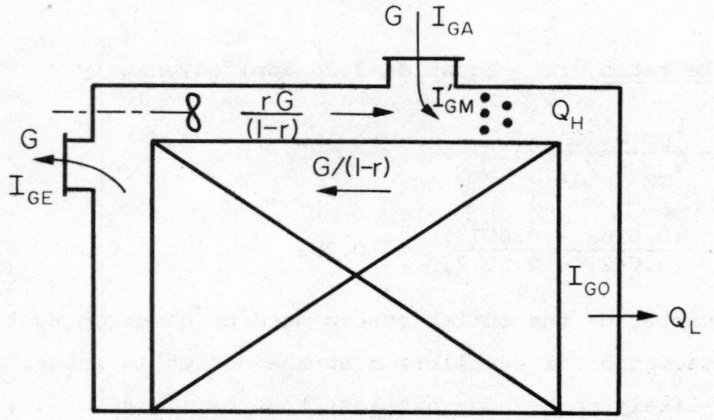

Fig. 3.12 Batch dryer

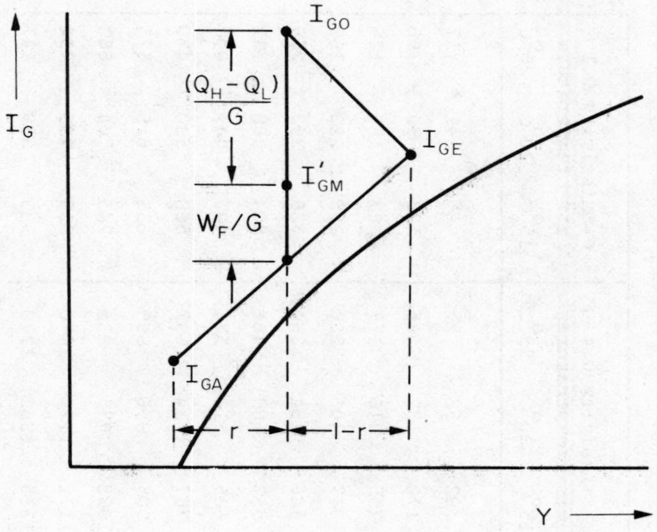

Fig. 3.13 Heat quantitities for a batch dryer on an enthalpy-humidity chart.

The fan-work Q_F is normally a small item. Kröll[10] gives the resistance to airflow in a simple drying chamber as about 20 Pa for an air velocity of 2 m s^{-1} over the drying goods. If the fan efficiency is 80 per cent, the work done for unit cross-section of the dryer is 20x2/0.8 = 50 W m^{-2}.

Heat losses are more significant, and depend considerably on the nature of the outside surface of the dryer, even at fairly low temperatures. The emissivity of freshly rolled sheet steel is 0.56, but is 0.88 when heavily oxidized, while galvanized sheet has a much lower emissivity of 0.2 .[17] At a temperature of 100°C, the heat loss from a galvanized vertical surface is about one half that of a rusted or painted surface of similar extent. Such a dull surface would lose about 1 kW m^{-2} of heat, and a reduction of the surface temperature to 50°C, say, by installing suitable lagging would eliminate about two-thirds of this loss. Values of the heat loss from vertical surfaces of varying emissivity are given in Table 3.1 for selected thermal conditions. These values are claimed to be also appropriate for very large bodies of irregular shape, since roughly as much surface faces upwards as downwards.

TABLE 3.1 TOTAL HEAT LOSS FROM VERTICAL SURFACES IN STILL AIR, W m^{-2}.

Calculated from data of Spiers[17]

Temperature of surface / °C	Emissivity 1.0 air temperature			Emissivity 0.7 air temperature			Emissivity 0.4 air temperature			Emissivity 0.1 air temperature		
	15°C	20°C	25°C	15°C	20°C	25°C	15°C	20°C	25°C	15°C	20°C	25°C
30	145	95	45	118	77	36	92	58	27	65	41	17
40	267	208	155	216	170	125	170	132	96	123	94	66
50	392	339	279	323	275	228	255	216	177	187	156	126
60	533	474	415	440	389	339	347	306	264	255	222	189
70	696	625	564	564	515	461	446	404	360	328	293	258
80	848	787	722	700	647	591	553	508	461	404	368	331
90	1023	960	896	844	789	734	665	618	574	485	447	408
100	1214	1143	1078	994	939	882	787	734	687	568	530	490
110	1405	1340	1274	1155	1099	1039	905	856	808	655	615	575
120	1615	1549	1482	1326	1266	1209	1035	985	935	745	704	662
130	1837	1771	1702	1504	1445	1386	1171	1120	1070	838	795	754
140	2071	2004	1935	1692	1633	1572	1313	1261	1270	933	889	847
150	2318	2250	2180	1889	1829	1765	1459	1408	1356	1031	988	944

The following worked example illustrates the overall effectiveness in using energy for a low-temperature, batch-drying process.

Example 3.5. A fluid-bed unit is used to dry wet bile acids produced from concentrated gall obtained as a meatworks' byproduct. The wet acid, containing 20 per cent moisture by mass, is charged to the dryer, and air is passed at the rate of 10 $m^3\ s^{-1}\ m^{-2}$ through the bed which has a cross-sectional area of 0.1 m^2. After ½ hour the averaged moistness of the bed has fallen to 2 per cent by mass. Estimate the fraction of the heat which is contained in the inlet air and used in drying when the inlet-air temperature is 52°C and the dry mass of solids charged is 100 kg.

At 52°C the air density is 1.086 kg m^{-3}, so the air used over the half-hour period is 1.086 x 10 x 0.1 x 1800 = 1955 kg.

Heat supplied = 1955 x 1 x 52 = 102 000 kJ

$$\text{Moisture evaporation} = 100\left[\frac{20}{80} - \frac{2}{98}\right] = 23 \text{ kg}$$

Enthalpy of saturated moisture vapour at 52°C is 2595 kJ kg^{-1}.

∴ Theoretical heating load = 23 x 2595 = 59 700 kJ

∴ Useful fraction = 59 700 / 102 000 = 0.585.

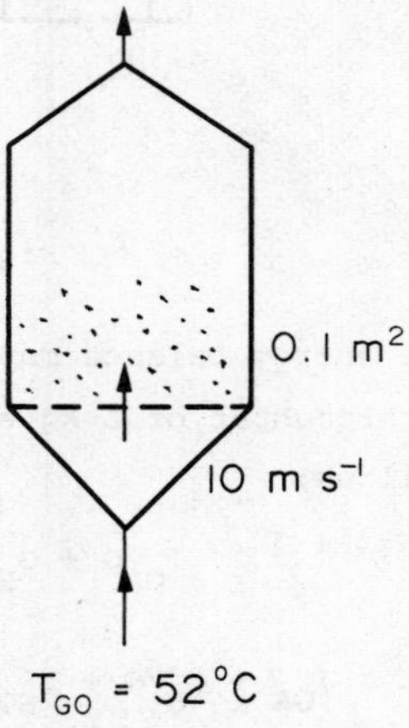

Fig.3.14 Fluid-bed dryer

At first, when the drying rates are the fastest, a somewhat greater fraction would be used, and towards the end of drying a somewhat lesser. This difference would show itself as an upward drift of the air-outlet temperature during each run.

Continuous drying. The elements of a typical continuous-drying installation are shown in Fig. 3.15 for the co-current movement of air and solids. Air is drawn in at an enthalpy I_{GA}, heated to raise its enthalpy and mixed with recycled exhaust air so that the humidity of the combined stream is I_{GO} at the inlet end of the dryer. When the solids progress in the opposite direction to the drying gas, then I_{GO} is the resultant air enthalpy at the solids outlet. The air humidifies on its passage through the dryer and emerges at the exhaust with enthalpy I_{GE}. The enthalpy of the solids changes from a value I_{SO} on entering to I_{SZ} on leaving.

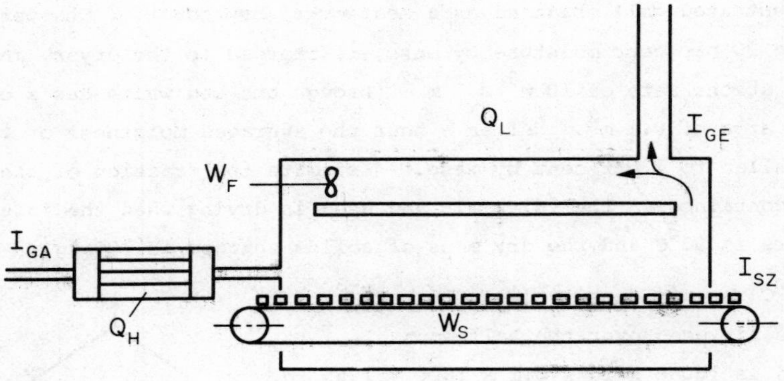

Fig. 3.15 Continuous dryer

The energy balance may be written for an intake of G kg s^{-1} dry gas and a throughput of L kg s^{-1} of drying goods (bone-dry solids basis) as follows:

$$I_{GA}G + Q_H + I_{SO}L + W_F + W_S - I_{GE}G - Q_L - I_{SZ}L = 0 \quad (3.24)$$

or $$I_{GA} + Q_H/G + I_{SO}L/G + W_F/G + W_S/G - I_{GE} - Q_L/G - I_{SZ}L/G = 0 \quad (3.25)$$

Since L/G normally ranges between 0.01 and 0.1, often the lumped solids-enthalpy term $(I_{SO} - I_{SZ})$ L/G in equation 3.25 is relatively minor. Values of Q_H are obtained from simple considerations of heat transfer, while estimates of the heat loss can be obtained from Table 3.1. Again, often the fan-work W_F is a small item, but the work extended in conveying the solids W_S may not be. The fan-work for a cross-circulated band dryer is of order 1 kW m^{-2} (band area), but on through-circulating the air through a perforated band the energy losses rise to about 2.5 kW m^{-2}.[15] The power needed to convey solids by suspending them in a fast airstream, as in an airlift dryer, is about 9 kW for each 1 m^3 s^{-1} employed.[16] On the other hand, however, the power taken by a large rotary dryer can be extensive, a dryer of 4 m diameter taking upwards of 100 kW. The power demands for some of the principal types of continuous dryers are listed by Lapple and co-authors.[12]

The various heat quantities involved may be displayed on an enthalpy-humidity chart, as depicted in Fig. 3.16.

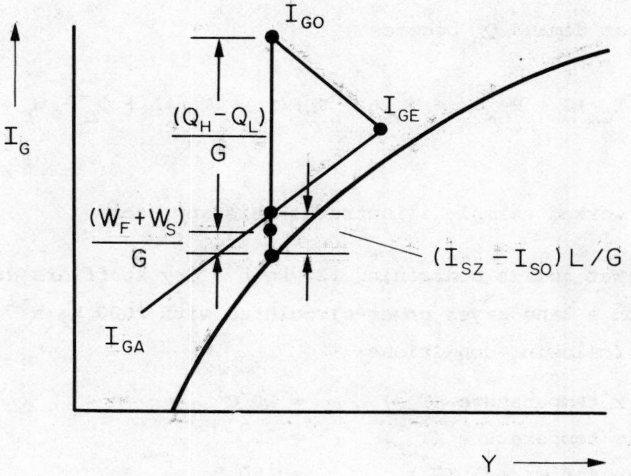

Fig. 3.16 Heat quantities for a continuous dryer on an enthalpy-humidity chart.

For calculational purposes, it is convenient to put equation 3.25 in a somewhat different form by neglecting the moisture brought in with the fresh air. With this assumption, the change in humid enthalpy of the gas across the dryer becomes:

$$I_{GE} - I_{GA} \simeq \bar{C}_{PG}(T_{GE} - T_{GA}) + H_{GW}G_V/G \qquad (3.26)$$

where T_{GE} and T_{GA} are the outlet and intake-air temperatures respectively and G_V is the rate at which moisture evaporates. With little error, one can take H_{GW} as the enthalpy of saturated moisture vapour at the air-outlet temperature T_{GE}. This assumption implies that the fraction of moisture driven off is heated to T_{GE} at which temperature it evaporates. Consequently the change in humid enthalpy of the moist solid is thus calculated by

$$I_{SZ} - I_{SO} \simeq (\bar{C}_S + \bar{C}_L X_Z)\ (T_{SZ} - T_{SO}) \qquad (3.27)$$

where X_Z is the moisture content of the solids being discharged from the dryer.

With these foregoing assumptions, equation 3.25 can now be rewritten as

$$-\bar{C}_{PG}(T_{GE} - T_{GA}) - H_{GW}G_V/G - (\bar{C}_S + \bar{C}_L X_Z)(T_{SZ} - T_{SO})\,L/G + (Q_H + W_F + W_S - Q_L)/G = 0 \qquad (3.28)$$

so that the heat demand Q_H becomes

$$Q_H = \bar{C}_{PG}(T_{GE} - T_{GA})G + H_{GW}G_V + (\bar{C}_S + \bar{C}_L X_Z)(T_{SZ} - T_{SO})L + Q_L - W_F - W_S \qquad (3.29)$$

The following worked example illustrates this approach.

Example 3.6 Wet solids containing 120 kg h^{-1} dry stuff are dried continuously on a band dryer cross-circulated with 2000 kg h^{-1} of dry air under the following conditions:

ambient-air temperature (T_{GA})	= 20°C
exhaust-air temperature (T_{GE})	= 70°C
evaporation (of water) (G_V)	= 150 kg h^{-1}
outlet-solids moisture content (X_Z)	= 0.25 kg kg^{-1}
inlet-solids temperature (T_{SO})	= 15°C
outlet-solids temperature (T_{SZ})	= 65°C
power demand ($W_F + W_S$)	= 5 kW
heat loss (Q_L)	= 18 kW

Estimate the heater load per unit mass of dry gas, and the fraction of this heat used in evaporating moisture.

Data

mean specific heat of dry air ($\bar{C}_{PG}$)	= 1.0 kJ $kg^{-1}K^{-1}$
enthalpy of saturated water vapour at T_{GE} (H_{GW})	= 2626 kJ kg^{-1}
mean specific heat of dry material ($\bar{C}_S$)	= 1.25 kJ $kg^{-1}K^{-1}$
mean specific heat of liquid moisture ($\bar{C}_L$)	= 4.18 kJ $kg^{-1}K^{-1}$

Flow rates. Now,

$$G = 2000/3600 = 0.556 \text{ kg s}^{-1}$$
$$L = 120/3600 = 0.0333 \text{ kg s}^{-1}$$
$$G_V = 150/3600 = 0.0417 \text{ kg s}^{-1}$$

Each term in equation 3.29 can now be evaluated in turn:-

(i) Dry-air enthalpy

$$\bar{C}_{PG}(T_{GE} - T_{GA})G = 1.0 \times (70-20) \times 0.556 = 27.8 \text{ kW}$$

(ii) Moisture-vapour enthalpy $H_{GW}G_V = 2626 \times 0.0417$ = 109.5 kW

(iii) Outlet-solids enthalpy $(\bar{C}_S + \bar{C}_L X_Z)(T_{SZ} - T_{SO})L$ = $[1.25 + (4.18 \times 0.25)](65 - 15) \times 0.0333 = 3.8$ kW

(iv) Heat loss Q_L = 18 kW

(v) Power demand $W_F + W_S$ = 5 kW

The heat load Q_H is the sum of items (i) to (iv) less item (v), namely

$$Q_H = 27.8 + 109.5 + 3.8 + 18 - 5 = 154.1 \text{ kW}.$$

The useful fraction of this heat is 109.5/154.1 = 0.710.

About two-thirds of the heat demand is used in vaporizing moisture, but often dryers use heat less effectively, and an inventory of mass and energy around a dryer can be a useful tool in the improvement of a drying process. The technique is illustrated by a recent survey[3] of the performance of a unit to dry coated paper by drawing the sheet through the chamber in festoons impinged with hot air. The balances clearly show that the thermal insulation was poor, heat losses accounting for almost one-third of the heat used. The outlet humidity was low, and the fraction of air returned to the fresh-air intake could have been greater for better conservation of heat. The tests also indicate that the paper was being needlessly overdried. The following worked example illustrates a similar appraisal of a rotary dryer.

Example 3.7 An uninsulated, countercurrent rotary dryer is being used to dry sugar-like crystals from a moisture content of 0.05 to 0.002 kg kg^{-1} at a dried product rate of 0.3 kg s^{-1}. The dryer is 1.2 m in diameter and 7 m long. Air at 15°C and 60% relative humidity is heated over steam-coils to 120°C before being admitted to the dryer and is exhausted from the dryer at 60°C. The solids are raised in temperature from 15°C to 70°C at discharge. The power required to drive the dryer is 8 D_s^2 (in kW), where D_s is the diameter of the shell; one third of the energy dissipated being transmitted to the shell through the girth-gear ring. Estimate the air and steam requirements. The barometric pressure is 100 kPa.

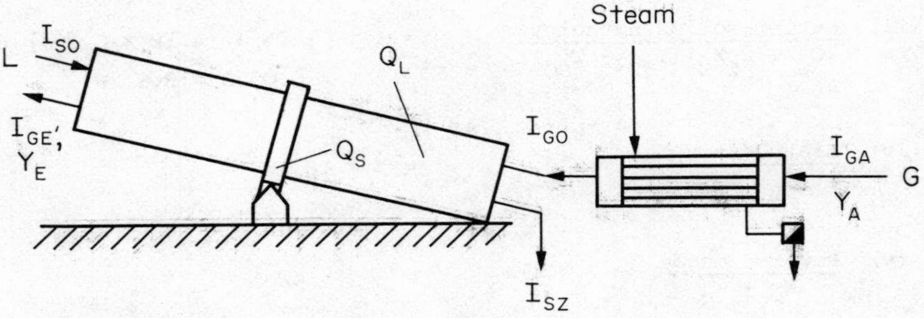

Fig. 3.17 Rotary-drying installation

(i) <u>inlet-air humidity</u>, Y_A

At 15°C, p_W^o = 1.704 kPa, so from equation 2.10

$$Y_A = \left(\frac{M_W}{M_G}\right)\frac{\psi p_W^o}{(P-\psi p_W^o)}$$

$$= \frac{0.6228 \times 0.6 \times 1.704}{100 - (0.6 \times 1.704)}$$

$$= 0.006\ 43 \text{ kg kg}^{-1}$$

(ii) <u>outlet-air humidity</u>, Y_E (unknown)

(iii) <u>inlet-air enthalpy</u>, I_{GO} (equation 2.22)

$$I_{GO} = \bar{C}_{PG}T_{GO} + [\bar{C}_{PW}T_{GO} + \Delta H_{VO}]Y_A$$

$$= (1.007 \times 120) + [(1.87 \times 120) + 2501] \times 0.00643$$

$$= 138.4 \text{ kJ kg}^{-1}$$

(iv) <u>outlet-air enthalpy</u>, I_{GE} (equation 2.22)

$$I_{GE} = \bar{C}_{PG}T_{GE} + [\bar{C}_{PW}T_{GE} + \Delta H_{VO}]Y_E$$

$$= (1.005 \times 60) + [(1.86 \times 60) + 2501]Y_E$$

$$= 60.3 + 2613\ Y_E \text{ kJ kg}^{-1}$$

(v) inlet-solids enthalpy, I_{SO} (equation 2.36).

Assume $\bar{C}_S = 1.3$ kJ kg^{-1} K^{-1} (sugar-like crystals), $\Delta H_W = 0$

$$I_{SO} = [\bar{C}_S + \bar{C}_{LW}X_O]T_{SO}$$

$$= [1.3 + (4.20 \times 0.05)] \times 15$$

$$= 22.6 \text{ kJ kg}^{-1}$$

(vi) outlet-solids enthalpy, I_{SZ} (equation 2.36).

$$I_{SZ} = [\bar{C}_S + \bar{C}_{LW}X_Z]T_{SZ}$$

$$= [1.3 + (4.18 \times 0.002)] \times 70$$

$$= 91.1 \text{ kJ kg}^{-1}$$

(vii) heat loss, Q_L.

The surface area of the shell is $\pi \times 1.2 \times 7 = 26.4$ m^2
The shell temperature will be close to the air temperature
i.e. mean shell temperature $\simeq (15 + 70)/2 = 42.5$
For an emissivity of 0.7, interpolation of the data given in Table 3.1 suggests a wall loss of 243 W m^{-2}.
i.e. $Q_L = 0.243 \times 26.4 = 6.4$ kW

(viii) frictional heat, Q_S.

$$Q_S = 8\, D_S^{\,2}/3$$

$$= 8 \times 1.2^2/3 = 3.8 \text{ kW}$$

(ix) dry solids flow, L.

$$L = 0.3/1.002 = 0.299 \text{ kg s}^{-1}$$

(x) dry gas flow, G (unknown).

moisture balance:

$$(Y_A - Y_E)G + (X_O - X_Z)L = 0$$

$$(0.00643 - Y_E)G + (0.05 - 0.002)\,0.299 = 0 \qquad \text{(I)}$$

enthalpy balance:

$$(I_{GO} - I_{GE}) + (I_{SO} - I_{SZ})L = -Q_S + Q_L$$

$$(138.4 - 60.3 - 2613Y_E)G + (22.6 - 91.1)\,0.299 = -3.8 + 22.3 \qquad \text{(II)}$$

Equations I and II become respectively

$$(0.00643 - Y_E)G + 0.01435 = 0$$

$$(78.1 - 2613Y_E)G - 23.1 = 0$$

whence one finds, on solving these expressions simultaneously,

$$G = 0.980 \text{ kg s}^{-1} \quad \text{and} \quad Y_E = 0.0209 \text{ kg kg}^{-1}$$

Since the humidities are everywhere less than 0.05, equation 2.22 is satisfactory for determining the humid enthalpy.

The ambient-air enthalpy I_{GA} is given by

$$I_{GA} = \bar{C}_{PG}T_{GA} + [\bar{C}_{PW}T_{GA} + \Delta H_{VO}]Y_A$$

$$= (1.00\text{x}15) + [(1.85\text{x}15) + 2501] \text{ x } 0.00643$$

$$= 31.3 \text{ kJ kg}^{-1}$$

Thus the heater load can now be evaluated:

$$Q_H = (I_{GO} - I_{GA})G$$

$$= (138.4 - 31.3) \text{ x } 0.989 = 102 \text{ KW}$$

If saturated steam at 135°C is used in the coils, then 2160 kJ become available for each kilogram of steam condensed, whence the theoretical steam consumption = 102/2160 = 0.0472 kg s^{-1}.

Since the moisture evaporated in the dryer is given by

$$G_V = 0.299(0.05 - 0.002) = 0.01435 \text{ kg s}^{-1}$$

the specific steam demand = 0.472/0.01435 = 3.29 kg kg^{-1}.

Thus over three as much steam has to be condensed as water is evaporated. Heat losses, at 6.4 x 100/102 = 6.3 per cent of the heater load, are not excessive even though the dryer is uninsulated.

3.4 Direct Heating

Dryers, which are to be worked above steam temperature (ca. 150°C), can be directly fired with any convenient fuel to yield hot combustion gases which can be passed over the drying goods. The fuel requirements to do this follow directly from the stoichiometry of the burning process. It is normally sufficient to consider that air is composed of a binary mixture containing 79 per cent (by volume) nitrogen and 21 per cent oxygen. Consider, for example, the complete combustion of methane to carbon dioxide. The stoichiometric equation is given by

$$CH_4 + 2O_2 \rightarrow CO_2 + 2H_2O$$

Therefore, 2 mol of oxygen are consumed in burning 1 mol of methane, or 2 x 100/21 = 9.52 mol of air per mole of fuel. The composition of the water-free combustion gases are thus:-

	mol/mol fuel	mol %
nitrogen	2 x 79/21 = 7.52	88.3
carbon dioxide	1	10.7
Σ	8.52	100.0

In a similar way, it is possible to derive the composition of the gas produced by burning any fuel of known molal composition.[(5)]

The enthalpies of carbon dioxide and air are little different at temperatures below 1500°C, as Table 3.2 shows. Thus the enthalpy of the combustion gas can often be approximated by that of air. The dry gas, produced by the complete combustion of methane for instance, has an enthalpy that is 1.0 per cent greater than air at 500°C, the difference rising with temperature to 1.9 per cent at 1500°C.

TABLE 3.2 ENTHALPY OF DRY GASES ($kJ\ kg^{-1}$) AT VARIOUS TEMPERATURES
Calculated from data of Mayhew and Rogers.[13]

temperature/ °C	N_2	CO_2	air	combustion gas, 88.3% N_2
500	530	514	517	522
1000	1128	1157	1092	1119
1500	1757	1860	1718	1750

At temperatures above 1500°C, the combustion of fuels is rendered incomplete because carbon dioxide and water vapour can dissociate. At 1700°C the degree of dissociation of carbon dioxide is 2.4 per cent in the combustion gases given off by burning heavy fuel-oil with the stoichiometric air requirement (no excess air). The water vapour is 0.51 per cent dissociated under these conditions.[18]

Mass changes. While the use of molal units is logical in cases of chemical change, it is sometimes more convenient to use mass units in estimating the composition of gases derived from burning fuels of complex, and often unknown, chemical composition, such as solid and liquid fossil fuels. There is a minimum mass of dry air g_A^o to burn unit mass of fuel, which yields ash a, water vapour g_v and dry gaseous products of combustion g_G. By mass balance, this minimum quantity of air is given by

$$g_A^o = g_G + g_V + a - 1 \tag{3.30}$$

The value of a is found from fuel analyses. The ash contents for commercial coals in the United Kingdom, for example, are largely independent of rank, and range as follows:[19]

large and ungraded coal	0.03 to 0.06
washed small coal	0.05 to 0.10
untreated small coal	0.12 to 0.18

The ash content of wood is negligibly small. Clearly, in the case of firing with oil or gas, the ash term does not appear.

The change in mass $(g_G - g_A^o)$ depends upon the fuel burnt.[9] The mean difference $(g_G - g_A^o)$ is 0.5 for solid fuels, zero for fuel-oils, - 0.8 for hydrogen-rich gases and + 0.8 for hydrogen-lean gases. The ratio g_G / g_A^o is normally very close to 1, as selected data in Table 3.3 indicate. The actual change in mass $(g_G - g_A)$ depends also on the excess air $(g_A - g_A^o) = g_A^+$ (say) as the flue gases are augmented by this quantity.

Water vapour arises from the hydrogen content of the fuel as well as the associated moisture therein. It follows that

$$g_V = M_W h/M_H + w \tag{3.31}$$

where M_W/M_H is the ratio of the molar mass of water to that of hydrogen and is numerically equal to 8.937, h is the fractional hydrogen content and w the fractional moisture content of the fuel respectively. The total moisture in the combustion gases derives from this added moisture and that brought in with the air; thus

$$(g_A^+ + g_G)Y_G = g_V + g_A Y_A \tag{3.32}$$

Re-arrangement of equation 3.32 yields an expression for the flue-gas humidity Y_G

$$Y_G = \frac{g_V}{(g_G + g_A^+)} + \frac{Y_A(g_A^o + g_A^+)}{(g_G + g_A^+)} \tag{3.33}$$

on noting that $g_A = g_A^o + g_A^+$. Since $g_G/g_A^o \simeq 1$, equation 3.33 reduces to

$$Y_G = Y_A + g_V/(g_G + g_A^+) \tag{3.34}$$

with sufficient accuracy for most calculational purposes.

On neglecting kinetic and surface energies associated with burning a fuel, even when finely dispersed, an enthalpy balance over the heating appliance yields for steady working the equation,

$$H_F + \Delta H_C + I_{GA}g_A = I_{GG}(g_G + g_A^+) + I_{Sa}a \qquad (3.35)$$

where

H_F is the enthalpy of the fuel,

ΔH_C is the heat of combustion (or calorific value),

I_{GA} is the humid enthalpy of the air drawn in,

I_{GG} is the humid enthalpy of the flue-gases,

and I_{Sa} is the humid enthalpy of the ash.

Normally, both H_F and $I_{Sa}a$ are negligibly small compared with other terms in equation 3.35. So this expression reduces to the equation

$$I_{GG} = \Delta H_C/(g_G + g_A^+) + I_{GA}g_A/(g_G + g_A^+) \qquad (3.36)$$

Further, we may assume $g_G/g_A^o \simeq 1$, so with little error equation 3.36 itself simplifies to the expression

$$I_{GG} = \frac{\Delta H_C}{g_G}\left[\frac{1}{1 + g_A^+/g_A^o}\right] + I_{GA} \qquad (3.37)$$

The ratio g_A^+/g_A^o is the excess air fraction. Unless the air is preheated, the inlet-air enthalpy is much smaller than the heat of combustion and so $I_{GG} \simeq \Delta H_C/g_G$ when little excess air is used. Tabulated in Table 3.3 are values of the <u>gross</u> calorific value ΔH_C which inherently accounts for the heat needed to vaporize the moisture evolved on combustion. Interestingly the value of the added specific enthalpy $\Delta H_C/g_G$ varies little from fuel to fuel, being 2.9 (± 2%) MJ kg^{-1} for coals, 3.1 (± 3%) MJ kg^{-1} for oils and 3.5 (± 5%) MJ kg^{-1} for hydrocarbon gases.

TABLE 3.3 VALUES OF COMBUSTION PARAMETERS FOR SELECTED FUELS

Principal data sources: Krischer[9] and Spiers[19].

Fuel	ash a	hydrogen h	water w	g_G/g_A^o	ΔH_C/ MJ kg^{-1}	$\Delta H_C/g_G$ /MJ kg^{-1}
coal, washed small	0.08	0.04	0.07	1.045	53.4	2.95
lignite	0.05	0.04	0.15	1.049	36.3	2.90
wood	trace	0.051	0.15	1.076	25.8	3.14
gas oil (diesel)	-	0.134	trace	0.989	42.4	3.28
fuel oil, medium	-	0.112	trace	0.992	39.9	3.17
methane	-	0.25	-	0.928	55.9	3.49
town's gas *	-	0.087*	-	0.947	40.7	3.60

* from a continuous vertical retort without steaming, with steaming h = 0.07

The following worked example illustrates the convenience of the foregoing approach.

Example 3.8. A medium fuel-oil is burnt with 15 per cent excess air, which is drawn in at 15°C and 60 per cent relative humidity. The hot flue-gases are blended with secondary air to obtain the gas-inlet temperature of 400°C in a spray dryer. If the dewpoint of the gases exhausted from the dryer is 60°C, what is the effective utilization of heat in the flue gases ?

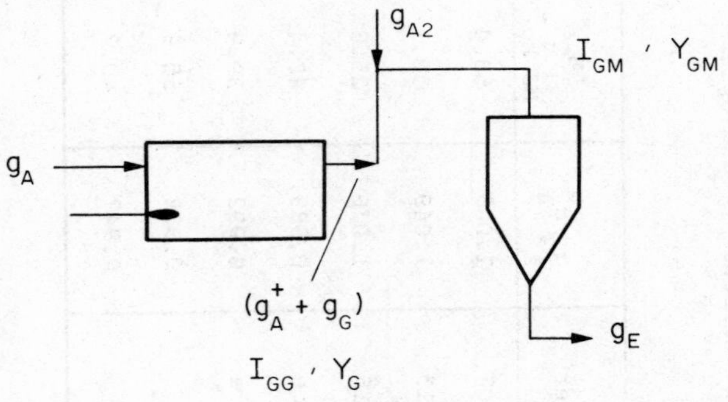

Fig. 3.18 Spray-drying installation

Basis. 1 kg of fuel burnt.

Data for medium fuel-oil are taken from Table 3.3.

Flue-gas, $(g_G + g_A^+)$.

Now, the dry-gas make is given by

$$g_G = \frac{\Delta H_C}{\Delta H_C/g_G} = \frac{39.9}{3.17} = 12.6 \text{ kg}$$

The stoichiometric air requirement thus becomes

$$g_A^o = \frac{g_G}{g_G/g_A^o} = \frac{12.6}{0.992} = 12.7 \text{ kg}$$

The excess air at 15 per cent excess is therefore

$$g_A^+ = \left(\frac{g_A^+}{g_A^o}\right) g_A^o = 0.15 \times 12.7 = 1.9 \text{ kg}$$

So the total dry flue gases are

$$g_G + g_A^+ = 12.6 + 1.9 = 14.5 \text{ kg}$$

Flue-gas humidity, Y_G .

From equation 3.31

$$g_V = 8.937\ h + w$$

$$= (8.937 \times 0.112) + 0 = 1.00\ kg$$

The fresh-air humidity (Y_A) is 0.00643 kg kg^{-1} (from example 3.7).

Since $g_G/g_A^o = 0.992$, equation 3.34 holds for the flue-gas humidity

$$Y_G = Y_A + g_V/(g_G + g_A^+)$$

$$= 0.00643 + 1.00/14.5$$

$$= 0.0754$$

(Note that 90 per cent of the humidity derives from the fuel being burnt)

Flue-gas enthalpy, I_{GG} .

The enthalpy of the excess air (I_{GA}) is 31.3 kJ kg^{-1} (from example 3.7). The added specific enthalpy ($\Delta H_C/g_G$) is 3.17 MJ kg^{-1}. Thus from equation 3.37

$$I_{GG} = \frac{\Delta H_C}{g_G} \frac{1}{(1+g_A^+/g_A^o)} + I_{GA}$$

$$= 3.17/(1 + 0.15) + 0.0313$$

$$= 2.79\ MJ\ kg^{-1}$$

Secondary-air balances.

Moisture balance:

$$(0.0754 \times 14.5) + (0.00643 \times g_{A2}) = Y_{GM}(14.5 + g_{A2}) \qquad I$$

Enthalpy balance:

$$(2.79 \times 14.5) + (0.0297 \times g_{A2}) = I_{GM}(14.5 + g_{A2}) \qquad II$$

Further, the mixture enthalpy I_{GM} can be determined by assuming that the thermophysical properties of the inlet gas to the dryer are the same as those of air, i.e. from equation 2.22 ,

$$I_{GM} = \bar{C}_{PG}T + Y[\bar{C}_{PW}T + \Delta H_{VO}]$$

$$= 1.025 \times 400/1000 + Y_{GM}[(1.939 \times 400) + 2501]/1000$$

$$= 0.41 + 3.277\ Y_{GM} \quad \text{MJ kg}^{-1} \qquad \text{III}$$

Simultaneous solution of equations I, II and III gives

$$g_{A2} = 77.1 \text{ kg}$$

$$Y_{GM} = 0.01735 \text{ kg kg}^{-1}$$

$$I_{GM} = 0.467 \text{ MJ kg}^{-1}$$

Evaporation load. The dewpoint of the outlet-gases is 60°C, and on the assumption that these behave as air, the outlet-gas humidity is 0.1547 kg kg^{-1}.

The dry gas intake into the dryer is $g_G + g_A^+ + g_{A2}$

$$= 12.6 + 1.9 + 77.1 = 91.6 \text{ kg}$$

$$\therefore \text{evaporation} = 91.6\ (0.1547 - 0.01735)$$
$$= 12.6 \text{ kg}$$

Heat utilization. The heat available is 39.9/12.6 = 3.167 MJ kg^{-1}. If moisture, assumed water, is evolved at a mean temperature of 100°C, 2.257 MJ are theoretically needed for each kilogram of moisture evaporated; therefore

$$\text{percentage heat utilization} = \frac{2.257 \times 100}{3.167} = 71.3\ \%$$

3.5 Heat Demand of an Ideal Dryer

The heat demanded by a dryer is defined through equation 3.29 in Section 3.3. However, it is convenient, as a basis for comparing alternative processes, to consider an idealized drying operation which represents the limiting behaviour of a real dryer. In this "ideal" dryer, all the moisture is driven off at the solids-inlet temperature which is assumed to be equal to the temperature of the ambient air; heating of the solids does not occur within the dryer; the dryer is perfectly adiabatic and there is no mechanical work. The solids are non-hygroscopic and thus the heat of moisture binding is zero.

Under these conditions, the heat demand stems from three requirements:

1. The heat needed to vaporize G_V kg of moisture, $G_V \Delta H_{VA}$.

2. The heat needed to raise the temperature of the moisture vapour from T_{GA} to T_{GE}, $G_V \bar{C}_{PW}(T_{GE} - T_{GA})$.

3. Heat needed to raise the temperature of the fresh air from T_{GA} to the outlet temperature T_{GE}, $G\,\bar{C}_{PG}(T_{GE} - T_{GA})$.

Thus the total heat needed Q_V is given by

$$Q_V = G_V \Delta H_{VA} + G(\bar{C}_{PG} + G_V \bar{C}_{PW}/G)(T_{GE} - T_{GA}) \qquad (3.38)$$

In terms of the specific heat demand (per unit mass of moisture evaporated) $\tilde{Q}_V$, equation 3.38 becomes

$$\tilde{Q}_V = Q_V/G_V = \Delta H_{VA} + \frac{G}{G_V}(\bar{C}_{PG} + G_V \bar{C}_{PW}/G)(T_{GE} - T_{GA}) \qquad (3.39)$$

Now the moisture evolved is directly related to the rise in humidity over the dryer, that is $G_V = G(Y_E - Y_A)$, so equation 3.39 can be rewritten as

$$\tilde{Q}_V = \Delta H_{VA} + [\bar{C}_{PG}/(Y_E - Y_A) + \bar{C}_{PW}]\,(T_{GE} - T_{GA}) \qquad (3.40)$$

In the case where the fresh air picks up all the needed heat (no recycle), the exhaust gases take out the quantity of heat (per unit evaporation)

$$\tilde{Q}_E = \Delta H_{VA} \left[\frac{\bar{C}_{PY}\big|_A^E \,(T_{GE} - T_{GA})}{\bar{C}_{PY}\big|_E^O \,(T_{GO} - T_{GE})} \right] \qquad (3.41)$$

where $\bar{C}_{PY}\big|_m^n$ is the mean humid heat between the conditions (T_m, Y_m) and (T_n, Y_n) and T_{GO} is the inlet-air temperature. Very approximately,

$$\bar{C}_{PY}\big|_A^E \simeq \bar{C}_{PY}\big|_E^O \qquad (3.42)$$

which holds strictly only when the drying conditions are very mild. Thus equation 3.41 may be approximated by

$$\tilde{Q}_E \simeq \Delta H_{VA}(T_{GE} - T_{GA})/(T_{GO} - T_{GE}) \qquad (3.43)$$

Finally, one notes that

$$\tilde{Q}_V = \Delta H_{VA} + \tilde{Q}_E \qquad (3.44)$$

so that

$$\tilde{Q}_V = \Delta H_{VA}[1 + (T_{GE} - T_{GA})/(T_{GO} - T_{GE})] \tag{3.45}$$

Equation 3.45 is a more convenient expression for the ideal heat demand than equation 3.40 when there is no air recycle.

It is also useful to have an explicit expression for the inlet-air temperature needed to sustain a specified operation. On combining equations 3.40 and 3.45 for $\tilde{Q}_V$, we find on eliminating common terms in ΔH_{VA} and $(T_{GE} - T_{GA})$

$$\bar{C}_{PG}(Y_E - Y_A) + \bar{C}_{PW} = \Delta H_{VA}/(T_{GO} - T_{GE}) \tag{3.46}$$

which may be re-arranged to give the desired relationship for T_{GO}

$$T_{GO} = T_{GE} + \Delta H_{VA}/[\bar{C}_{PG}/(Y_E - Y_A) + \bar{C}_{PW}] \tag{3.47}$$

The ideal heat demand is plotted in Fig. 3.19 for fresh air at 10°C and a relative humidity of 0.8 at a pressure of 100 kPa in terms of the relative humidity and temperature of the outlet-gas stream. These hygrothermal conditions fix the exhaust-gas humidity Y_E (through equation 2.10) and thus equation 3.40 can be solved for the specific heat demand $\tilde{Q}_V$. Cross-plotted on Fig. 3.19 are contours of the inlet-gas temperature calculated through equation 3.47.

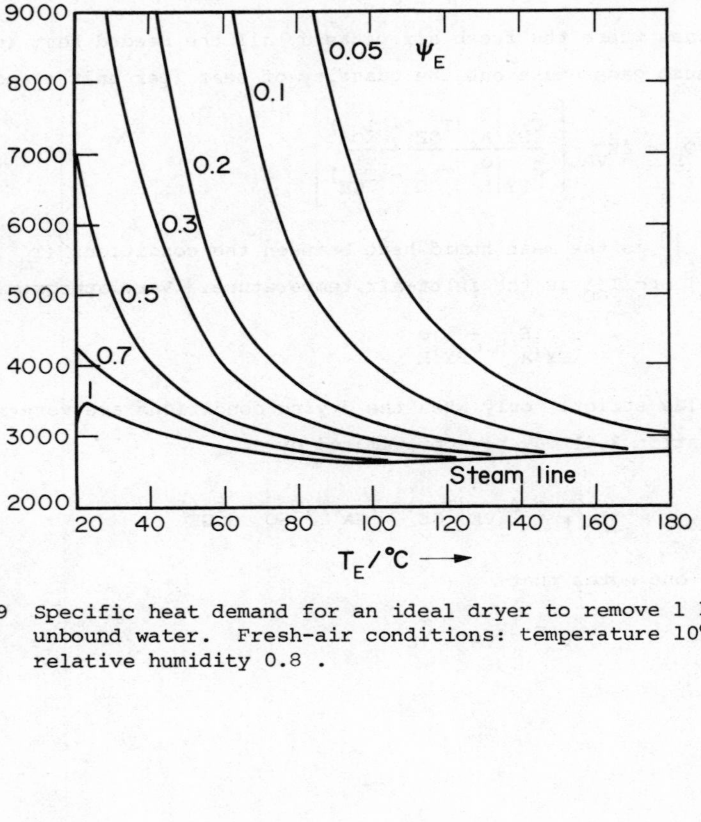

Fig.3.19 Specific heat demand for an ideal dryer to remove 1 kg unbound water. Fresh-air conditions: temperature 10°C, relative humidity 0.8 .

The feasible operating region is represented by the area above the irregularly shaped bound corresponding to saturated-air conditions. The above diagram shows how the thermal economy can be improved by raising the temperature of the inlet air or by permitting a damper exhaust. However, operation at higher temperatures will increase heat losses and be limited by the heating facilities available. Working with a moister exhaust is limited by the need to prevent dew forming on cool surfaces. The options for thermal economy are thus not without restraint.

The usefulness of the foregoing ideas in assessing the differences in process strategies is illustrated by the following worked examples.

<u>Example 3.9</u> Two drying operations are to be compared. In method 1, the air leaves the dryer at a temperature of 110°C and a dewpoint of 60°C. In method 2, 20 per cent less air is used, but the exit temperature is 10°C higher. Which method requires the greater minimum energy demand ? The ambient conditions are: temperature 15°C, humidity 0.0088 kg kg^{-1}, pressure 100 kPa.

At 60°C, the saturated-air humidity is 0.1547 kg kg^{-1} at 100 kPa. From thermophysical tables, ΔH_{VA} is 2466 kJ kg^{-1} for T_{GA} = 15°C, $\bar{C}_{PG}$ = 1.01 kJ kg^{-1} K^{-1} and $\bar{C}_{PW}$ = 1.87 kJ kg^{-1} K^{-1}.

<u>Method 1.</u> From equation 3.40,

$$\tilde{Q}_V = \Delta H_{VA} + [\bar{C}_{PG}/(Y_E - Y_A) + \bar{C}_{PW}](T_{GE} - T_{GA})$$

$$= 2466 + [1.01/(0.1547 - 0.0088) + 1.87]\ (110-15)$$

$$= 3301 \text{ kJ kg}^{-1}$$

<u>Method 2.</u> $\Delta Y = (0.1547 - 0.0088)/0.8 = 0.1824$ kg kg^{-1}

$$T_{GE} = (110 + 10) = 120^{\circ}\text{C}$$

$$\therefore \quad \tilde{Q}_V = 2466 + [1.01/0.1824 + 1.87]\ (120 - 15)$$

$$= 3244 \text{ kJ kg}^{-1}$$

The second method has a smaller energy demand due to the smaller quantity of air being heated despite the higher exit temperature. However, the heat loss may also be higher if the wall temperature of the dryer reflects this higher temperature.

Example 3.10 What are the inlet-air temperatures, in the absence of air recycle, for the two operations considered in Example 3.9 ?

Method 1. From equation 3.47

$$T_{GO} = T_{GE} + \Delta H_{VA}/[\bar{C}_{PG}/(Y_E - Y_A) + \bar{C}_{PW}]$$

$$= 110 + 2466/[1.01/(0.1547 - 0.0088) + 1.87]$$

$$= 390^{\circ}C$$

Method 2. $T_{GO} = 120 + 2466/[1.01/0.1824 + 1.87]$

$$= 453^{\circ}C$$

Thus the air-inlet temperature must be $63^{\circ}C$ higher at the lesser air-rate to achieve the required drying. The inlet temperatures are fairly high and it would be more convenient to resort to isothermal operation with heaters placed inside the drying chamber.

3.6 Entropy-humidity Diagram

Like all practical separation processes, drying is irreversible in a thermodynamic sense. The degree of this irreversibility determines the excess consumption of energy over the theoretical amount of energy needed to separate moisture from its host material. A measure of this irreversibility is the increase of a property we call entropy. The increase in entropy dS for an infinitesimally small amount of heat dQ added reversibly is given by the expression

$$dS = dQ/T \tag{3.48}$$

where T is the absolute temperature of the surroundings. In an irreversible process, an index of the excess energy is thus $T\Delta S$, where ΔS is the entropy increase during the observed process. Sometimes the quantity $T\Delta S$ is called the lost work. It is that work which would have been accomplished in a reversible process, but is lost forever once the actual process is carried out.

Consider a moist gas. The change in entropy on heating from temperature T_1 to T_2 follows from equation 3.48:

$$S_2 - S_1 = \int_{T_1}^{T_2} \frac{dQ}{T} = \int_{T_1}^{T_2} \frac{C_{PY} dT}{T} \tag{3.49}$$

From the definition of humid enthalpy (equation 2.20), equation 3.49 can be rewritten as

$$S_2 - S_1 = \int_{T_1}^{T_2} \left[\frac{\partial I_G}{\partial T}\right]_P \frac{dT}{T} \tag{3.50}$$

Thus it is possible to construct an entropy-humidity diagram from the temperature coefficients of humid enthalpy $(\partial I_G/\partial T)_P$ taken from the the enthalpy-humidity chart. An absolute value for the entropy at a particular temperature can be found by Gibb's rule for an ideal vapour-gas mixture. For an elaboration of this point and further details about drawing up entropy diagrams, the reader is referred to the procedures outlined by Bošnjaković[2]. An entropy-humidity diagram for moist air is sketched in Fig.3.20, on which is superimposed the state path (1 → 2) for a simple drying process.

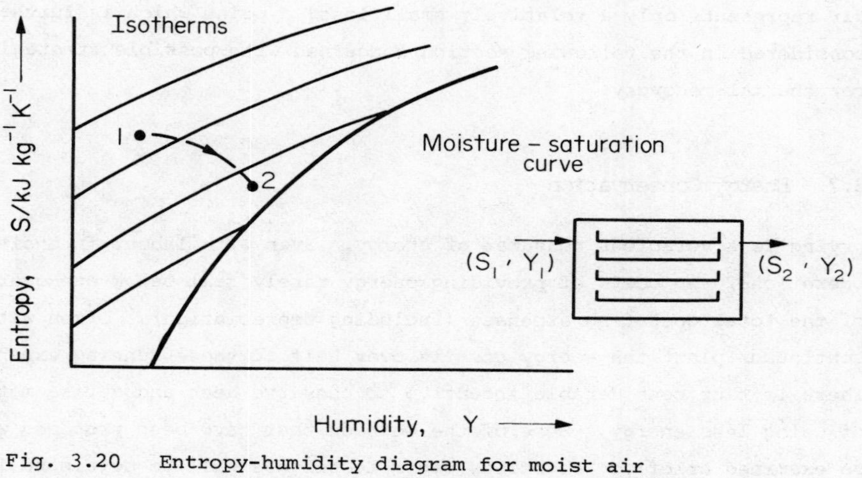

Fig. 3.20 Entropy-humidity diagram for moist air

The increased heating load due to the "wastefulness" of the drying process can be found in the following way. If the extra work were obtained in a reversible heat-engine (Carnot cycle), then the ratio of useful work at temperature T_A to the net energy supplied from an external source at T_S is $T_S/(T_S - T_A)$. It follows then that the extra heat demanded by the process is given by

$$\Delta Q = \left[\frac{T_S}{T_S - T_A}\right] \cdot T_A \Delta S \tag{3.51}$$

where ΔS is the associated increase in entropy.

The entropy increase $-\Delta S_V$ on removing unit mass of moisture in the dryer is given by

$$-\Delta S_V = \frac{S_2 - S_1}{Y_2 - Y_1} - \frac{\Delta S_L}{\Delta X} \tag{3.52}$$

where $-\Delta S_L$ is the specific entropy increase of the moist solids on passing through the dryer and ΔX is the associated change in the moisture content. The value of $-\Delta S_V$ (and thus the irreversibility) becomes smaller as the inlet conditions (S_1, Y_1) lie nearer to the moisture-saturation curve, as Fig. 3.20 shows. Further, one finds from equation 3.51 that the increase in heating load ΔQ has a similar order of magnitude to that of the actual heat required in drying itself. Thus the escape of the warm, humid gases from the dryer to the outside air represents only a relatively small loss, a point which is further considered in the following section concerned with possible strategies for thermal recovery.

3.7 Energy Conservation

Drying is a voracious consumer of energy. Even with labour-intensive operations, the costs of providing energy rarely fall below one-quarter of the total operating expenses (including depreciation). Often with continuous plant the energy cost is over half of these running expenses. There is thus considerable incentive to conserve heat and devise methods of using less energy. Some of the methods that have been proposed will be examined briefly. Doubtless other techniques will be developed in the near future as operators seek to trim costs.

1. Offgas heat recovery. The most obvious way of regaining some of the energy expended in drying is to warm up the fresh air with the hot humid gases being discharged. Nevertheless, there is some doubt that this course is the most advantageous policy, as it was observed in the previous section that the waste heat in the offgases was of minor consequence compared with the thermodynamic irreversibilities of the

drying process. Moreover, a practical obstacle to the use of this waste heat lies in the high moisture content of the airstream. This point is illustrated by the following worked example.

Example 3.11 Air of dewpoint 60°C is being expelled from a dryer at a temperature of 80°C. Fresh air is drawn in at 15°C and 60 per cent relative humidity. It is proposed to preheat the fresh air to 55°C, allowing the exhaust gases to cool to their dewpoint. How much of the available heat can be used in this way ?

Basis. 1 kg dry air.

The offgases have a humidity of 0.1547 kg kg^{-1} (saturation humidity at 60°C). The mean specific heat of dry air from 60°C to 80°C is 1.007 kJ kg^{-1} K^{-1} and of water vapour is 1.876 kJ kg^{-1} K^{-1} ; ΔH_V at 60°C is 2358 kJ kg^{-1}. The enthalpy of the offgases is available as:

1. Sensible heat. [1.007 + (1.876 x 0.1547)] (80-60) = 25.9 kJ kg^{-1}
2. Latent heat. 0.1547 x 2358 = 364.8 kJ kg^{-1}

$\therefore$ total available heat = 25.9 + 364.8 = 390.7 kJ kg^{-1}.

The mean specific heat for air over the temperature range 15 to 55°C, on neglecting the water content, is 1.005 kJ kg^{-1} K^{-1}.

$\therefore$ heat recovered = 1.005 x (55 - 15) = 40.2 kJ K^{-1}

or 40.2 x 100/390.7 = 10.2% of the available heat.

Clearly, as this example demonstrates, direct heat exchange is of little use unless all the evolved heat of the condensing moisture vapour can be taken up in some way.

In industries which use copious quantities of warm water, the heat available could be abstracted by direct contact with cool feedwater. The wet scrubbing of the dryer's exhaust gases also takes out any fine airborne particles arising from the handling of dusty products. The process then has some attractions, particularly if the operation reconstitutes the feed material, as in the case of scrubbing dust-laden exhausts from milk-powder plants to recover a liquid milk. Alternatively, if direct contact between the offgas and the feed slurry can be tolerated, the feed can be preconcentrated to some extent by evaporation.

The next example illustrates the production of hot water by recovering the waste heat in the outlet gases from a pulp dryer.

Example 3.12 Air leaves a pulp dryer at 100°C and with a dewpoint of 60°C. This airstream is to be co-currently contacted with recycled process water at 45°C to produce hot water at 60°C. If the pulp enters the dryer with a moisture content of 1.5 kg kg^{-1} and leaves with a moisture content of 0.15 kg kg^{-1}, how much water can be treated per tonne of dried pulp ? The inlet-air humidity is 0.0088 kg kg^{-1}.

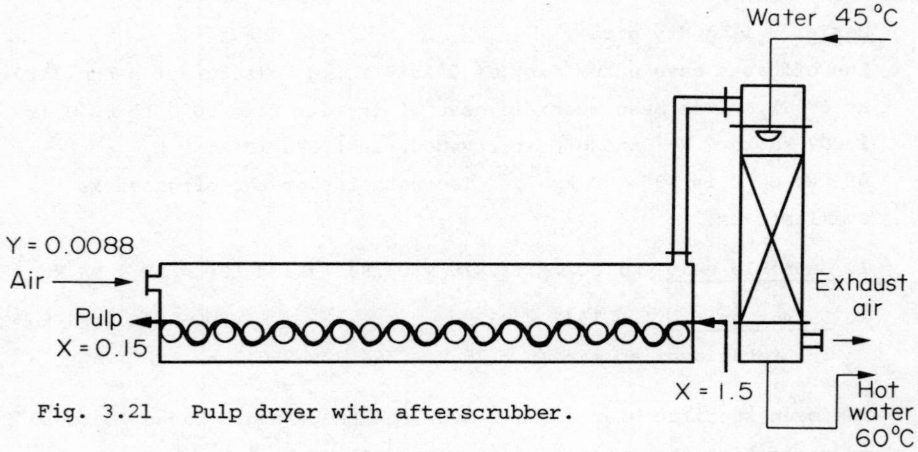

Fig. 3.21 Pulp dryer with afterscrubber.

The offgas is saturated at 60°C and the corresponding humidity is 0.1547 kg kg^{-1}.

The saturated vapour enthalpy at 60°C is 2609 kJ kg^{-1} and the vapour superheat from 60°C to 100°C is 1.88x40 = 75 kJ kg^{-1}; the dry-gas superheat is 1.008x40 = 40.3 kJ kg^{-1}.

∴ Available enthalpy per unit mass of dry gas

$$= 40.3 + 0.1547[2609 + 75] = 455.5 \text{ kJ}$$

Enthalpy uptake in liquid per unit mass = 4.18 x (60-45) = 62.7 kJ

∴ Hot water treated = 455.5/62.7 = 7.26 kg/kg dry air.

$$\text{Air/pulp ratio} = \Delta X/\Delta Y = (1.5 - 0.15)/(0.1547 - 0.0088) = 9.25 \text{ kg/kg dry pulp matter}$$

Water treated per unit mass of pulp product

$$= 7.26 \times 9.25 \times 1.15 = 77.2 \text{ kg kg}^{-1}$$

The density of water at 45°C is 0.990 t m^{-3}

∴ water treated = 77.2/0.990 = 78 m^{3}t^{-1}

Alternatively, the outlet-gas stream could interchange its available heat with another stream provided the flow of the latter were much greater. Such a stream might ventilate a second and larger drying chamber working at a relatively low temperature. One possible arrangement is shown in Fig.3.22

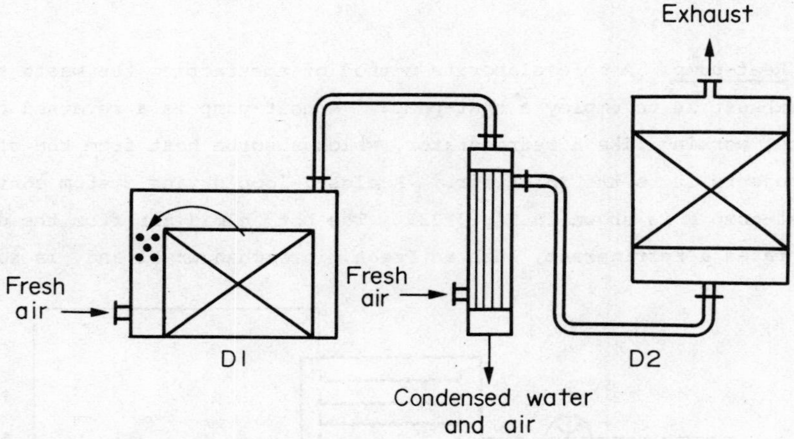

Fig. 3.22 Dual drying system

Air from the main dryer D1 is passed to an air-air exchanger which warms up a much greater quantity of fresh air which then is used in a secondary predryer D2. The exchanger, from which moisture from the primary airstream condenses as dew, must be designed to be easily cleaned, particularly if the stuff in the first dryer tends to be fluffy or dusty. A dual dryer of this type has been built in Switzerland for drying grasses, vegetables and fruit having a total specific heat demand of 2300 kJ kg^{-1} water evaporated, a saving of about one-third of the heating needs for a standard arrangement[11] (see Fig.3.19).

However, there are circumstances, particularly if the dryer is directly fired and the offgas temperature is thus relatively high, when direct heat recovery may be worthwhile. Unconventional recuperative devices of moderate cost may be more suitable for this purpose than standard air-to-air exchangers. One such device is the energy-recovery or heat wheel, which consists of a thermal-storage disc rotating between counterflowing airstreams. Another is the heat-pipe, an evacuated tube internally lined with a wick wetted with a suitable fluid. Heat transfer between the hot and cold ends of the pipe takes place by evaporation and subsequent condensation of this fluid along the pipe. The heat-pipe is especially attractive as there are no moving parts and cross-contamination of the airstreams is avoided.

2. Heat-pump. A more elaborate method of abstracting the waste heat in the exhaust is to employ a heat-pump. A heat-pump is a reversed heat-engine, working like a refrigerator, which absorbs heat from the offtake and rejects it to the inlet air. A closed-loop drying system containing a heat-pump P is shown in Fig.3.23. The hot, humid air from the dryer evaporates a refrigerant, such as Freon, in exchanger E1 and, in so doing,

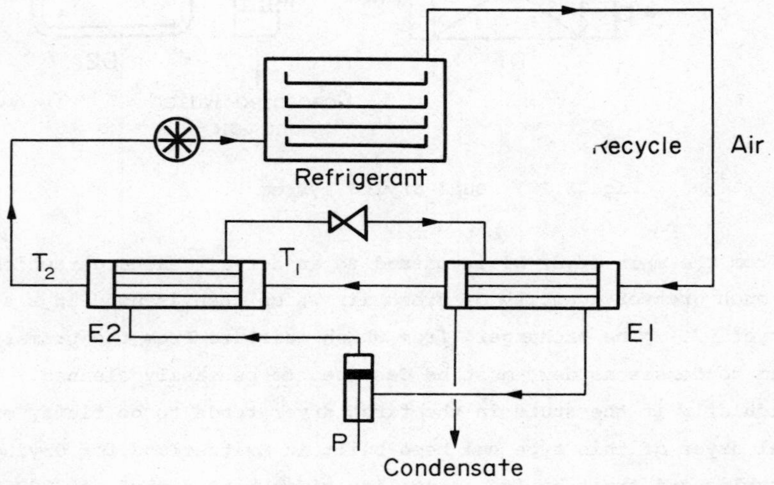

Fig. 3.23 A closed-loop dryer employing a heat-pump

some moisture condenses and is withdrawn. For a closed-loop system at steady state, this dew represents the moisture evaporated in the dryer. The refrigerant vapour is compressed by the heat-pump P to such a pressure that the vapour may be subsequently condensed in exchanger E2 in which the recycled air to the intake of the dryer is heated up again. The heat-pump may be also be placed within the dryer itself to act as a dehumidifier for the air being recirculated inside.(5)

The theoretical coefficient of performance for a reversible (Carnot) heat-engine operating between temperatures T_1 and T_2 is $T_2/(T_2 - T_1)$. For $T_2 = 350$ K and $T_1 = 330$ K, the coefficient of performance is thus $350/(350-330) = 17.5$. Therefore a 1 kW heat-pump will theoretically deliver 17.5 kJ s^{-1} of heat to the air intake. The actual coefficient of performance will be much smaller than the theoretical value as this reversible, Carnot-cycle operation cannot be achieved in practice. The heat-pump may need about triple the minimum energy input, and thus about 6 kW of thermal energy might be delivered for each kW of mechanical energy dissipated.

At present, the use of heat-pumps is limited by the upper working temperatures for dehumidifiers[6] and the size of compressors[7] currently available. Heat-pumps appear particularly suitable for use with those drying chambers which are worked batchwise to dry products that demand careful control of the hygrothermal conditions throughout the operation. The economic feasibility of using heat-pumps depends upon the relative costs of providing electrical and thermal energy, these costs having become more closely equal over recent years. The gain in temperature from T_1 and T_2 depends upon the rise needed to get dehumidification of the air for satisfactory operation of the dryer. Thus the moisture loading in the dryer governs the extent to which a heat-pump will be of use.

3. <u>Vapour recompression</u>. Evaporators are usually built in the form of a series of interconnected chambers in which the vapour from one vessel is used as the heating medium of its neighbouring chamber downstream. By this means, it is not uncommon to evaporate upwards of 3 kg of water for each kg of steam used. By contrast, the steam economy of the rotary dryer considered in Example 3.7 is 1/3.29 = 0.304 kg kg^{-1}, an order of magnitude less. If it were possible to dry materials in an atmosphere of superheated steam, and use the compressed offgas stream as the heating medium, then the thermal economy of a multiple-effect evaporator could be approached.

Superheated-steam drying has, in fact, been known for a long time,[4] but applications have been hindered by lack of suitable equipment and the corrosive conditions encountered.[8] Indeed, in some cases, drying under these conditions can be advantageous. An example may be seen in the drying of timber with superheated steam, which appears to be less severe on the material than drying under a conventional kiln schedule.[14] A general arrangement is shown in Fig.3.24 for drying a stack of timber boards in a kiln by recompressing the vapour offtake.

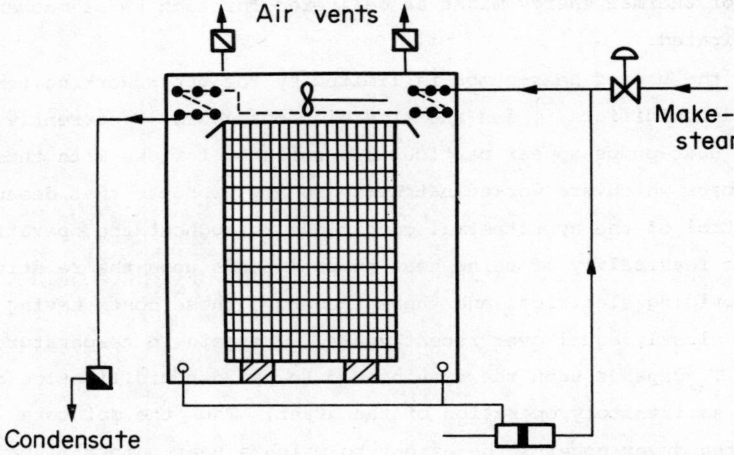

Fig.3.24 Thermal economy by vapour recompression. Arrangement for kiln-seasoning timber.[14]

To obtain a temperature difference across the heating coils, the saturation temperature of the compressed vapour must exceed the temperature in the kiln. For instance, if the desired working temperature is 130°C, then the discharge pressure of the compressor must be greater than the saturation pressure (270 kPa) of steam at 130°C. Saturation pressures rise very rapidly with temperature above 150°C, and thus working at such temperatures becomes increasingly costly and less attractive.

By way of example, Miller[14] compares the energy requirements of vapour-recompression drying with conventional drying methods for seasoning softwood boards in a batch kiln. In the former case, the energy needed to remove the moisture from a given charge is about 2.5 times the energy consumed, an economy which is about fourfold better than the conventional case. The electrical energy used in compression is about one-quarter of the total energy demand, the balance being supplied by the make-up steam.

REFERENCES

1. Bošnjaković, F. and P.L. Blackshear. "Technical Thermodynamics", p.8-13, Holt, Rinehart and Winston, New York (1965).
2. Bošnjaković, F. and P.L. Blackshear, _ibid_. p.307.
3. Dryden, I. G. C. (ed.), "The Efficient Use of Energy", p. 171-4, IPC Science and Technology, Guildford (1975).
4. Hausbrand, E. "Drying by Means of Air and Steam", Chapter 5, Scott and Greenwood, London (1901).
5. Himmelblau, D.M., "Basic Principles and Calculations in Chemical Engineering", 3/e, Chapter 2, Prentice-Hall, Englewood Cliffs N.J. (1974).
6. Hodgett, D. L., Efficient drying using heat pump, _Chem.Engr_. 311, 510-2 (1976).
7. Jebson, R. S., Possible Uses of Heat Pumps in New Zealand Industry, _NZIE Proc.Tech.Groups_, 1, 2 (CH), 105-111 (1975).
8. Kollman, F.P.P., High Temperature Drying, _Forest Prod.J_., _11_, 508-15 (1961).
9. Krischer, O. "Die wissenschaftlichen Grundlagen der Trocknungstechnik", 2/e p.32-46, Springer, Berlin/Göttingen/ Heidelberg (1962).
10. Kröll, K., "Trockner und Trocknungsverfahren", p.40-1, Springer, Berlin/Göttingen/Heidelberg (1959).
11. Kröll, K., _ibid_. p.59.
12. Lapple, W.C., W.E. Clark and E.C. Dybal, Drying design and costs, _Chem.Eng_., _62_(11), 186-200 (1955).
13. Mayhew, Y.R. and G.F.C. Rogers, "Thermodynamics and Transport Properties of Fluids", 2/e, p.16-7, Blackwell, Oxford (1972).
14. Miller, W.R., Vapour recompression drying - a feasibility study, N.Z.F.S., Forest Res.Inst., Forest Prod. Rep. Timber Drying 3, (unpublished)(1975).
15. Nonhebel, G. and A.A.H. Moss, "Drying of Solids in the Chemical Industry", p.142, Butterworths, London (1971).
16. Nonhebel, G. and A.A.H. Moss, _ibid_. p.237.
17. Spiers, H.M., "Technical Data on Fuel", 6-e, p.59-69, British Nat. Cte., World Power Conference, London (1962).
18. Spiers, H.M., _ibid_. p.102.
19. Spiers, H.M., _ibid_. p.296.

PLATE 3. A continuous fluid-bed dryer fitted to the outlet cone of a spray dryer for skimmilk. The two-stage drying results in lower energy and investment costs compared with spray-drying alone. [Anhydro A/S, with permission.]

Chapter 4

HUMIDIFICATION

4.1 Saturation Processes

It has been remarked before that a universal feature of the behaviour of every dryer is the progressive humidification of the air from the intake to the outlet. The extent to which the air becomes saturated with moisture governs the course of drying for the goods inside the unit. Suppose on humidification the state conditions of the air change from (I_{G1} , Y_{G1}) to (I_{G2} , Y_{G2}). Then it follows from equation 3.12 that these conditions can be related to the enthalpy associated with unit mass of moisture added (I_L) by the expression

$$\frac{I_{G2} - I_{G1}}{Y_{G2} - Y_{G1}} = I_L \tag{4.1}$$

There are three limiting cases of interest: changes at constant enthalpy, humidity and temperature. The state paths for these humidification processes are shown in Fig. 4.1.

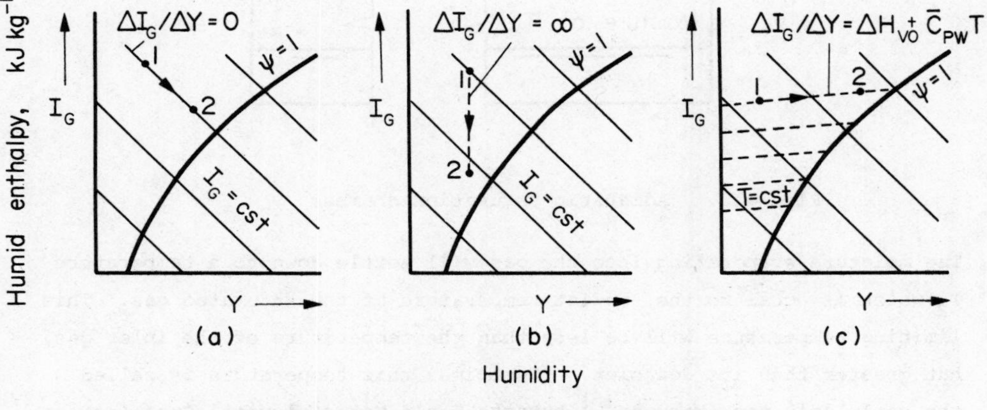

Fig. 4.1 Humidification at (a) constant enthalpy, (b) constant humidity and (c) constant temperature.

1. Isenthalpic change. (Fig. 4.1a). In this case, $\Delta I_G/\Delta Y_G = 0$ and thus from equation 4.1, I_L must be zero. The added moisture,

having zero enthalpy, is liquid at the enthalpy-datum temperature of $0^{o}C$.

2. Isohumid change. (Fig. 4.1b). Since there is no increase in humidity in this limit, $\Delta I_G/\Delta Y_G = \infty$ and the state path is a vertical line on the enthalpy-humidity diagram. This situation corresponds to pure heating or cooling, and is approached by "humidifying" with infinitely hot moisture.

3. Isothermal change. (Fig. 4.1c). This case can be attained only by adding saturated or superheated moisture vapour at the given temperature. It follows that I_L is the moisture-vapour enthalpy H_{GW} :

$$I_L = H_{GW} = \Delta H_{VO} + \bar{C}_{PW}T \tag{4.2}$$

where T is the operating temperature.

4. Adiabatic change. More commonly a gas is saturated under closely adiabatic conditions. Consider the saturation of a gas in an infinitely long and perfectly insulated chamber, as sketched in Fig.4.2.

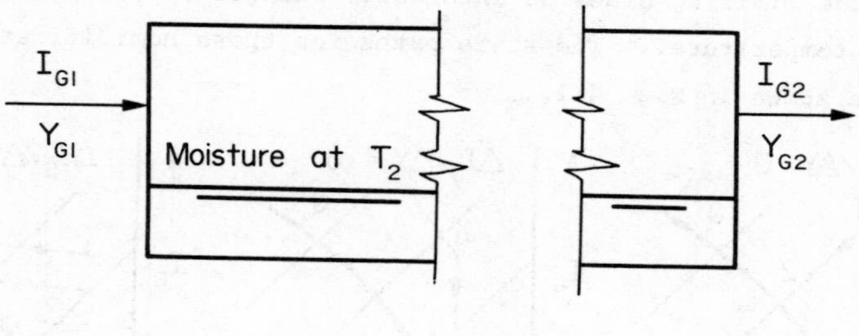

Fig. 4.2. Adiabatic-saturation chamber

The moisture evaporating into the gas will settle down to a temperature T_2 which is equal to the outlet temperature of the saturated gas. This limiting temperature will be less than the temperature of the inlet gas, but greater than its dewpoint. Sometimes this temperature is called the cool-limit temperature,[6] but the World Meteorological Organization[29] has adopted the term thermodynamic wet-bulb temperature. The use of this latter term is understandable in the case of air-water mixtures, as the limiting temperature T_2 is very close to that recorded by a thermometer having a wetted sensor and placed in a moving airstream. However, as we shall see, the recorded temperature can be appreciably different from the limiting temperature T_2. Therefore, the simpler descriptive term, adiabatic-saturation temperature (T_{GS}), is less confusing and is to be preferred.

An enthalpy balance for unit mass of dry gas passing through the adiabatic-saturation chamber (Fig. 4.2) yields the expression

$$I_{G1} + (Y_{G2} - Y_{G1})\, I_L = I_{G2} \tag{4.3}$$

where I_L is the enthalpy of the moisture added to the air on vaporizing unit mass of moisture. It follows that

$$I_L = + H_{LW} \tag{4.4}$$

where H_{LW} is the enthalpy of liquid moisture at temperature T_2. Re-arrangement of equations 4.3 and 4.4 gives

$$\frac{I_{G2} - I_{G1}}{Y_{G2} - Y_{G1}} = + H_{LW} \tag{4.5}$$

The adiabatic-saturation path, corresponding to equation 4.5, is plotted in Fig. 4.3.

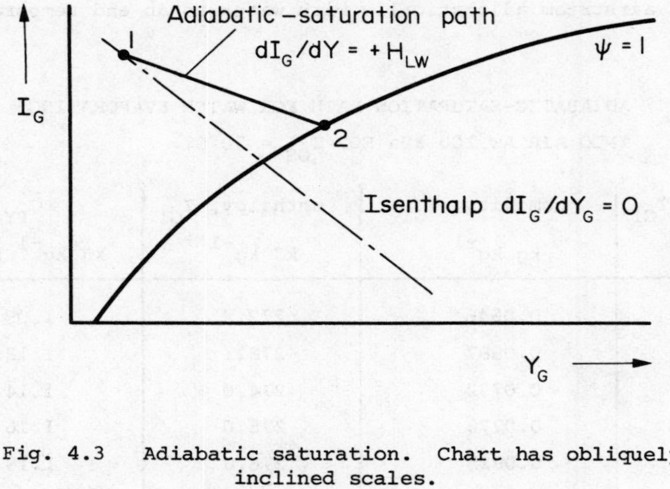

Fig. 4.3 Adiabatic saturation. Chart has obliquely inclined scales.

4.2 Adiabatic-saturation Temperature

Equation 4.5 in the previous section can be re-written in terms of the adiabatic-saturation conditions (I_{GS}, Y_{GS}) as

$$(I_{GS} - I_{G1}) = H_{LW}(Y_{GS} - Y_{G1}) \tag{4.6}$$

Now the difference in humid enthalpies may be related to the corresponding temperature difference by incorporating the humid heat, which is defined in equation 2.21 as the heat capacity of unit mass of dry gas with its associated moisture vapour; that is

$$(I_{GS} - I_{G1}) = \bar{C}_{PY}(T_{GS} - T_{G1}) + H_{GW}(Y_{GS} - Y_{G1}) \tag{4.7}$$

where H_{GW} is the enthalpy of saturated moisture vapour at a temperature T_{GS}. The humid heat is evaluated at a mean temperature $(T_{GS} + T_{G1})/2$ and at the inlet humidity Y_{G1}. Combination of equations 4.6 and 4.7 yields

$$\frac{Y_{GS} - Y_{G1}}{T_{GS} - T_{G1}} = \frac{\bar{C}_{PY}}{(H_{GW} - H_{LW})} = -\frac{\bar{C}_{PY}}{\Delta H_{VS}} \tag{4.8}$$

where ΔH_{VS} is the latent heat of vaporization of moisture at T_{GS}.

Table 4.1 below enumerates for illustrative purposes the state path for saturating an airstream adiabatically with water to an end temperature T_{GS} of 50°C.

TABLE 4.1 ADIABATIC-SATURATION PATH FOR WATER EVAPORATING INTO AIR AT 100 kPa FOR T_{GS} = 50°C.

temperature, T_{G1} °C	humidity, Y_{G1} kg kg^{-1}	enthalpy, I_{G1} kJ kg^{-1}	$\bar{C}_{PY}$ kJ kg^{-1} K^{-1}
100	0.0646	272.2	1.09
90	0.0687	273.1	1.12
80	0.0732	274.0	1.14
70	0.0778	275.0	1.16
60	0.0825	276.0	1.19
50	0.0875	277.0	-

The values of $\bar{C}_{PY}$ are seen to depend on the starting point for the humidification only to a minor extent, ranging between 8 and 19 per cent higher than the heat capacity of dry air at the same temperature. Thus the adiabatic-saturation path (equation 4.7), when plotted on Grosvenor's humidity chart[10] with axes of humidity and temperature, is essentially a straight line of slope $-\bar{C}_{PY}/\Delta H_{VS}$, as shown in Fig. 4.4. The mean slope $-\bar{C}_{PY}/\Delta H_{VS}$ becomes larger as the adiabatic-saturation temperature T_{GS} rises.

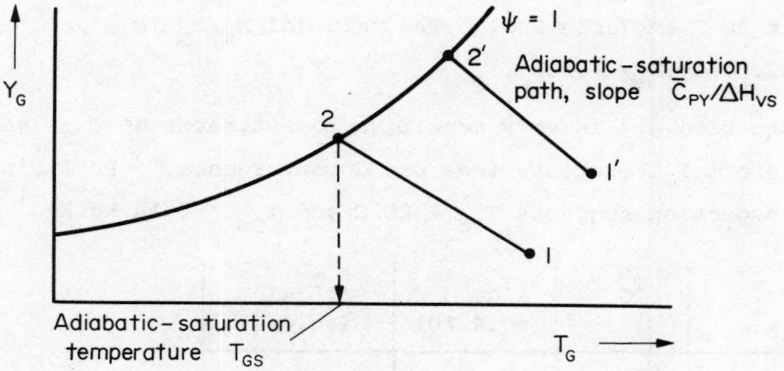

Fig. 4.4 Grosvenor chart illustrating the adiabatic-saturation process.

Equation 4.7 can be solved for the adiabatic-saturation temperature T_{GS}, given the initial set of conditions (T_{G1}, Y_{G1}), provided one has an expression for the saturation humidity in terms of temperature. Vapour-pressure data for organic materials have already been given in graphical form in Fig. 2.2, and saturation humidities for water substance are located in the Appendix. Alternatively, the data may be fitted to an expression of the type

$$\ln p_W^o = -a/T + b \tag{4.9}$$

where a and b are "best-fit" constants. An extensive listing of such constants may be found in the Handbook of Chemistry and Physics.[27] It follows from equation 2.8 that the saturation humidity Y_{GS} is given by

$$Y_{GS} = \left(\frac{M_W}{M_G}\right)\frac{1}{[P\exp(a/T_{GS}-b) - 1]} \tag{4.10}$$

Equations 4.7 and 4.10 can be solved iteratively. Very precise fits of vapour-pressure data for a number of substances are possible by putting $T\ln p_W^o$ in the form of a Chebyshev polynomial in temperature.[2]

Example 4.1. Calculate the adiabatic-saturation temperature for methanol evaporating into air containing 0.1 kg methanol per kg dry air at an inlet temperature of 60°C at 100 kPa total pressure.

Data. The constants a and b in equation 4.10 are respectively 4608 and 18.26 when p_W^o is the saturated-vapour pressure in kPa and T the temperature in K.

For an initial estimate, $\bar{C}_{PY}$ is taken to be 1.1 kJ kg^{-1}K^{-1} and ΔH_V is taken to be 1200 kJ kg^{-1} from the available value at 20°C in Table 2.5. The value of M_W/M_G is 32.04/28.9 = 1.109 .

Equation 4.7 is very sensitive to estimates of T_{GS}, and an informal iteration gives rapid convergence. Preliminary inspection suggests $T_{GS} \simeq 20$°C and $Y_{GS} \simeq 0.14$ kg kg^{-1}

n	T_G^n	Y_{GS} (eq.4.10)	T_{calc}^n (eq.4.7)
1	20	0.144	12
2	18	0.127	30
3	19	0.135	21.3
4	19.1	0.1376	19.0

Since ΔH_V was taken at a temperature of 20°C, no further correction is needed in this regard.

At T_{GS} = 19.1°C the value of $\bar{C}_{PY}$ for the methanol-air mixture is given by

$$\bar{C}_{PY} = \bar{C}_{PG} + \bar{C}_{PW}Y_G$$

$$= 1.004 + (0.761 \times 0.1)$$

$$= 1.080 \text{ kJ kg}^{-1}\text{K}^{-1}$$

This value is close to the first estimate.

For calculations by computer, a satisfactory algorithm can be obtained by decrementing temperature along the adiabatic-saturation path from the known conditions (T_{G1} , Y_{G1}), testing whether the moisture-saturation curve has been overstepped or not, and finally using a step-halving procedure to close onto the adiabatic-saturation point to the required temperature approach.

The adiabatic-saturation process is closely followed in many cases, so it is common practice to show the state paths to reach specified adiabatic-saturation temperatures on humidity charts. These adiabatic-saturation curves, having a uniform slope of + H_{LW} (eq. 4.5), decline linearly on those enthalpy-humidity diagrams plotted with oblique axes. The difference in slope between the adiabatic-saturation contours and the isenthalpic lines is small, so for this reason it is often easier to follow adiabatic-saturation curves on the Grosvenor chart which has co-ordinates of humidity and temperature, (see Fig. 4.4). Some Grosvenor charts have slightly distorted scales to render the adiabatic-saturation curves straight and parallel. While such a construction is convenient for the purposes of interpolation, clearly the axes are not linearly related and the Lever Rule does not hold.

4.3 Vapour Diffusion

The rate at which moisture vapour can escape from a damp surface into the bulk of an airstream flowing over it clearly sets a limit to the maximum drying rate under a prescribed set of hygrothermal conditions. The outward movement of vapour goes through a region where the temperature varies from place to place, but the motion is surprisingly insensitive to such differences in temperature. Under most drying conditions, one may assume that properties such as density and diffusion coefficients remain constant over the whole field at some mean-temperature value.[30]

Let us consider the transport of moisture vapour from a plane surface 1, adjacent to the place where the moisture is evaporating, to another plane surface 2 in the bulk gas-stream where the humidity is less. Over the region considered the physical properties may be regarded as being uniform.

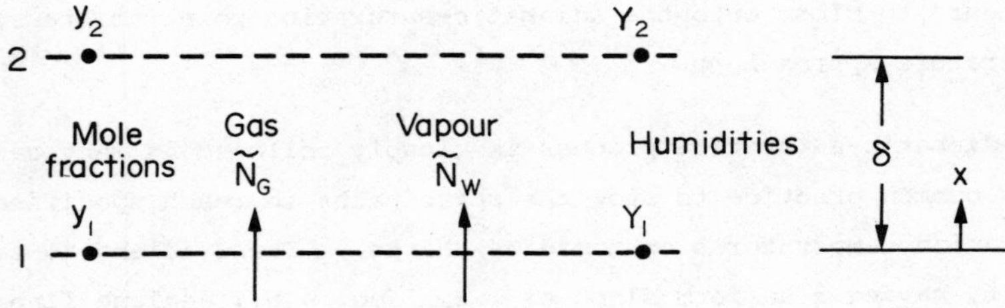

Fig. 4.5 Vapour diffusion

The rate at which the vapour moves relative to the medium is given by Fick's law, namely

$$J_W = -\mathcal{D}_{WG}\frac{dc_W}{dx} \tag{4.11}$$

where J_W is the diffusion flux of moisture vapour (mol $m^{-2}s^{-1}$),
c_W is the molar concentration of moisture (mol m^{-3}),
and $\mathcal{D}_{WG}$ is a "constant" of proportionality, the diffusion coefficient (m^2s^{-1}).

For practical purposes we are more interested in the rate relative to a fixed surface, say the dryer's wall. The moisture efflux is then the sum of the diffusion flux and moisture transported by the total motion. This latter quantity is given by the product of the total flux $(\tilde{N}_W + \tilde{N}_G)$ and the mole fraction of moisture vapour contained therein, c_W/c, where c is the mean molar or "total" concentration. Thus the moisture-transfer flux from surface to surface is given by

$$\tilde{N}_W = J_W + c_W(\tilde{N}_W + \tilde{N}_G)/c \tag{4.12}$$

On collecting terms for this flux $\tilde{N}_W$, we find

$$\tilde{N}_W = \frac{J_W + c_W\tilde{N}_G/c}{(1 - c_W/c)} \tag{4.13}$$

In some situations, the flux $\tilde{N}_G$ is a significant quantity, particularly in cases when air jets impinge on a surface or air is blown through open fabrics. Nevertheless, $\tilde{N}_G$ is often negligible for transport at <u>right-angles</u> to the moist surface, since experiments[26] have shown that most of the fall in vapour concentration occurs within a thin (and quiescent) zone close to the evaporating surface. Under such conditions, equation 4.13 becomes

$$N_W = J_W/(1 - c_W/c) \qquad (4.14)$$

which, on the insertion of the expression (eq.4.11) for Fick's law, yields

$$\tilde{N}_W = \frac{-\mathcal{D}_{WG}}{(1-c_W/c)} \cdot \frac{dc_W}{dx} \qquad (4.15)$$

When the mean molal concentration is independent of composition (the case for an ideal gas), equation 4.15 can be re-written in a more convenient form in terms of mole fractions, $y = c_W/c$:

$$\tilde{N}_W = \frac{-c\,\mathcal{D}_{WG}}{(1-y)} \cdot \frac{dy}{dx} \qquad (4.16)$$

Mass-transfer coefficients. An expression for the steady diffusion of moisture vapour is readily obtained from equation 4.16 by integration over the distance δ between the considered surfaces (see Fig. 4.5). Whence we find

$$\tilde{N}_W = \left[\frac{c\mathcal{D}_{WG}}{\delta}\right] \ln \frac{(1-y_2)}{(1-y_1)} \qquad (4.17)$$

The quantity $[c\,\mathcal{D}_{WG}/\delta]$ has units of flux, and represents the conductance for the transport of moisture. This conductance is usually called the mass-transfer coefficient F. The logarithmic term is the driving force, which is clearly not equal to the difference in the moisture-vapour concentration between the surfaces.

In drying calculations, it is more convenient to use humidities, which are vapour/gas mass fractions, rather than mole fractions of moisture vapour. For an ideal gas, it follows from equation 2.7 that

$$Y = \frac{m_W}{m_G} = \left(\frac{M_W}{M_G}\right) \frac{c_W}{(c-c_W)} \qquad (4.18)$$

so, on introducing the mole fraction of moisture vapour, $y = c_W/c$, we find

$$Y = \left(\frac{M_W}{M_G}\right) \frac{y}{(1-y)} \qquad (4.19)$$

from which we get the useful identity

$$(1-y) \equiv D/(D+Y) \tag{4.20}$$

where $D = M_W/M_G$. On back-substitution of this identity into equation 4.17 together with the mass-transfer coefficient F,

$$F = c\mathscr{D}_{WG}/\delta \tag{4.21}$$

we obtain the following expression for the moisture-vapour flux

$$\tilde{N}_W = F \ln \frac{(D+Y_1)}{(D+Y_2)}, \quad Y_1 > Y_2 \tag{4.22}$$

This expression holds for the flux expressed in molal units (e.g. mol $m^{-2}s^{-1}$). Mass units are sometimes more convenient, and the corresponding expression becomes

$$N_W = FM_W \ln \frac{(D+Y_1)}{(D+Y_2)} \tag{4.23}$$

It is also interesting to discern in what way the flux depends upon the simple <u>humidity potential</u>, $(Y_1 - Y_2)$. Expansion of the logarithmic driving force yields the relationship

$$N_W = FM_G\left[\frac{D}{D+Y_1}\right]\left[\frac{(D+Y_1)\ \ln\{1 + \frac{(Y_1-Y_2)}{(D+Y_2)}\}}{(Y_1 - Y_2)}\right](Y_1 - Y_2) \tag{4.24}$$

$$= (FM_G) \cdot \phi_1 \cdot \phi_2 \cdot (Y_1 - Y_2) \tag{4.25}$$

When the humidity potential is small, that is $(Y_1-Y_2) \ll (D+Y_2)$, the term in the second square bracket (ϕ_2) is almost unity, and

$$N_W \simeq (FM_G) \cdot \left[\frac{D}{D+Y_1}\right] (Y_1 - Y_2) \tag{4.26}$$

$$= (FM_G) \cdot \phi_1 \cdot (Y_1 - Y_2) \tag{4.27}$$

Moreover, should the humidity level itself be very small, that $Y_1 \ll D$, then the term in the first square bracket of equation 4.24 (ϕ_1) is also almost one, and

$$N_W \simeq (FM_G)\ (Y_1 - Y_2) \tag{4.28}$$

Under these circumstances alone, then, is the moisture-vapour efflux directly proportional to the humidity difference driving the transport.

Because of this linearity under mild drying conditions, many authorities still use another mass-transfer coefficient, following the precedent set by Lewis.[15] This coefficient is defined by the ratio of the flux to the humidity potential:

$$K = N_W/(Y_1 - Y_2) \tag{4.29}$$

Kinetic data for drying are still frequently expressed in terms of this coefficient, rather than the less humidity-dependent conductance F. Nevertheless, in calculations concerned with conditions of intensive drying, it is probably more satisfactory to convert the field data into the F-type conductances which are less sensitive to humidity levels. By comparing equations 4.23 and 4.29, we find that

$$K = FM_G\left[\frac{D}{D+Y_1}\right]\left[\frac{(D+Y_1)\ \ln\{1 + \dfrac{(Y_1-Y_2)}{(D+Y_2)}\}}{(Y_1 - Y_2)}\right] \tag{4.30}$$

which can be re-arranged to yield the identity

$$F \equiv \frac{K(Y_1-Y_2)}{M_W\ \ln\{1 + \dfrac{(Y_1-Y_2)}{(D+Y_2)}\}} \tag{4.31}$$

As anticipated, the right-hand side of this expression takes a limiting value of K/M_G at low humidities and small humidity potentials.

Example 4.2 Acetone vapour is diffusing through air. Estimate the percentage change in the mass-transfer coefficient K when the humidity conditions are changed as set out below:

conditions	Y_1	Y_2
old	0.1	0.02
new	0.5	0.15

The molar mass ratio is given by

$$D = M_W/M_G = 0.05808/0.02897 = 2.005$$

Under the first set of conditions

$$F = \alpha_I K_I/M_W$$

where $$\alpha_I = \left[\frac{(Y_1 - Y_2)}{\ln\{1 + \frac{(Y_1 - Y_2)}{(D+Y_2)}\}} \right]_I \qquad \text{(eq.4.31)}$$

$$= \frac{(0.1 - 0.02)}{\ln\{1 + \frac{(0.1-0.02)}{(2.005+0.02)}\}}$$

$$= 2.065$$

Under the second set of conditions

$$F = \alpha_{II} K_{II}/M_W$$

where $$\alpha_{II} = \left[\frac{(Y_1 - Y_2)}{\ln\{1 + \frac{(Y_1 - Y_2)}{(D+Y_2)}\}} \right]_{II}$$

$$= \frac{(0.5 - 0.15)}{\ln\{1 + \frac{(0.5 - 0.15)}{(2.005+0.15)}\}}$$

$$= 2.326$$

$$\therefore \quad \frac{K_{II}}{K_I} = \frac{\alpha_I}{\alpha_{II}} = \frac{2.065}{2.326} = 0.888$$

i.e. The K-type coefficient would appear to be 12 per cent less under the second set of conditions compared with that for the first.

This calculation illustrates the importance of the "proper" choice of mass-transfer coefficients. An excellent review of the range for selection is given by Spalding.[21] In this work, the modified conductance (FM_G), which will be given the symbol K_o, will be used for the following reasons:

1. The quantity K_o is the ratio of the vapour flux to the humidity potential under mild drying conditions.
2. This coefficient arises as a natural grouping of transport parameters (eq. 4.17).
3. The coefficient can be readily used with humidity differences as a measure of the driving force.

To summarize: we express the rate of moisture transfer through unit area, the vapour "flux", by an expression of the form

$$N_W = K_o\phi\Delta Y \qquad (4.32)$$

where ΔY is the humidity difference driving the movement and ϕ is a lumped factor equal to the product of $\phi_1\phi_2$ of equation 4.24.

4.4 Humidity-potential Coefficient

The parameter ϕ in equation 4.32 will be called the humidity-potential coefficient. In the limit of very mild drying conditions, this coefficient takes a value of 1 when the vapour flux becomes directly proportional to the humidity potential ΔY. Inspection of equation 4.24 shows that this coefficient ϕ departs from its limiting value as both the humidity potential and the humidity level itself rise. The first influence is that of high mass-transfer rates, the second is that due to the "inappropriateness" of the choice of the humidity difference as the driving force for the vapour efflux.

The magnitude of the coefficient ϕ is given by those terms in equation 4.24 in square brackets, namely

$$\phi = \left[\frac{D}{D+Y_1}\right]\left[\frac{(D+Y_1 \ln\left\{1 + \dfrac{(Y_1-Y_2)}{(D+Y_2)}\right\}}{(Y_1 - Y_2)}\right] \qquad (4.33)$$

$$= \phi_1 \cdot \phi_2 \qquad (4.34)$$

in which ϕ_1 relates to the effect of humidity level and ϕ_2 to that of humidity difference. The parameter ϕ_1 is always less than 1, whereas ϕ_2 is always greater than 1 for any finite humidity difference. The latter coefficient takes its largest value for the limiting condition of vapour flow into a perfectly dry gas ($Y_2 = 0$). Let this limiting value be ϕ_2^o, which is readily found from equations 4.33 and 4.34 to be given by

$$\phi_2^o = \frac{(D+Y_1)}{Y_1} \ln \{1 + Y_1/D\} \tag{4.35}$$

Clearly, ϕ will range between the limits of ϕ_1 (zero humidity potential) and ϕ_2^o (maximum humidity potential). These limiting values of ϕ are plotted in Fig. 4.6 for the case of water vapour diffusing through air from a zone where the humidity ranges from 0.01 to 1 kg kg^{-1}.

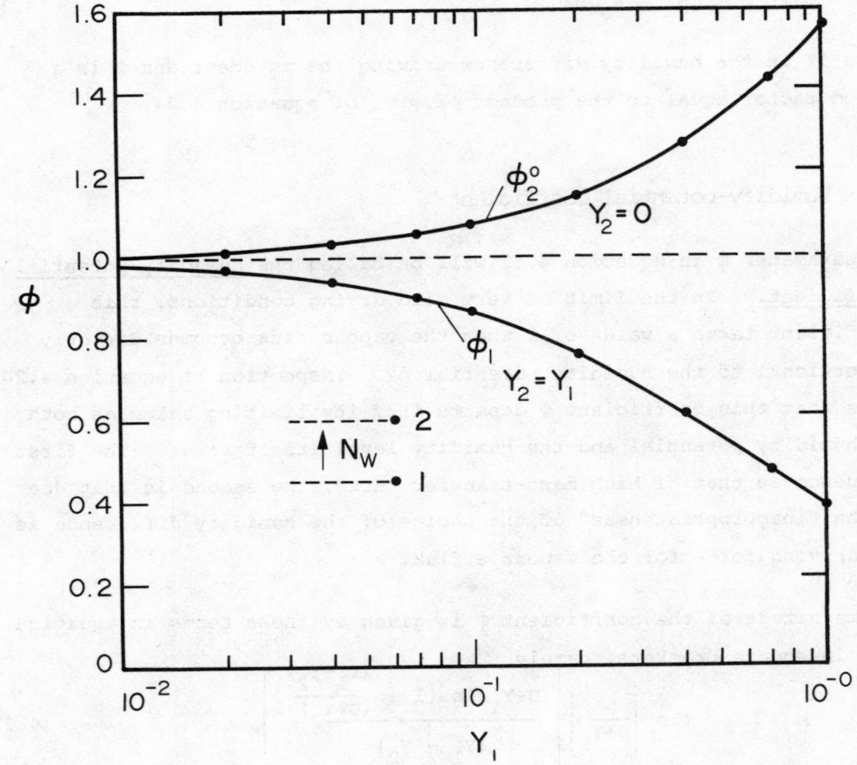

Fig. 4.6 Humidity-potential coefficients for water vapour diffusing through air.

The air in dryers is usually extremely humid. Indeed, the lost work for drying becomes less as the process is operated closer to air-saturation conditions. Thus often the humidity-potential coefficient can be approximated by the zero-flux limit,[2] that is

$$\phi \simeq \phi_1 = D/(D+Y_1) \tag{4.36}$$

Further, if Y_1 remains constant everywhere in the dryer, then the humidity-potential coefficient is also almost invariant and the vapour flux becomes directly proportional to the humidity difference <u>irrespective</u> of the humidity level. Such conditions occur in the adiabatic drying of very wet stuff.

<u>Example 4.3</u> Evaluate the humidity-potential coefficients ϕ , ϕ_1 and ϕ_2^o for the flow of acetone vapour through air between humidity levels $Y_1 = 0.25$ and $Y_2 = 0.20$.

$$D = M_W/M_G = 2.005$$

$$\phi_1 = \frac{D}{(D+Y_1)} = \frac{2.005}{(2.005 + 0.25)} = 0.889$$

$$\phi_2 = \frac{(D+Y_1) \ln \{1 + \frac{(Y_1-Y_2)}{(D+Y_2)}\}}{Y_1 - Y_2}$$

$$= \frac{(2.005 + 0.25) \ln \{1 + \frac{(0.25 - 0.20)}{(2.005+ 0.20)}\}}{(0.25 - 0.20)}$$

$$= 1.011$$

$$\phi = \phi_1\phi_2 = 0.889 \times 1.011 = 0.899$$

$$\phi_2^o = \frac{(D+Y_1) \ln \{1 + Y_1/D\}}{Y_1}$$

$$= \frac{(2.005 + 0.25) \ln \{1 + 0.25/2.005\}}{0.25}$$

$$= 1.060$$

In this case ϕ_2 scarcely deviates from unity and the approximation $\phi \simeq \phi_1 = D/(D+Y_1)$ is good enough.

4.5 Evaporation from a Moist Surface

So far we have considered the flux of moisture vapour between two specified surfaces in the gas phase. In practice, we take one surface to lie in the bulk of the gas stream and the other to be in the air-skin next to the exposed surface of the stuff being dried. Suppose this surface is thoroughly sodden so that the moisture exerts its full vapour pressure p_W^o . The partial pressure of moisture vapour in this skin must also be p_W^o and it follows that the corresponding humidity is given by

$$Y_1 = Y_S = \left(\frac{M_W}{M_G}\right) \frac{p_W^o}{(P-p_W^o)} \tag{4.37}$$

which is the same as equation 2.8 . Should the material be hygroscopic and have a relative humidity $\psi = p_W/p_W^o$, then the surface humidity becomes

$$Y_1 = Y_S = \left(\frac{M_W}{M_G}\right) \frac{\psi p_W^o}{(P-\psi p_W^o)} \tag{4.38}$$

which is equation 2.10 .

Let us consider the layer of moisture-laden gas next to a wet surface. Throughout this layer there will be changes in temperature and humidity as shown in Fig. 4.7 .

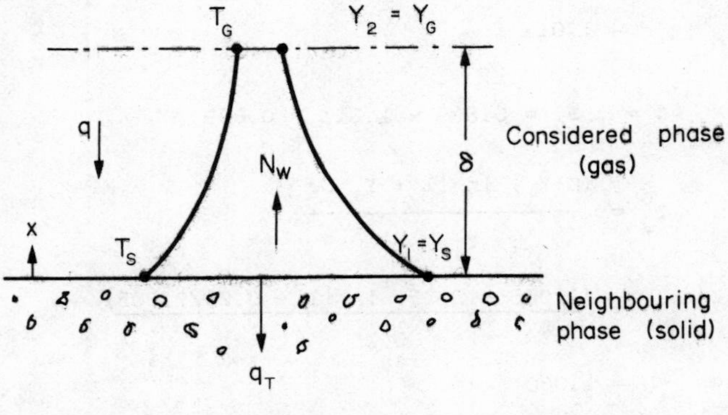

Fig. 4.7 Humidity and temperature profiles near a wet surface

The nett heat flux reaching the surface is the difference between that conducted through the air film due to the temperature difference across it and the enthalpy convected away by the movement of moisture, that is:

$$q = \lambda \frac{dT}{dx} - \bar{C}_{PG}(T-T_S)N_W \tag{4.39}$$

On integrating equation 4.39 between the limits x = 0 (surface) and x = δ (bulk), one finds

$$N_W \bar{C}_{PG}(T_G-T_S)/q = e^{N_W\bar{C}_{PG}\delta/\lambda} - 1 \tag{4.40}$$

Now the ratio λ/δ is called the heat-transfer coefficient h which may be compared with the definiton of the mass-transfer coefficient $K_o = (cM_G)\mathcal{D}_{WG}/\delta$. Equation 4.40 can be recast to incorporate this coefficient h :

$$q = h(T_G - T_S)\left[\frac{E}{\exp E-1}\right] \tag{4.41}$$

where $E = N_W\bar{C}_{PG}/h$

The term in square brackets is sometimes called the Ackermann correction [1] ϕ_E and accounts for the influence of the moisture-vapour efflux on the transfer of heat to the surface.

Example 4.4 Evaluate the Ackermann correction for a cross-circulated dryer for which $h = 50\ W\ m^{-2}\ K^{-1}$ and $N_W = 0.065\ g\ m^{-2}\ s^{-1}$. (These conditions correspond to gentle drying in an oven with an air velocity of about $1\ m\ s^{-1}$ over the goods, a bulk-air temperature of 90°C and a surface-air temperature of 50°C.)

We may take $\bar{C}_{PG} = 1\ J\ g^{-1}K^{-1}$ (note units), so that

$$E = N_W\bar{C}_{PG}/h$$

$$= 0.065 \times 1/50 = 0.0013$$

$$\phi_E = E/(\exp E-1)$$

$$= 0.0013/(\exp 0.0013 - 1)$$

$$= 0.9994$$

Clearly, the influence of evaporation on heat transfer at these drying rates is negligible. Even if the temperature drop doubles, corresponding to a bulk-air temperature of 130°C, the Ackermann correction changes only slightly to 0.9987. The correction is thus normally fairly small, except when drying under extreme conditions, and usually can be ignored.

The total heat-transfer q_T into the neighbouring solid phase is given by the difference of the flux reaching the surface q and the enthalpy taken away by evaporation, that is

$$q_T = q - N_W \Delta H_{VS} \tag{4.43}$$

On inserting equation 4.41 for q and equation 4.32 for N_W with $\Delta Y = Y_S - Y_G$, one gets

$$q_T = h(T_G - T_S)\left[\frac{E}{\exp E - 1}\right] - K_o \phi (Y_S - Y_G)\, \Delta H_{VS} \tag{4.44}$$

Equation 4.44 contains two transport coefficients, h and K_o, which are interrelated through the analogies between transport processes. The simplest form of the analogy, which assumes turbulent flow of the fluid over the swept surface, is that suggested by Chilton and Colburn:[9]

$$j_M = \frac{K_o}{G_T} \cdot Sc^{2/3} = \beta j_H = \beta \cdot \frac{h}{\bar{C}_P G_T} \cdot Pr^{2/3} \tag{4.45}$$

where

G_T is the <u>total</u> gas flow per unit cross-section (kg $m^{-2}s^{-1}$),

Sc is the Schmidt number,

Pr is the Prandtl number,

β is a coefficient close to unity.

The Schmidt and the Prandtl numbers are dimensionless transport criteria which each depend upon the appropriate ratio of boundary-layer thicknesses, as listed in Table 4.2. Since these layers are of similar thickness in gas-transport processes, all the listed transport criteria have values of similar order close to 1.

TABLE 4.2 DIMENSIONLESS CRITERIA FOR MOISTURE TRANSPORT

Symbol	Name	Definition	Ratio of boundary-layer thicknesses
Pr	Prandtl	$\bar{C}_p \mu/\lambda$	heat transfer : skin friction
Sc	Schmidt	$\mu/\rho\mathcal{D}$	moisture transfer : skin friction
Le	Lewis *	$\lambda/\bar{C}_p\rho\mathcal{D}$	moisture transfer : heat transfer
Lu	Luikov	$\mathcal{D}\bar{C}_p\rho/\lambda$	heat transfer : moisture transfer

$\bar{C}_p$ = heat capacity, μ = dynamic viscosity, λ = thermal conductivity, ρ = density, $\mathcal{D}$ = diffusivity.

*Footnote: Lewis, in his original paper[15] in 1922, observes that the quantity $h/K\bar{C}p$ is roughly a constant for a given system. This quantity, which is essentially $Sc/Pr^{2/3}$ from equation 4.45, is usually called the Lewis number in air-conditioning practice.[24] Further, to add to the confusion, a recent mass-transfer text[20] has defined the Lewis number as Pr/Sc. The definition adopted in this book is more universal in the process-engineering literature,[4, 7, 25] but perhaps it is wiser to side-step the issue and use the Luikov number instead.

Experimentally the coefficient β has been found to be about 1. The mean value of this coefficient is given as 1.1 for the evaporation of water and ethanol and 1.0 for benzene and carbon tetrachloride in the experiments of Heertjes and Ringens[12] who studied the evaporation of these liquids into an airstream flowing around a small porous slab wrapped in rayon gauze and saturated with the liquid. Other data[16] for the evaporation of water into the turbulent flow of air, helium and Freon-12 over a 50 mm long wetted wick confirm the value 1.1. The appearance of the coefficient β has been ascribed to differences in the way the mean velocity should be computed in heat and in mass transfer. This view[5] results in the following expression for β:

$$\beta = \frac{M_S \ln(\bar{M}_S/\bar{M}_G)}{\frac{(M_W - M_G)D}{(D+Y_S)} \cdot \ln\left[\frac{(D+Y_S)}{(D+Y_G)}\right]} \qquad (4.46)$$

where $\bar{M}_j$ is the mean molar mass at the subscripted location.

Example 4.5 Find the value of the velocity coefficient β for acetone vaporizing into air with $Y_S = 0.25$ and $Y_G = 0.20$.

In this case,

$$M_W = 0.05808 \text{ kg mol}^{-1}$$

$$M_G = 0.02897 \text{ kg mol}^{-1}$$

$$D = M_W/M_G = 0.05808/0.02897 = 2.005$$

The acetone mole fractions follow from the humidities

$$y_S = \frac{Y_S}{(D+Y_S)} = \frac{0.25}{(2.005 + 0.25)} = 0.1109$$

$$y_G = \frac{Y_G}{(D+Y_G)} = \frac{0.20}{(2.005 + 0.20)} = 0.0907$$

The corresponding mean molar masses can now be calculated:

$$\bar{M}_S = M_W y_S + M_G(1 - y_S)$$

$$= (0.05808 \times 0.1109) + 0.02897\ (1 - 0.1109)$$

$$= 0.03220$$

$$\bar{M}_G = M_W y_G + M_G(1 - y_G)$$

$$= (0.05808 \times 0.0907) + 0.02897\ (1 - 0.0907)$$

$$= 0.03161$$

From equation 4.46

$$\beta = \frac{\bar{M}_S \ln(\bar{M}_S/\bar{M}_G)}{\dfrac{(M_W - M_G)D}{(D+Y_S)} \cdot \ln\left[\dfrac{(D+Y_S)}{(D+Y_G)}\right]}$$

$$= \frac{0.0322 \ln 0.03220/0.03161}{\dfrac{(0.05808 - 0.02897) \times 2.005}{(2.005 + 0.25)} \ln\left[\dfrac{(2.005 + 0.25)}{(2.005 + 0.20)}\right]}$$

$$= 1.034$$

This value lies within the range of experimental data. The correction is minor and within the uncertainty of the transfer coefficients themselves.

Since heat-transfer coefficients are more numerous than the corresponding mass-transfer data, it is preferable to eliminate the less certain mass-transfer coefficients K_o whenever possible. Equation 4.45 can be rewritten to give an explicit expression for K_o, namely :

$$K_o = \frac{\beta h}{\bar{C}_P} \cdot \left(\frac{Pr}{Sc}\right)^{2/3} = \frac{\beta h}{\bar{C}_P} \cdot Lu^{2/3} \qquad (4.47)$$

where the Luikov number Lu is $\mathcal{D}\bar{C}_p\rho/\lambda$. The Luikov number for liquids evaporating into air at room conditions varies from 1.170 for water to 0.314 for a heavy organic solvent such as m-xylene.[28] Since the heat capacity of air is about 1 J $g^{-1}K^{-1}$, it follows that the heat-transfer coefficient in W $m^{-2}K^{-1}$ has roughly the same numerical magnitude as the mass-transfer coefficient in units of g $m^{-2}s^{-1}$ when the moisture is water and the gas air. For example, under conditions when the heat-transfer coefficient is 50 W $m^{-2}K^{-1}$ and the humidity potential 0.005 g g^{-1}, the maximum evaporation, and thus drying rate, is therefore 50 x 0.005 = 0.25 g $m^{-2}s^{-1}$. One might expect to get such a drying rate in a commercial four-rack tray dryer.[17]

Equation 4.47 may be invoked to eliminate the mass-transfer coefficient from equation 4.44 for the total heat flux q_T ; the final result is

$$q_T = h\left[(T_G - T_S)\left[\frac{E}{\exp E - 1}\right] - \frac{\beta}{\bar{C}_P} Lu^{2/3} \phi(Y_S - Y_G)\Delta H_{VS}\right] \qquad (4.48)$$

A particularly important case occurs when $q_T = 0$, and this circumstance is explored in the next section.

4.6 Psychrometry

One of the commonest applications of the foregoing theory in Section 4.5 relates to the long-standing use of a thermometer with a wet sensor to measure humidities. The so-called wet-bulb thermometer is an ordinary thermometer, the temperature-sensing element of which is covered by a porous cloth or sheath kept wet. A stream of air or gas is maintained over the wetted element, either through any existing motion, or by creating a draught when the gas is still, to permit steady evaporation of the liquid from the exposed surface. When both a wet-bulb and dry-bulb (unwetted) element are incorporated in the same instrument, the unit is called a psychrometer.

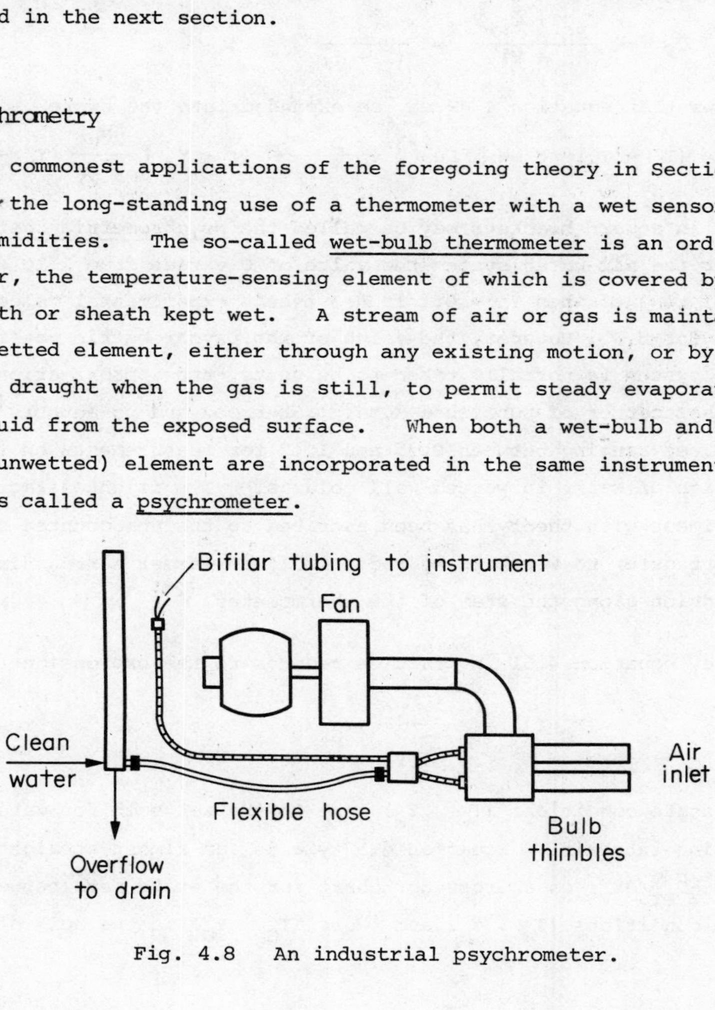

Fig. 4.8 An industrial psychrometer.

Under steady evaporative conditions, the total heat flux q_T into the wet bulb is zero and equation 4.48 for this case becomes

$$(T_G - T_S) = -\beta\ Lu^{2/3}\ \phi .\ \frac{\Delta H_{VS}}{\bar{C}_P}\ (Y_G - Y_S) \qquad (4.49)$$

when the small Ackermann correction is ignored.

Further, whenever the wick is wetted with water and the surrounding gas is air, a remarkably simple result occurs. On noting that $\phi = D/(D+Y_S)$ for small driving forces and that the heat capacity $\bar{C}_P$ may be related to the humid heat $\bar{C}_{PY}$ by the identity

$$\bar{C}_P = \frac{\bar{C}_{PG} + \bar{C}_{PW} Y}{(1 + Y)} = \frac{\bar{C}_{PY}}{(1 + Y)} \qquad (4.50)$$

It follows that equation 4.49 may be expanded into the expression:

$$(T_G - T_S) = -\left[\beta Lu^{2/3} \frac{D}{(D + Y_S)} (1 + Y_S)\right] \frac{\Delta H_{VS}}{\bar{C}_{PY}} (Y_G - Y_S) \qquad (4.51)$$

The term in square brackets may be called the <u>psychrometric coefficient</u> σ. For the air-water system the value of σ varies from 1.10 when $Y_S = 0.01$ to 1.05 when $Y_S = 0.1$ if Heertjes's experimental value of 1.0 is taken for β. However, the value of the psychrometric coefficient for <u>this system</u> is normally taken to be unity, the approximation being within the scatter of published data. Sherwood and co-authors[20] quote values ranging between 0.95 and 1.12 for measurements on the evaporation of water in wetted-wall columns. The slight disagreement of experiment with theory has been ascribed to the unaccounted effect of heat transfer to wet bulb by radiation from warmer surroundings or by conduction along the stem of the thermometer.[20]

Therefore, equation 4.51 in practice reduces to the expression

$$(T_G - T_W) = -\frac{\Delta H_{VS}}{\bar{C}_{PY}} (Y_G - Y_W) \qquad (4.52)$$

for the state conditions (T_W , Y_W) next to the wet bulb for water evaporating into air. Equation 4.52 yields an almost straight line of slope $-\bar{C}_{PY}/\Delta H_{VS}$ on a Grosvenor chart for the state path between the wet-bulb conditions (T_W , Y_W) and those (T_G , Y_G) in the bulk of the air. In this form, the equation for the wet-bulb depression (T_G-T_W) is identical with the equation for the adiabatic-saturation change (T_G-T_{GS}), equation 4.7 . This convenient quirk was first formally pointed out by Carrier,[8] who, however, noted that the idea then (1911) was already nearly a century old but had never been properly developed because of imperfections in the experimental data to substantiate the equivalence. Thus readings of the dry-bulb and wet-bulb temperature yield the humidity of the bulk of the air by following the appropriate adiabatic-saturation path from surface state A to the bulk state B, as shown in Fig. 4.9 .

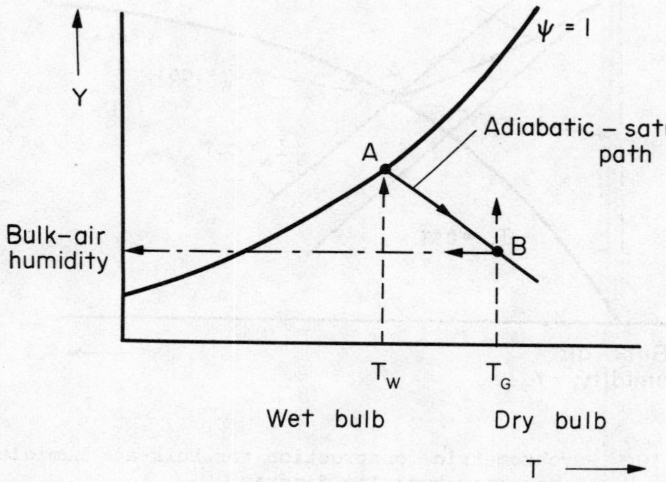

Fig. 4.9 Psychrometric construction for bulk-air humidity.

Equation 4.52 may also be generated by a rough enthalpy balance between that reaching the surface by heat transfer and that convected away by evaporation

$$\bar{C}_{PY}(T_G-T_W) = -\Delta H_{VS}(Y_G-Y_W) \qquad (4.53)$$

Equations 4.52 and 4.53 are essentially recast forms of equation 4.8, which is derived from the more fundamental expression (eq. 4.6) that incorporates humid enthalpies. In the notation of equation 4.53, this expression for the difference in humid enthalpies may be written as

$$\frac{(I_W-I_G)}{(Y_W-Y_G)} = H_{LW} \qquad (4.54)$$

where H_{LW} is the enthalpy of the liquid moisture at the wet-bulb temperature T_W. Thus the enthalpy-humidity chart can also be used for a similar psychrometric construction to that presented in Fig. 4.9. Care should be taken to avoid confusing the adiabatic-saturation line with an isenthalpic line which has a closely similar slope on those enthalpy-humidity charts with obliquely inclined axes. This construction is shown in Fig. 4.10.

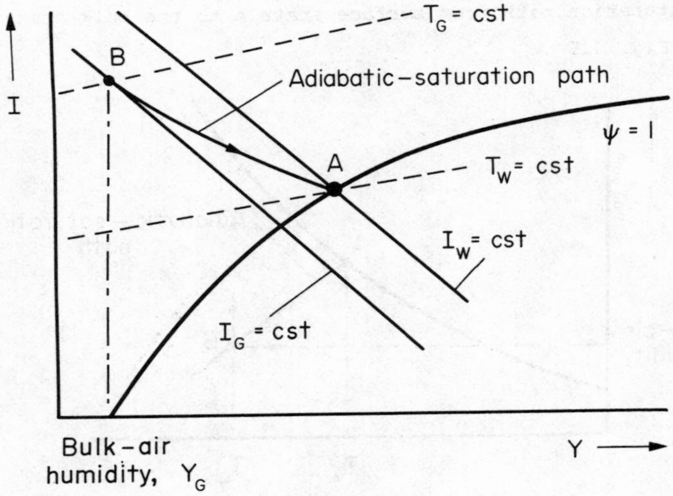

Fig. 4.10 Psychrometric construction for bulk-air humidity on enthalpy-humidity diagram.

Although the principles of psychrometry have been known for nearly two hundred years, the subject is still being researched, mainly because of experimental uncertainties in practice. For example, Thelkeld[(24)] calculates how the heat received by the thermometric sensor, and thus the indicated wet-bulb temperature, depends upon the size of the wet bulb, the temperature and speed of the air past the device. Even with a bulb of only 2.5 mm diameter, the deviation in reading from the true wet-bulb temperature can be significant for airspeeds below 1 m s^{-1} with an air temperature of 50°C (see Fig. 4.11). The effect becomes larger with increasing air temperature and bulb size.

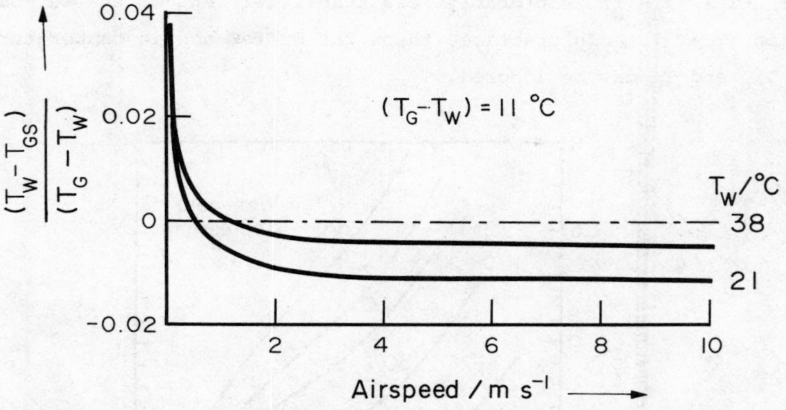

Fig. 4.11 Errors in temperature reading of a wet-bulb thermometer 2.5 mm diameter.

For these reasons commercial psychrometers often have radiation shields over the sensing bulb across which a draught of air is directed from an inbuilt blower. Somewhat surprisingly, perhaps, the wet-bulb reading is little affected should the element be shadowed in the wake of another element nearby.[31] Further, although the liquid should be free of contaminants that would depress the vapour pressure, the surface may sustain <u>microscopically small</u> dry spots without influencing the indicated temperature.[23] As these spots enlarge, then the indicated temperature begins to approach the dry-bulb value.

Even for the air-water system, the wet-bulb temperature is never <u>precisely</u> equal to the adiabatic-saturation temperature. Although the Luikov number for evaporation into dry air is almost independent of temperature, the Luikov number for air saturated with water vapour rises with saturation temperature, due to the presence of moisture, from 1.171 at 10°C to 1.232 at 60°C.[14] Thus one might expect the psychrometric coefficient σ to depart noticeably from unity when the atmosphere is very humid. On the other hand, the calculations of Ashworth and Keey[3] show that the difference between the so-called thermodynamic and the actual wet-bulb temperature falls with increasing surface temperature, being identically equal at the boiling point when the gradient $\Delta Y/\Delta T$ becomes infinite. These calculations are graphically displayed in Fig.4.12. Under most commercial drying conditions, the temperature ratio $(T_{GS}-T_W)/(T_G-T_W)$ is probably less than 1 per cent of the wet-bulb depression (T_G-T_W). In practice, then, the difference in temperature between T_{GS} and T_W may be ignored.

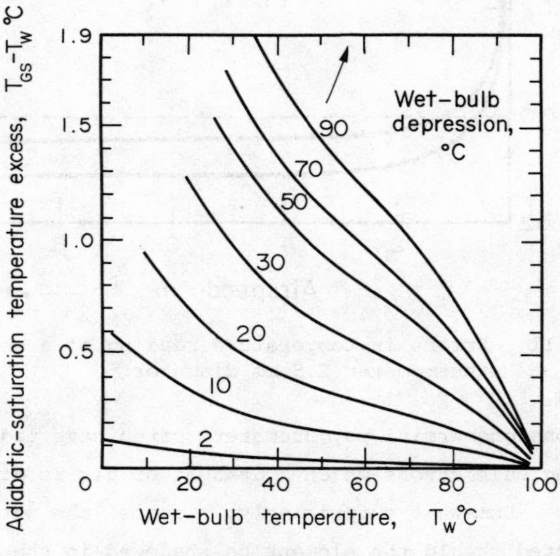

Fig. 4.12 The difference between the adiabatic-saturation and wet-bulb temperature for water evaporating into air.

Non-aqueous systems. For systems, other than water evaporating into air, the psychrometric coefficient is evaluated as

$$\sigma = \left[\frac{Lu}{Lu_{AW}}\right]^{2/3} \qquad (4.55)$$

where Lu_{AW} is the limiting Luikov number for the air-water system, taken to be 1.170. Luikov numbers for some solvents evaporating into air are given in Table 4.3 below.

TABLE 4.3 VALUES OF THE LUIKOV NUMBER (Lu) FOR LIQUIDS EVAPORATING INTO AIR AT 25°C. (28)

Liquid	Lu = Pr/Sc	Liquid	Lu = Pr/Sc
Water	1.170	toluene	0.379
methanol	0.730	carbon tetrachloride	0.375
propanol	0.541	ethylene tetrachloride	0.328
benzene	0.410	chlorobenzene	0.325
ethyl acetate	0.388	m-xylene	0.314

The temperature variation of Lu is small and may be neglected to a first approximation.[19]

Equation 4.55 assumes that

$$\beta \cdot \frac{D}{(D + Y_S)} \cdot (1 + Y_S) = 1$$

which is a very good approximation for practical values of D and Y_S. Equation 4.52 for the wet-bulb depression now becomes

$$(T_G - T_W) = -\sigma \cdot \frac{\Delta H_{VS}}{\bar{C}_{PY}} \cdot (Y_G - Y_W) \qquad (4.57)$$

and the corresponding equation for the enthalpy change (cf eq.4.54) is

$$(I_G - I_W) = -\sigma\, H_{LW} (Y_G - Y_W) \qquad (4.58)$$

Thus the slopes of the wet-bulb depression lines are σ times steeper than those of the adiabatic-saturation contours on the enthalpy-humidity diagram. On the Grosvenor chart, where the humidity is plotted as ordinate, the wet-bulb lines fall more steeply by a factor of 1/σ, cutting off the adiabatic-saturation lines at an angle $\cos^{-1}(1/\sigma)$ as shown in Fig. 4.13. On this basis, it is possible to use existing humidity charts, cross-plotted with adiabatic-saturation curves, for psychrometric analyses of evaporating non-aqueous liquids.

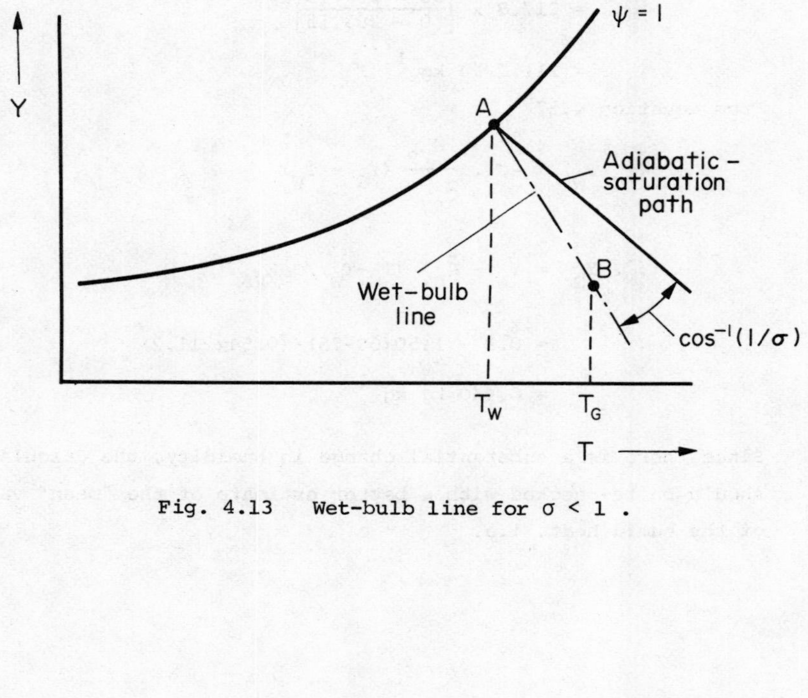

Fig. 4.13 Wet-bulb line for σ < 1 .

Humidity charts for solvents such as benzene, carbon tetrachloride and toluene evaporating into air are given by Perry.[18] These charts are based on a "radiation coefficient" β of 1.06 and psychrometric coefficients σ of 0.54, 0.57 and 0.49 respectively for these liquids. Some lines of constant wet-bulb temperature are cross-plotted over the adiabatic-saturation contours.

Example 4.6 Calculate the bulk-air humidity when carbon tetrachloride from a soaked bobbin at a temperature of 25°C evaporates into air at 60°C.

If steady conditions hold, then the indicated temperature of the surface will be that of the surrounding airskin,

i.e. $T_W = T_S = 25^\circ C$.

From Perry's data, cited above, $\sigma = 0.54$.

The humid heat can be calculated at the known surface conditions, $T_W = 25^\circ C$, $Y_W = 0.9$ kg kg^{-1} (From Perry[18] Fig.20-12, p 20-7).

From Table 2.5, $\bar{C}_W = 0.552$ kJ kg^{-1}K^{-1} and $\bar{C}_G = 1.005$ kJ kg^{-1}K^{-1}

$$\therefore \bar{C}_{PY} = 1.005 + (0.552 \times 0.9) = 1.50 \text{ kJ kg}^{-1}\text{K}^{-1}.$$

Table 2.5 gives ΔH_{VO} as 217.8 kJ kg^{-1} at $T_o = 0^\circ C$. The critical temperature is 283.15°C, so from the cube-root rule

$$\Delta H_{VS} = \Delta H_{VO}\left[\frac{T_S - T_c}{T_o - T_c}\right]^{1/3}$$

$$= 217.8 \times \left[\frac{25 - 283.15}{0 - 283.15}\right]^{1/3}$$

$$= 211.2 \text{ kJ kg}^{-1}$$

From equation 4.57

$$(T_G - T_W) = -\sigma \cdot \frac{H_{VS}}{\bar{C}_{PY}} (Y_G - T_W)$$

$$\therefore Y_G = Y_W - \bar{C}_{PY}\ (T_G - T_W)/\sigma\Delta H_{VS}$$

$$= 0.9 - 1.50(60-25)/(0.54 \times 211.2)$$

$$= 0.440 \text{ kg kg}^{-1}$$

Since there is a substantial change in humidity, the calculation should be re-checked with a better estimate of the "mean" value of the humid heat, i.e.

$$\bar{C}_{PY} = 1.005 + 0.552 \times (0.9 + 0.440)/2$$
$$= 1.37 \text{ kJ kg}^{-1}\text{K}^{-1}$$

and the new estimate of the air humidity becomes

$$Y_G = 0.9 - 1.37(60 - 25)/(0.54 \times 211.1)$$
$$= 0.480 \text{ kg kg}^{-1}$$

a third iteration gives $Y_G = 0.475$ kg kg^{-1}.

Interpolation between the contours of wet-bulb temperature in Fig.20-12 yields a value of $Y_G = 0.52$ kg kg^{-1}. The difference is ascribable to the marked change in gradient of the wet-bulb contours which become less steep with diminishing humidity as $\bar{C}_{PY}$ itself becomes smaller.

<u>Example 4.7</u> Re-work example 4.6 from first principles (eq.4.49).

We have

$$(T_G - T_S) = -\beta \, Lu^{2/3} \, \phi \, . \, \frac{\Delta H_{VS}}{\bar{C}_{PY}} \, . \, (Y_G - Y_S)$$

(i) <u>Velocity coefficient</u> β .

$M_W = 0.1538$ kg mol^{-1} and $M_G = 0.02897$ kg mol^{-1}

$$\therefore D = M_W/M_G = 0.1538/0.02897 = 5.309$$

The results of example 4.6 will be used to estimate the mole fractions:

$$y_G = \frac{Y_G}{(D+Y_G)} = \frac{0.475}{(5.309 + 0.475)} = 0.082$$

and

$$y_S = \frac{Y_S}{(D+Y_S)} = \frac{0.9}{(5.309 + 0.9)} = 0.145$$

The mean molar masses then become:

$$\bar{M}_G = M_W y_G + M_G(1 - y_G)$$
$$= (0.1538 \times 0.082) + 0.02897 \times (1 - 0.082)$$
$$= 0.03921$$

$$\bar{M}_S = M_W y_S + M_G(1 - y_S)$$

$$= (0.1538 \times 0.145) + 0.02897 \times (1 - 0.145)$$

$$= 0.04707$$

From equation 4.46

$$\beta = \frac{\bar{M}_S \ln (\bar{M}_S/\bar{M}_G)}{\dfrac{(M_W - M_G)D}{(D+Y_S)} \ln\left[\dfrac{(D+Y_S)}{(D+Y_G)}\right]}$$

$$= \frac{0.04707 \ln (0.04707/0.03921)}{\dfrac{(0.1538 - 0.02897)}{(5.309 + 0.9)} \times 5.309 \ln\left[\dfrac{(5.309 + 0.9)}{(5.309 + 0.475)}\right]}$$

$$= 1.137$$

This compares with an experimental "radiation coefficient" of 1.06 .

(ii) Boundary-layer ratio Lu.

A value of 0.375 is found from Table 4.3 .

(iii) Humidity-potential coefficient ϕ

$$\phi_1 = \frac{D}{(D+Y_S)} = \frac{5.309}{(5.309 + 0.9)} = 0.855$$

$$\phi_2 = \frac{(D+Y_S) \ln \left[1 + \dfrac{(Y_S - Y_G)}{(D+Y_G)}\right]}{(Y_S - Y_G)}$$

$$= \frac{(5.309 + 0.9) \ln \left[1 + \dfrac{(0.9 - 0.475)}{(5.309 + 0.475)}\right]}{(0.9 - 0.475)}$$

$$= 1.036$$

$$\phi = \phi_1 \times \phi_2 = 0.855 \times 1.036 = 0.886$$

(iv) Heat capacity $\bar{C}_P$

This is calculated for the bulk-air conditions, for which the mass fraction of moisture vapour is given by

$$m_G = y_G(M_W/\bar{M}_G)$$

$$= 0.082 \times 0.1538/0.03921$$

$$= 0.322$$

[Alternatively, this could be evaluated from the humidity itself directly, since $m_G \equiv Y/(1+Y)$ and $m_G = 1/(1+Y)$]

$$\therefore \bar{C}_P = 1.005(1 - 0.322) + (0.552 \times 0.322)$$

$$= 0.859 \text{ kJ kg}^{-1}\text{K}^{-1}$$

(v) <u>Air humidity</u> Y_G

On re-arranging equation 4.49, we find

$$Y_G = Y_S - \frac{\bar{C}_P Lu^{-2/3}}{\beta\ \phi\ \Delta H_{VS}}\ (T_G - T_S)$$

$$= 0.9 - \frac{0.859 \times 0.375^{-2/3}}{1.136 \times 0.886 \times 211.2} \times (60-25)$$

$$= 0.628$$

A second iteration gives $Y_G = 0.637$. This latter value is 22.5% greater than that taken from Perry's chart, and some 34% greater than that predicted by the simplified analysis (eq.4.57). These differences highlight the uncertainties of psychrometric estimations for non-aqueous liquids at moderate humidity levels.

4.7 Evaporation with Heating

There are a number of situations where the heat received by a surface is not counterbalanced by the heat removed by way of evaporation. Such cases occur in the drying of materials over steam-heated rolls, in the drying of loops of thin sheets, wet on one side only, and in the later stages of drying when moisture-free patches may appear at the exposed surface. Heat transfer takes place over the whole surface, whereas moisture transfer can only arise from the wetted area.

Let us consider a unit exposed area, over which a fraction f_W is wet. Suppose thermal equilibrium exists at the surface which has a temperature T_S everywhere. Equation 4.48 for the heat flux, on neglecting the

Ackermann correction and introducing the psychrometric coefficient σ, becomes

$$q_T = h[(T_G-T_S) + f_W \cdot \frac{\sigma \Delta H_{VS}}{\bar{C}_{PY}} (Y_G-Y_S)] \quad (4.59)$$

$$= q_C + q_W \text{, say.} \quad (4.60)$$

These heat fluxes may take similar or different signs depending upon the nature of the process. A negative heat flux signifies a heat loss from the surface; a positive flux a heat gain.

The relative magnitude of these fluxes may be displayed on an enthalpy-humidity diagram by defining the quantities:

$$I_C = \frac{\bar{C}_{PY}}{h} \cdot q_C = \bar{C}_{PY}(T_G-T_S) \quad (4.61)$$

$$I_W = \frac{\bar{C}_{PY}}{h} \cdot q_W = f_W \sigma \Delta H_{VS}(Y_G-Y_S) \quad (4.62)$$

and $$I_T = \frac{\bar{C}_{PY}}{h} \cdot q_T = I_C + I_W \quad (4.63)$$

In Fig. 4.14, point G on the isotherm T_G represents the state conditions of the bulk of the air, and S at the intersection of the isotherm T_S with the saturation curve those of the airskin next to the wet surface. The vertical line GF represents I_C through equation 4.61. Point P divides the isothermal line FS in the ratio $f_W\sigma:(1-f_W\sigma)$. Graphically ΔH_{VS} represents the difference in slopes between that of the isotherm (H_{GW}) and that of the adiabatic-saturation contour (H_{LW}). Thus a line of slope H_{LW} from P to the extension of line FG to Q will yield the quantity I_W as the line segment FQ through equation 4.62. The quantity I_T is the vector sum of FG (+ I_C) and FQ (- I_W).

In the case where $f_W\sigma = 1$, P coincides with S and Q with G, so QP(GS) becomes the adiabatic-saturation line for a temperature T_S. Further, $- I_W = I_C$ and $I_T = 0$, the inwards and outwards flow of heat counter-balancing.

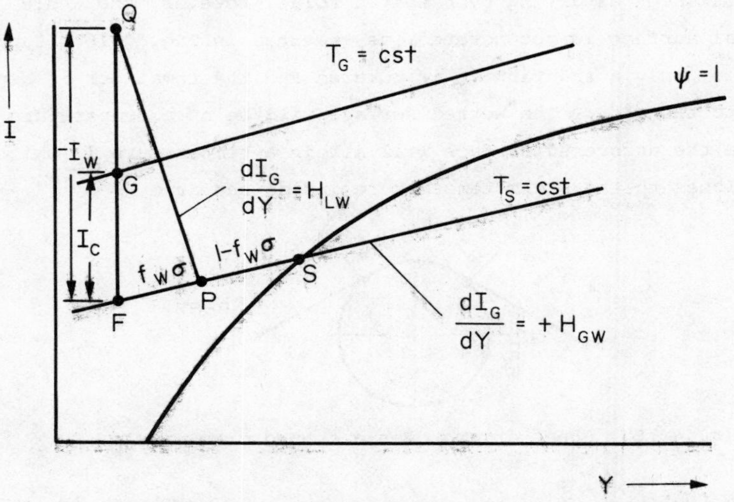

Fig. 4.14 Evaporation with heating for $f_W\sigma < 1$ and $q_W < 0$

Another special case arises when both q_C and q_W are negative, for instance when slurries are dried over a steam-heated cylinder. Usually the moisture involved is water, so $f_W\sigma = 1$ should the drum's entire surface be covered. Under these conditions, P and S coincide on the moisture-saturation curve, but G is now located below F on the diagram. The total enthalpy difference I_T is negative, as there is a net heat loss:

$$- I_T = - I_C - I_W \tag{4.64}$$

The diagram for this situation is shown in Fig. 4.15.

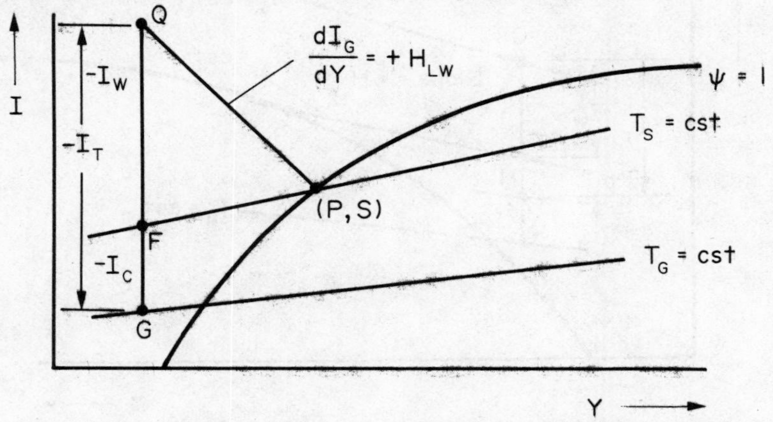

Fig. 4.15 Evaporation from a surface heated from below for $f_W\sigma = 1$.

In most instances of drying over heated rolls, however, the whole cylindrical surface is not covered, as sketched in Fig. 4.16. Suppose, for example, only a fraction f_W is covered and the remainder is exposed directly to the air. The wetted surface will be at a temperature T_W, say, while the uncovered surface will attain a temperature T_S which will be much closer to the steam temperature inside the drum.

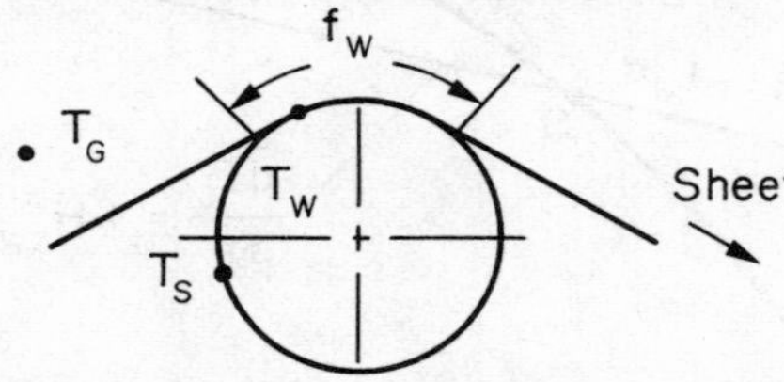

Fig. 4.16 Sheet drying over a heated drum.

It follows that

$$q_C = h\,[f_W(T_G - T_W) + (1-f_W)(T_G - T_S)] \quad (4.65)$$

and

$$-I_C = -\frac{\bar{C}_{PY}}{h}\,q_C = \bar{C}_{PY}[f_W(T_G - T_W) + (1-f_W)(T_G - T_S)] \quad (4.66)$$

By geometric considerations, the quantities on the right-hand side of equation 4.66 can be represented by

$$HF = \left(\frac{FP}{FS} \times GF\right) + \left(\frac{PS}{FS} \times GQ\right) \quad (4.67)$$

where the letters refer to the corresponding line segments in Fig.4.17.

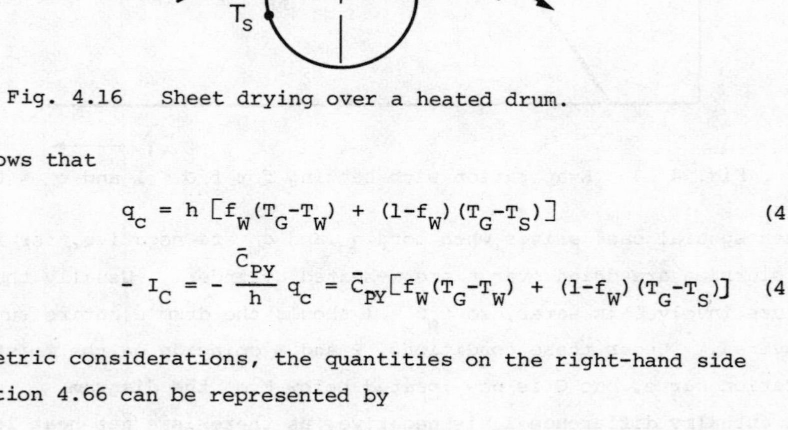

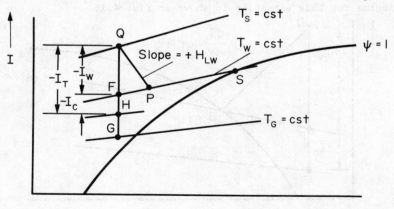

Fig.4.17 Evaporation from a partially covered surface heated from below for $f_W < 1$ and $\sigma = 1$. The line segment HF is calculated through eq.4.67.

This construction is still approximate for the conditions which can occur in the drying of slurries over heated drums when there may be significant conducted heat being transmitted along the shafts to the bearings.

4.8 Humidification in Dryers

The foregoing ideas may be collated to provide information about humidity and temperature profiles in drying equipment.[13, 22] The drying surface, if thoroughly sodden, will reach the wet-bulb temperature shortly after entering the dryer, whereas the air will progressively approach an adiabatic-saturation temperature as the gas picks up moisture. This path to saturation can be complex should heat be either added or withdrawn from the air, which is often the case in commercial drying installations.

Consider the evaporation from a infinitesimally short zone of length dz within a dryer, as sketched in Fig. 4.18.

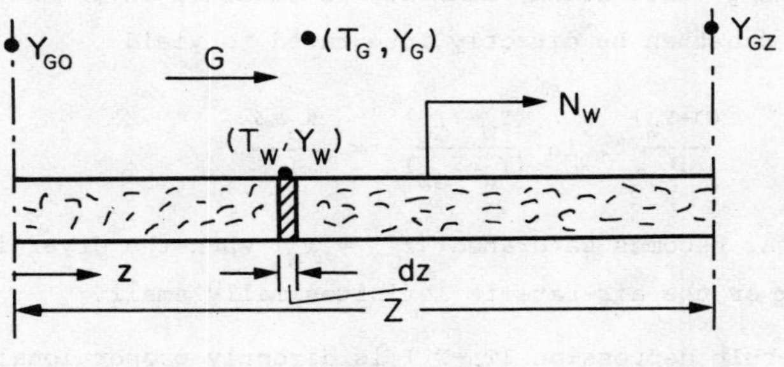

Fig. 4.18 Humidification in a dryer

The rate of humidification of the air is equal to the moisture-transfer rate from this zone:

$$G dY_G = K_o \phi (a dz)(Y_W - Y_G) \qquad (4.68)$$

where G is understood as the dry-air flow per unit cross-sectional area of the dryer and a is the exposed surface of the drying goods per unit volume.

Re-arrangement of equation 4.67 yields

$$\int_{Y_{GO}}^{Y_{GZ}} \frac{dY_G}{(Y_W-Y_G)} = \int_{o}^{Z} \frac{K_o \phi a dz}{G} \tag{4.69}$$

To a first approximation, the variation of the mass-transfer coefficient K_o with distance may be ignored and ϕ may be set equal to $D/(D+Y_W)$. Under these circumstances equation 4.69 yields

$$\frac{1}{D}\int_{Y_{GO}}^{Y_{GZ}} \frac{(D+Y_W)dY_G}{(Y_W-Y_G)} = \frac{K_o aZ}{G} \tag{4.70}$$

Note that this result is expressed in dimensionless form, the right-hand side of the expression being the non-dimensional "extent" of humidification.

Equation 4.70 is general for any humidification path, but for an adiabatic change of state when $\sigma = 1$ there is a particularly simple result. Under these conditions the wet-bulb temperature is constant throughout (as long as the surface remains fully wetted) and thus Y_W the corresponding (saturation) humidity is likewise invariant. Equation 4.70 can then be directly integrated to yield

$$\frac{(D+Y_W)}{D} . \ln \frac{(Y_W-Y_{GO})}{(Y_W-Y_{GZ})} = \frac{K_o aZ}{G} \tag{4.71}$$

The outgoing gas becomes saturated ($Y_{GZ} = Y_W$) when the dryer is infinitely long or the air-rate is infinitesimally small.

Since the wet-bulb depression (T_G-T_W) is directly proportional to the humidity potential (Y_W-Y_G), as demonstrated by equation 4.52 for instance, it follows that

$$\frac{(T_{GZ}-T_W)}{(T_{GO}-T_W)} = \frac{(Y_{GZ}-Y_W)}{(Y_{GO}-Y_W)} = \exp - \left[\frac{D}{(D+Y_W)} . \frac{K_o aZ}{G} \right] \tag{4.72}$$

Thus the temperature falls off exponentially with distance from the air inlet.

1. <u>Adiabatic evaporation ($\sigma = 1$)</u>. Whenever the psychrometric coefficient is one, the cool limit for the saturated air is the wet-bulb temperature of the evaporative surface. The temperature of the air falls exponentially from its inlet value to this limiting temperature, as sketched in Fig. 4.19 .

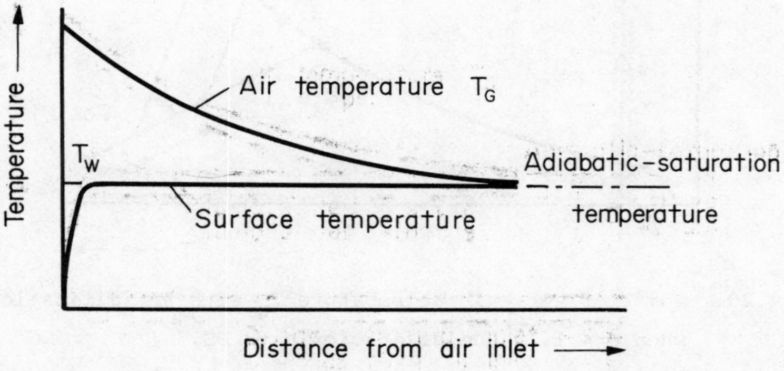

Fig. 4.19 Temperature changes in adiabatic saturation ($\sigma = 1$).

2. <u>Adiabatic evaporation ($\sigma < 1$)</u>. For systems other than water evaporating into air, the psychrometric coefficient will be less than one. At a given bulk-air humidity, the wet-bulb temperature will be higher than the adiabatic-saturation temperature, since the wet-bulb contours have a steeper ascent than the adiabatic-saturation curves. The temperature difference becomes progressively smaller as the air becomes damper, until the temperatures coincide at saturation. This shift in wet-bulb temperature is readily illustrated on a Grosvenor chart (Fig. 4.20), from which the temperature profiles of Fig. 4.21 can be deduced. When $\sigma = 1$, $\cos^{-1}(1/\sigma) = 0$, so that the wet-bulb and adiabatic-saturation contours become identical, as we expect for this condition.

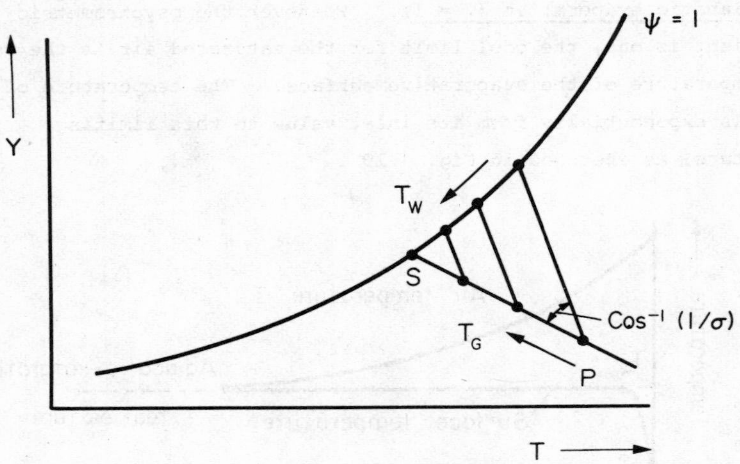

Fig.4.20 Shift of wet-bulb temperature T_W with humidification when $\sigma < 1$. Humidification path PS.

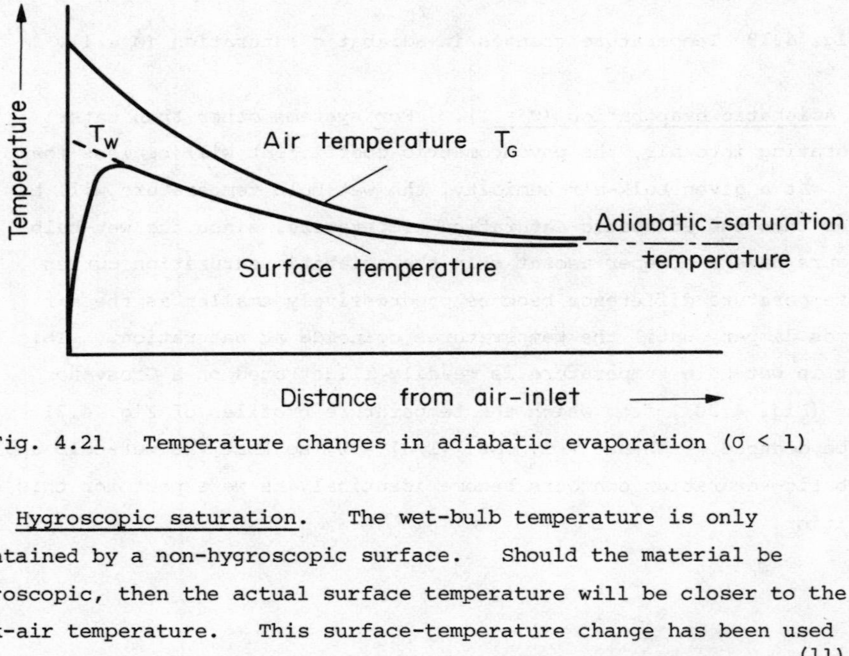

Fig. 4.21 Temperature changes in adiabatic evaporation ($\sigma < 1$)

3. Hygroscopic saturation. The wet-bulb temperature is only maintained by a non-hygroscopic surface. Should the material be hygroscopic, then the actual surface temperature will be closer to the bulk-air temperature. This surface-temperature change has been used as the basis of a moisture-control technique for nearly dry goods.[11] It follows that the ultimate temperature of the air saturated by hygroscopic material will be different from the adiabatic-saturation value, and an enthalpy balance for the gas-solid system suggests that it is less. However, there is no simple way of calculating this effect.

Fortunately, in practice most dryers are worked with a counterflow of goods and air so that the dampest air is in contact with the wettest material, which almost always may be regarded as being non-hygroscopic.

4. Isothermal evaporation. To reduce the extent to which the drying air must be preheated, some installations have internal heaters. Consider such an arrangement in which the internal heating is disposed to maintain a constant temperature in the air. The humidification path PS is now a vertical line on the Grosvenor chart, and the corresponding wet-bulb temperatures rise along the envelope RS as the gas becomes more humid, as Fig. 4.22 shows.

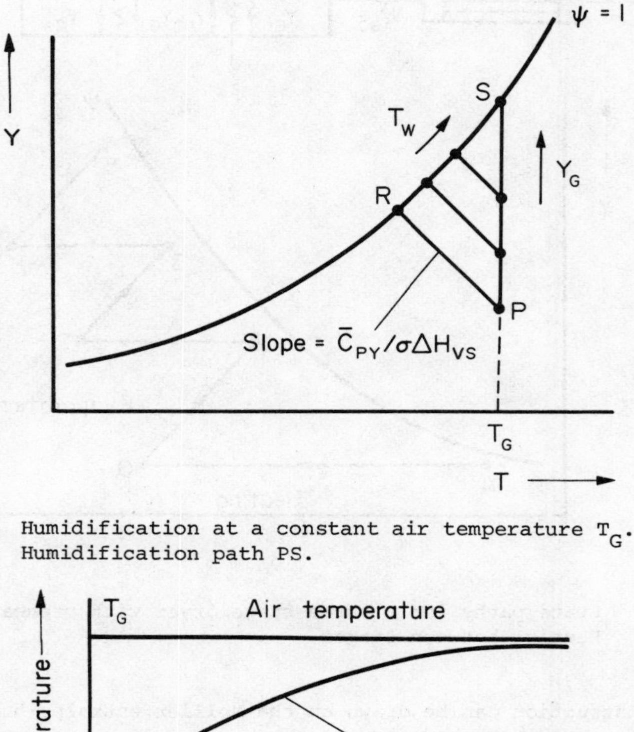

Fig.4.22 Humidification at a constant air temperature T_G. Humidification path PS.

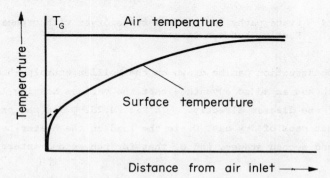

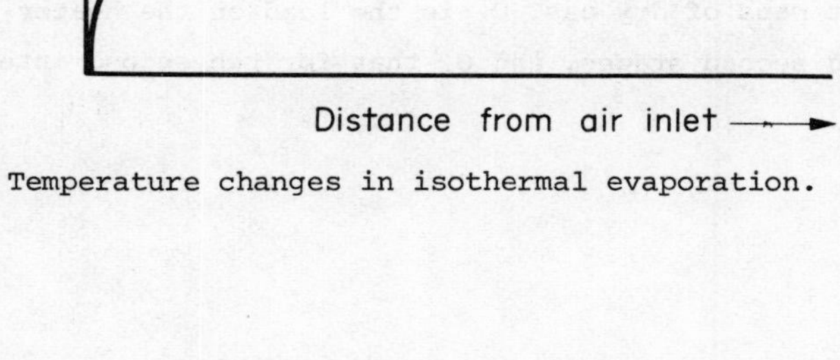

Fig. 4.23 Temperature changes in isothermal evaporation.

5. Evaporation with discontinuous heating. The heating may be installed in a dryer in blocks, so the evaporation takes place in a series of essentially adiabatic stages between the heating elements. Often this is a more convenient arrangement than providing continuous internal heaters, and also it gives greater flexibility in tuning drying operations. The state changes for a three-stage dryer with preheater are illustrated on a Grosvenor chart in Fig. 4.24. Each heating process is represented by a horizontal line on this diagram, while each evaporative stage is adiabatic for a perfectly insulated dryer.

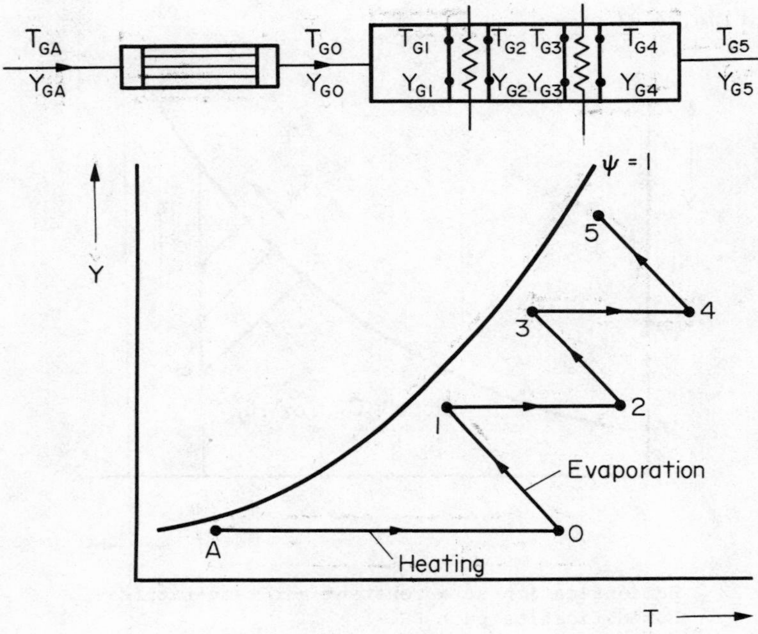

Fig. 4.24 State paths for a three-stage dryer with preheater. Heating between stages.

A similar construction can be drawn on the Mollier enthalpy-humidity chart, which has an added advantage that the various heating loads can be read off the diagram directly. In Fig. 4.25, Q_o is the preheating load per unit mass of dry gas, Q_1 is the load on the heater between the first and second stages, and Q_2 that for the second interstage heater.

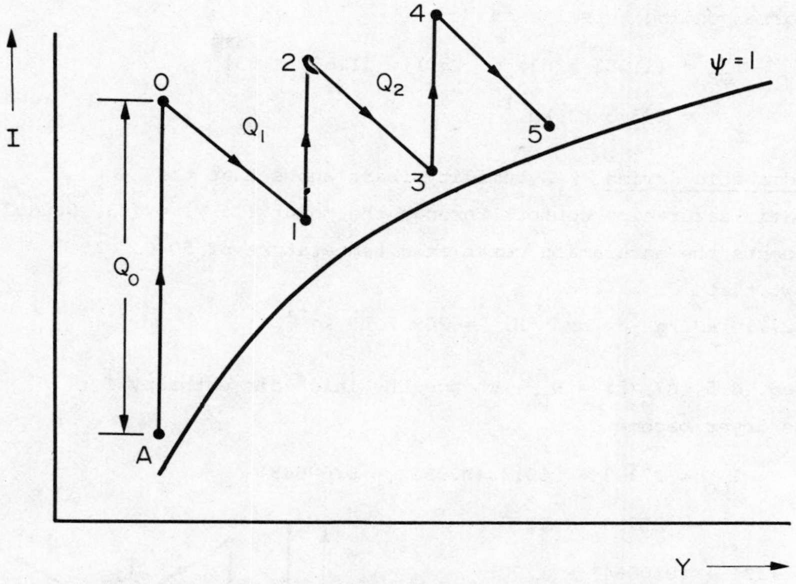

Fig.4.25 State paths for a three-stage dryer with preheater. Heating between stages.

Example 4.8. Outside air at 15°C and 60% relative humidity is drawn into a dryer and discharged at 63°C and 50% relative humidity. Calculate the state paths and heating requirements per unit mass of moisture evaporated for the following arrangements:

(a) adiabatic drying with preheating to the required air-inlet temperature;

(b) isothermal drying with preheating to 63°C;

(c) two-stage adiabatic drying with preheating and interstage heating to 100°C.

The air pressure is 100 kPa.

Inlet-air conditions. The inlet-air humidity is 0.00643 (from example 3.7); its enthalpy is

$$I_{GA} = (1.004 \times 15) + 0.00643\,[2501 + (1.86 \times 15)]$$
$$= 31.3 \text{ kJ kg}^{-1}$$

Outlet-air conditions. The humidity follows from equation 2.10. At 63°C the vapour pressure p_W^o of water is 22.85 kPa; therefore

$$Y = \left(\frac{M_W}{M_G}\right) \frac{\psi p_W^o}{(P - \psi p_W^o)} = \frac{0.622 \times 0.5 \times 22.85}{(100 - (0.5 \times 22.85))}$$
$$= 0.0802 \text{ kg kg}^{-1}$$

The corresponding enthalpy is

$$I_{GE} = (1.005 \times 63) + [2501 + (1.865 \times 63)]$$
$$= 273.3 \text{ kJ kg}^{-1}$$

(a) <u>Adiabatic drying</u>. A humidity chart shows that the adiabatic-saturation contour through the point (T, Y) = (63, 0.0802) intersects the saturation curve at a temperature of 50°C. It follows that

$I_W = 277.0 \text{ kJ kg}^{-1}$ and $H_{LW} = 209.3 \text{ kJ kg}^{-1}$.

From eq. 4.5, $\Delta I_G/\Delta Y = H_{LW}$, whence the inlet-air enthalpy I_{GO} to the dryer becomes

$$I_{GO} = 273.3 - 209.3\ (0.0802 - 0.00643)$$
$$= 257.8 \text{ kJ kg}^{-1}$$

$$\bar{C}_{PY} \simeq 1.01 + (0.00643 \times 1.90)$$
$$= 1.02 \text{ kJ kg}^{-1} \text{ K}^{-1}$$

From equation 2.23 for the humid enthalpy one has

$$T_{GO} = (I_{GO} - \Delta H_{VO} Y_{GO}) / \bar{C}_{PY}$$
$$= 257.8 - (2501 \times 0.00643)/1.02$$
$$= 237^{\circ}\text{C}$$

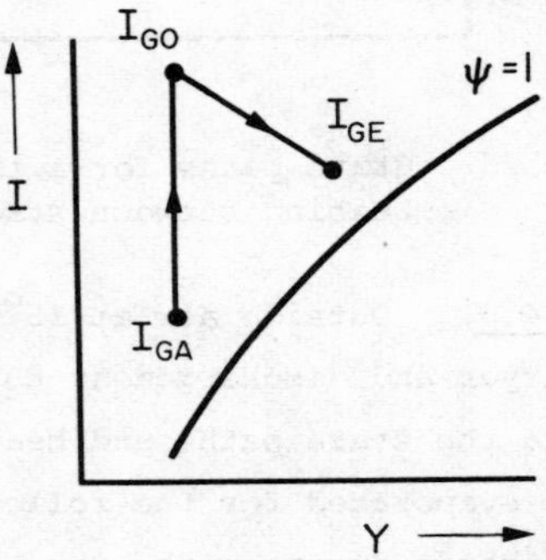

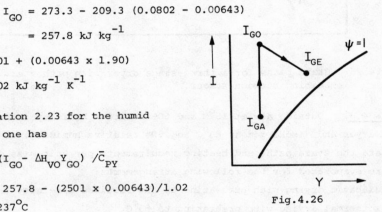

Fig.4.26
Adiabatic drying

Alternatively, one notes from equation 4.8

$$T_{GO} = T_W + \Delta H_{VS}\ (Y_W - Y_{GO}) / \bar{C}_{PY}$$
$$= 50 + 2382\ (0.0875 - 0.00603)/1.02$$
$$= 239^{\circ}\text{C}$$

The difference in estimates illustrates the small errors incurred by using eq.4.8 over a large temperature difference.

$$\text{Heating load} = \frac{I_{GO} - I_{GA}}{(Y_{GW} - Y_{GA})} = \frac{257.8 - 31.3}{(0.0802 - 0.00643)}$$
$$= 3070 \text{ kJ kg}^{-1}$$

(b) Isothermal drying. Under these conditions, the heating load becomes

$$\frac{(I_{GE} - I_{GA})}{(Y_{GE} - Y_{GA})} = \frac{(273.3 - 31.3)}{(0.0802 - 0.00643)}$$

$$= 3280 \text{ kJ kg}^{-1}$$

Thus (3280 - 3070) = 210 kJ extra heat is needed for each kilogram of moisture evaporated.

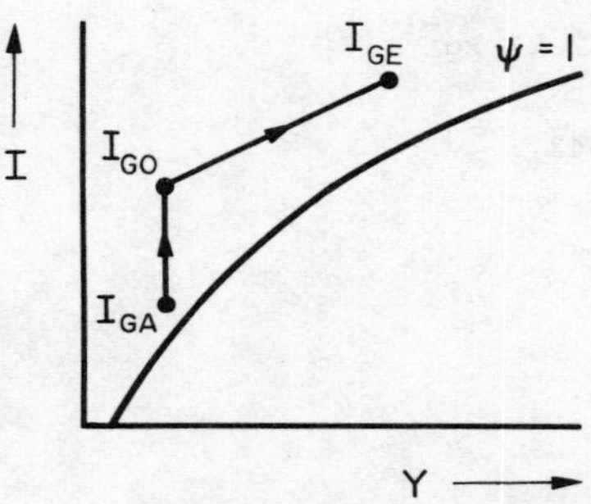

Fig. 4.27 Isothermal drying

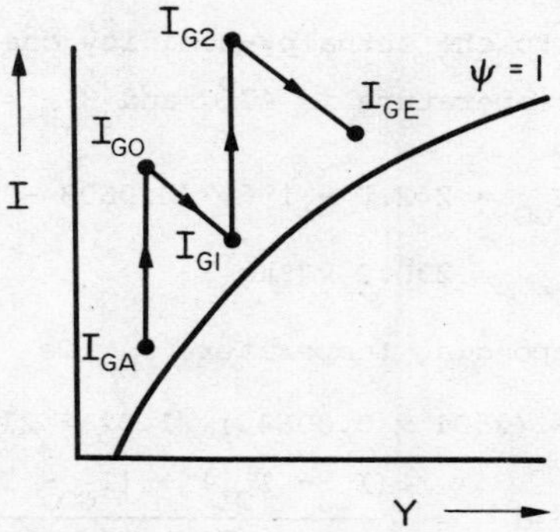

Fig.4.28 Two-stage drying

(c) Two-stage drying. For the second stage we assume that

$$\bar{C}_{PY} \simeq 1.01 + (0.06 \times 1.88)$$

$$= 1.12 \text{ kJ kg}^{-1} \text{ K}^{-1}$$

From eq. 4.8

$$\frac{Y_W - Y_{G2}}{T_W - T_{G2}} = -\frac{\bar{C}_{PY}}{\Delta H_{VS}}$$

$$\therefore \quad Y_{G2} = Y_W + \bar{C}_{PY}(T_W - T_{G2})/\Delta H_{VS}$$

$$= 0.0875 + 1.12(63 - 100)/2382$$

$$= 0.0701 \text{ kg kg}^{-1}$$

A second iteration gives $\bar{C}_{PY} = 1.14 \text{ kJ kg}^{-1} \text{K}^{-1}$ and $Y_{G2} = 0.0698 \text{ kg kg}^{-1}$.

Thus, $I_{G2} = 273.3 - 209.3(0.0802 - 0.0698)$

$$= 271.1 \text{ kJ kg}^{-1}$$

It follows from Fig. 4.28 that the outlet temperature from the first stage has no fixed value, except that it cannot exceed an upper bound of 100°C, the lower the temperature, the lower the temperature of the air from the preheater. There is also a lower bound corresponding to saturated-air conditions. Let the outlet temperature be 60°C, then

$$I_{G1} = (1.005 \times 60) + 0.0698 \left[2501 + (1.865 \times 60)\right]$$
$$= 242.7 \text{ kJ kg}^{-1} .$$

Reference to the enthalpy-humidity chart shows that the new wet-bulb temperature is 47°C and $H_{LW} = 196.7$ kJ kg^{-1}.

$$\therefore I_{GO} = 242.7 - 196.7(0.0698 - 0.00643)$$
$$= 230.2 \text{ kJ kg}^{-1}$$

The corresponding temperature T_{GO} is

$$230.2 - (2501 \times 0.00643) /1.02 = 210°\text{C}$$

$$\text{Heating load} = \frac{(I_{G2} - I_{G1}) + (I_{GO} - I_{GA})}{(Y_{GE} - Y_{GA})}$$
$$= \frac{(271.1 - 242.7 + (230.2 - 31.3)}{(0.0803 - 0.00643)}$$
$$= 3077 \text{ kg kg}^{-1}$$

The heating load is the same as the single-stage adiabatic case, within the error inherent in the calculation. Such a result might have been anticipated, as the enthalpy change for an adiabatic process should be independent of the route taken. The advantage of the two-stage process lies in reducing the preheater temperature from 237°C to 210°C.

REFERENCES

1. Ackermann, G., Wärmeübergang und molekulare Stoffübertragung an gleichen Feld, VDI Forschungsh. 382, 1-16 (1937).
2. Ambrose, D., J. F. Comsell and A. J. Davenport, The use of Chebyshev polynomials for the representation of vapour pressures between the triple point and the critical point, J. Chem.Thermodynamics, 2, 283-294 (1970).
3. Ashworth, J. C. and R. B. Keey, The evaporation of moisture from wet surfaces, Chem.Eng.Sci., 27, 1797-1806 (1972).
4. Beek, W. J. and K. M. K. Muttzall, "Transport Phenomena", p. 293, Wiley Interscience, London (1975).

5. Bedingfield, G. M. and J. B. Drew, Analogy between heat transfer and mass transfer, a psychrometric study, Ind.Eng.Chem., 42, 1164-1173 (1950).
6. Bosnaković, F. and P. L. Blackshear, "Technical Thermodynamics", p. 31-3, Holt, Rinehart and Winston, New York (1965).
7. Brauer, H. "Stoffaustausch einschließlich chemischer Reaktionen", p. 48, Verlag Sauerländer, Aarau/Frankfurt a. M. (1971).
8. Carrier, W.H., Rational psychrometric formulae, their relation to the problems of meteorology and of air-conditioning, Trans.ASME, 33, 1005-1053 (1911).
9. Chilton, J.H. and C.P. Colburn, Mass transfer (absorption) coefficients, Prediction from data on heat transfer and fluid friction, Ind.Eng.Chem., 26, 183-7 (1934).
10. Grosvenor, W.M., Calculations for dryer design, Trans.AJChE, 1, 184-202 (1907).
11. Harbert, F.C., Automatic control of drying processes moisture measurement and control by the temperature difference method, Chem.Eng.Sci., 29, 888-890 (1974).
12. Heertjes, P.M. and W.P. Ringens, The j_H and j_D factor of air use for drying, Chem.Eng.Sci., 5, 226-231 (1955).
13. Keey, R.B., "Drying Principles and Practice", p.174-6, Pergamon, Oxford (1972).
14. Kusuda, J., Calculation of the temperature of a flat-plate wet surface under adiabatic conditions with respect to the Lewis relation, in "Humidity and Moisture", 1, p.29, Reinhold, New York (1965).
15. Lewis, W.K., The evaporation of a liquid into a gas, Mech.Eng., 44, 445-6 (1922).
16. Lynch, E.J. and C.R. Wilke, Effect of fluid properties on mass transfer in the gas phase, AJChE Journal, 1, 1-19 (1955).
17. Nonhebel, G. and A.A.H. Moss, "Drying of Solids in the Chemical Industry", p.78, Butterworths, London (1971).
18. Perry, H.E., ed., "Chemical Engineers' Handbook", 5/e, p.20-7-20-8, McGraw Hill, New York (1973).
19. Reid, R.C. and T.K. Sherwood, "The Properties of Gases and Liquids", 2/3, p.460, 510, McGraw Hill, New York (1966).
20. Sherwood, T.K., R.L. Pigford and C.R. Wilke, "Mass Transfer", p.262, McGraw Hill, New York (1975).
21. Spalding D.B., "Convective Mass Transfer", p.66-72, Arnold, London (1963).
22. Strumiłło, C., "Podstowy Teorii i Techniki Suszenia", p.134-7, Wydawictwa Nauk.Tech., Warszawa (1975).
23. Suzuki, M. and S. Maeda, On the Mechanism of drying of granular beds. Mass transfer from a discontinuous source, J.Chem.Eng.Japan, 1, 26-31 (1968).
24. Threlkeld, J.L., "Thermal Environmental Engineering", 2/e, p.198-205, Prentice-Hall, New York (1970).
25. Treybal, R.E., "Mass Transfer Operations", 2/e, p.34, McGraw Hill, New York (1968).
26. Treybal, R.E., ibid p. 39.
27. Weast, R.C., ed., "Handbook of Chemistry and Physics", 48/e, p.D140-2, Chemical Rubber Co., Cleveland Ohio (1967/8).
28. Wilke, C.R. and D.T. Wasan, A new correlation for the psychrometric ratio, AIChE - IChE Symp.Series, 6, 21-6 (1965).
29. World Meteorological Organization, Tech. Regulations Vol.1 General", 2/e. WMO 49 BD 2, Geneva, Switzerland (1959).
30. Wylie, R.G., The effect of variability of the fluid properties in laminar convective heat and mass transfer processes, Section 4.1, p.17-23, 1 Australasian Heat & Mass Transfer Conf., Monash, Melbourne, Australia (1973).
31. Wylie, R.G., Interpretation of the wet-shadowing experiment in a basic study of the psychrometric system, Section 5.1, p.17-24, ibid.

PLATE 4. A twin-valve lock for discharging powder from a dryer worked under vacuum. [Luwa-SMS GmbH, with permission.]

Chapter 5

PROCESS OF DRYING

5.1 Ideal Moist Solids

Moisture is often bound to its host substance in multifarious ways, as outlined in Chapter 2. Attempts to describe the manner whereby moisture is dislodged and evaporates for many common materials involve formidable problems of analysis, and are usually tractable only for constant drying conditions. However, the process conditions almost always vary from place to place in a dryer and, in the case of batch drying, they change with time as well. It is thus useful to describe a body of comparatively simple structure, through which the movement of moisture can be analysed or experimentally modelled in a straightforward way, so that the drying behaviour can be predicted for conditions more representative of those in commercial equipment. This procedure, although very rough in a quantitative sense, nevertheless provides a number of important clues about drying behaviour in general and strategies for process operation.

Consider a porous body in which the void spaces are greater than 1 μm in extent. In these circumstances, moisture merely fills the cavities without any significant attachment to the solid skeletal frame. The body is non-hygroscopic, as the moisture always exerts its full vapour pressure, even to the end of drying. We shall call such a body an ideal moist solid. Soaked beds of inert particles such as glass spherules form ideal moist solids, for example. Shallow particulate beds have commonly been used for laboratory investigations of drying, despite their atypical behaviour towards the end of drying, since these materials are uniform in composition and experimentally reproducible.

5.2 Simple Model Systems

1. Two-pore model. The simplest possible version of an ideal moist solid is a two-pore system of joined capillaries of unequal diameter. When the pores are completely full, the body is totally sodden. As drying takes

place, as soon as any superficial moisture has been driven off, moisture begins to evaporate preferentially from the wider pore. The moisture level here remains constant, whereas that in the narrower pore falls, since moisture from one limb to the other supplies the evaporative loss. This process continues as long as this capillary action can be sustained. After a certain "critical" condition has been reached, both menisci begin to withdraw into the body. Moisture then evaporates from sites below the visible surface of the material.

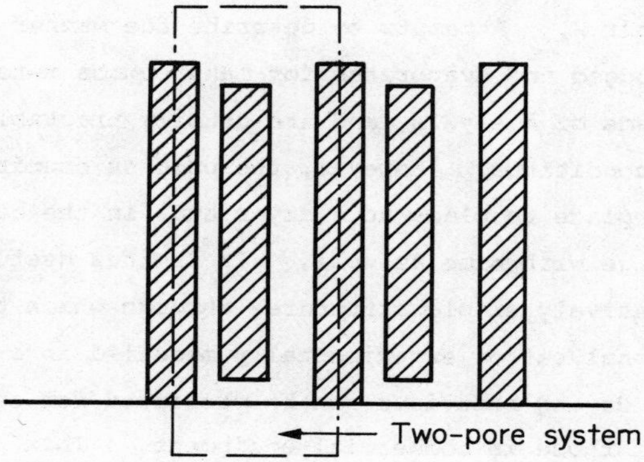

Fig. 5.1 A two-pore system

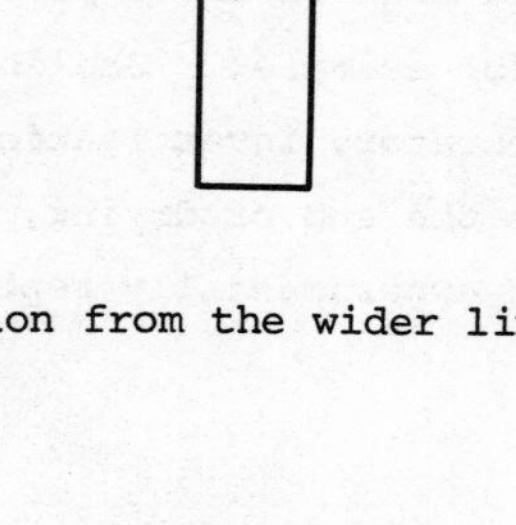

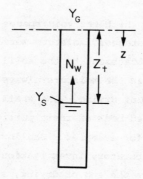

Fig. 5.2 Evaporation from the wider limb of a two-pore system.

As long as moisture evaporates from the exposed surface, the loss of moisture over unit surface is given by equation 4.32, namely

$$N_W = \alpha K_o \phi(Y_S - Y_G) \quad (5.1)$$

where α is a coefficient, slightly smaller than 1, and relates to the fractional area of the body covered by the capillary openings. At first, there is evaporation from both the wider and narrower pores, but α becomes somewhat smaller with time as the meniscus retreats in the narrower pore to feed the other by capillary action. Once the meniscus withdraws into the larger pore, the evaporation rate begins to fall off markedly. The rate of evaporation may be related to the diffusional path for the moisture vapour in this larger pore, if the evaporation, now very minor, from the smaller pore is neglected. On using equation 4.20 for the humidity, the rate of fall of the meniscus is given by

$$\frac{d}{d\tau}(\rho_L z) = N_V = \frac{c\,\mathcal{D}_{WG}}{z} \ln \frac{(D+Y_S)}{(D+Y_G)} \quad (5.2)$$

Equation 5.2, on integration for an elapsed time t, yields the distance fallen Z_t in that time:

$$Z_t = \left[\frac{2c\,\mathcal{D}_{WG}t}{\rho_L} \ln \frac{(D+Y_S)}{(D+Y_G)}\right]^{\frac{1}{2}} \quad (5.3)$$

Thus the level, and therefore the moisture content, falls as the square root of the time. On the other hand, the evaporation rate is inversely proportional to the level (eq.5.2), so the rate diminishes quadratically as the pore dries out.

These considerations lead us to expect that the rate-of-drying curve for a two-pore system would be composed of two sections: one wherein the rate remains almost constant while capillary action takes place; the other for which the rate falls appreciably with reducing moisture content. The postulated curve is shown in Fig. 5.3.

This mode of drying behaviour has been witnessed frequently in some of the earliest laboratory drying experiments.[2, 27, 29] This observation has led to the formal division of drying periods into a "constant-rate" period (corresponding to ABC) and a "falling rate" period (corresponding to CD). The average moisture content of the body at the transition

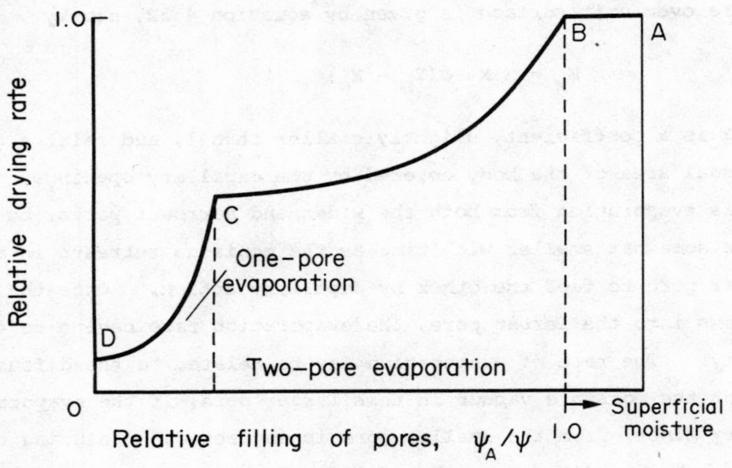

Fig. 5.3 Drying out of a two-pore system

between these periods is called the critical point or the critical moisture content. The term "constant-rate" period is somewhat misleading, as the initial drying rates are rarely entirely constant, and the term initial drying period is to be preferred. Nevertheless, the drying rates do sometimes only change to a small degree over this period, and so approximate in magnitude to the rate of evaporation from a free liquid surface under the same drying conditions, as illustrated in the following table.

TABLE 5.1 MAXIMUM EVAPORATION RATES FROM MATERIALS IN A PAN WITH CROSSFLOWING AIR AT 3 m s^{-1} AND 50°C (30)

Material	Rate of evaporation / kg m^{-2} h^{-1}	Material	Rate of evaporation / kg m^{-2} h^{-1}
water	0.27	brass turnings	0.24
whiting pigment	0.21	fine sand	0.20 - 0.24
brass filings	0.24	clays	0.23 - 0.27

2. Uniform bed. For experimental purposes, a bed of closely graded, inert particles such as glass beads provides a convenient model system. If such a bed is thoroughly sodden, the moisture is distributed continuously through the free spaces between touching particles. Moisture, when threading through the bed in this way, is said to be in the funicular state. Clearly this is a prerequisite for capillary motion to take place. On drying the bed, this funicular state appears to collapse suddenly, when the moisture shrinks into rings at the waists within the assembly of particles. The moisture is now said to be in the pendular state. The onset of this pendular state defines the critical point of transition, when capillary motion ceases, and the falling-rate period of drying begins.

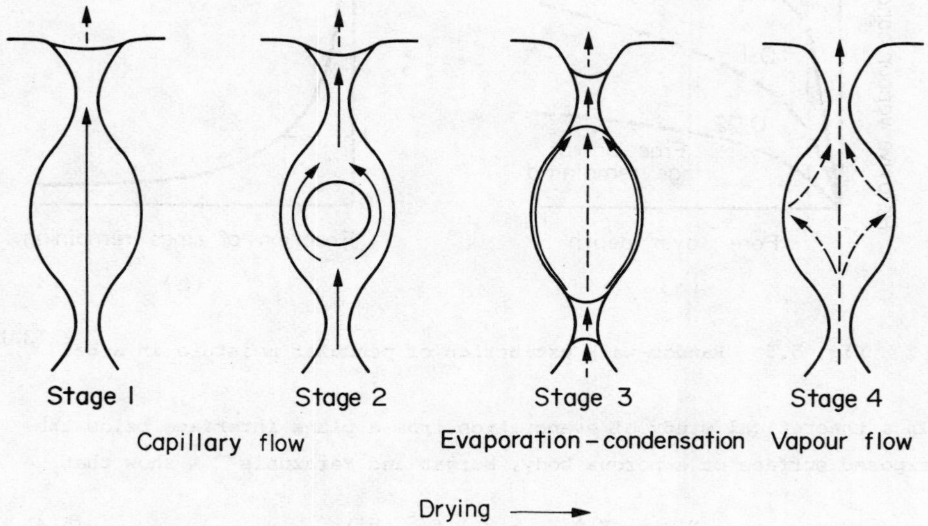

Fig. 5.4 Moisture at various stages in the drying out of a porous material.

A possible sequence of events in the dessication of a single cavity in the bed is shown in Fig. 5.4 above. The sequence in which the various rings of moisture are destroyed is open to speculation, but one reasonable postulate[(32)] is based on a random distribution of such rings in a porous bed. There is a probability p that moisture evaporated at one location will recondense at another site and a probability (1-p) that the moisture will be able to escape to the surface. These probabilities will be a function of the depth of the evaporative site below the surface. Results for a "random walk" of these rings within a square lattice are shown in Fig. 5.5. By extrapolating the normalized moisture-vapour concentration to 1 in Fig. 5.5(a), it is possible to estimate the apparent depth of the evaporative interface ζ. The value of this recession ζ does not exceed one-layer depth until about 80 per cent of the rings have been broken up for the case when the probability that a ring will shift position by one layer is 1/2.

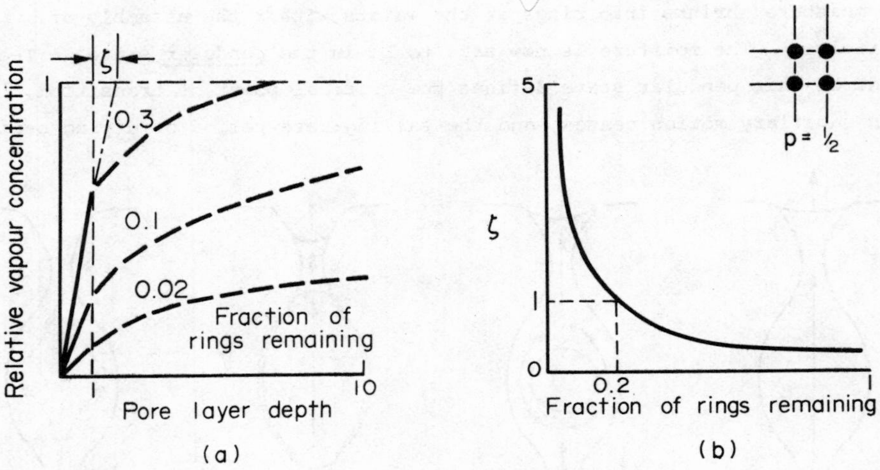

Fig. 5.5 Random-walk extinction of pendular moisture in a bed.[32]

In a theoretical study of evaporation from a plane interface below the exposed surface of a porous body, Morgan and Yerazunis[19] show that

$$N_V = K_o \phi (Y_W - Y_G) f(\xi, Bi) \qquad (5.4)$$

where $f(\xi, Bi)$ is the relative rate of drying ($f = 1$ when the interface is at the surface ($\xi=1$)),

ξ is the relative depth of the evaporative plane below the surface,

and Bi is the mass-transfer Biot number, the relative conductance of the air boundary layer to moisture compared with that through the porous mass when dry.

The reduction in normalized evaporation rate $f(\xi, Bi)$ with position ξ of a receding evaporative interface is shown in Fig. 5.6 for 12.7 mm thick beds of polystyrene, glass and steel spheres.

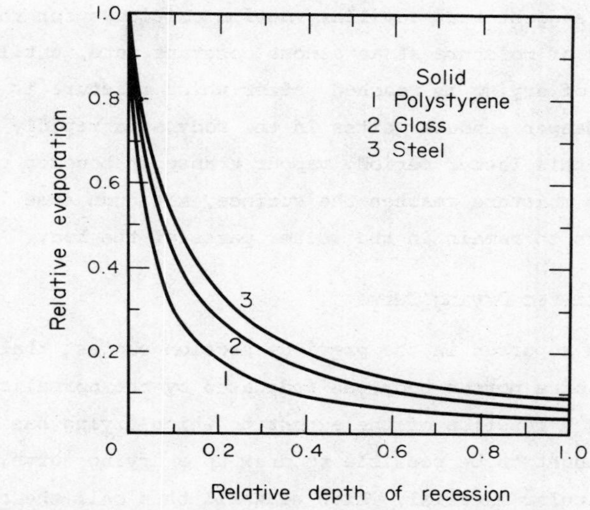

Fig. 5.6 Evaporation from below the surface of a bed of particles 12.7 mm thick.(19)

By fitting equation 5.4 to data obtained in the drying out of a 12.7 mm thick layer of glass beads of 100 μm in diameter, the apparent evaporative plane is shown to reside close to the surface until 80 per cent of the moisture has vaporized, as illustrated in Fig. 5.7, a result that is consistent with the random-walk analysis.

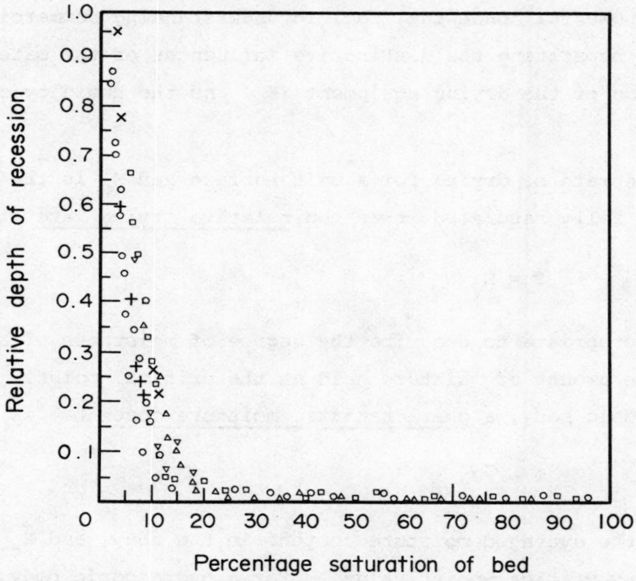

Fig.5.7 Variation of apparent depth of evaporative plane with mean moisture content of drying beds of glass beads. (19)

These analyses suggest that capillary motion continues for the removal of the majority of moisture at an almost constant rate, until the critical point of drying is reached, after which moisture is lost from progressively deeper pendular sites in the body at a rapidly diminishing rate. During this latter period, vapour transport becomes the dominant process whereby moisture reaches the surface, although some liquid-phase movement appears to remain in the wetter parts of the body.[37]

5.3 Characteristic Drying Curve

The experiments reported in the previous section suggest that the relative ease of drying of a porous body, as indicated by the normalized drying rate, is simply a function of the extent to which drying has occurred. Therefore, it ought to be possible to draw up a drying curve, "characteristic" of a particular material, which we shall thus call the characteristic drying curve. If such a curve can be derived, then the rate of drying over a unit exposed surface is given by

$$N_V = f\, K_o\, \phi(Y_W - Y_G) \tag{5.5}$$

where f is a dimensionless function of the moisture content which takes account of the thermophysical properties of the material being dried. This formulation, first used by Van Meel[39] to analyse batch drying, provides a powerful conceptual tool in understanding commercial drying practice by separating the distinctive influences of the material (f), the operation of the drying equipment (K_o) and the humidity conditions, $\phi(Y_W - Y_G)$.

If N_V is the rate of drying for a unit surface and N_W is the rate when the body is fully saturated, then the relative drying rate f is clearly

$$f = N_V/N_W \tag{5.6}$$

It seems appropriate to describe the degree of moistness of a body in terms of the amount of moisture held at the critical point. For a non-hygroscopic body, a characteristic moisture content Φ is defined by

$$\Phi = \bar{X}/\bar{X}_{cr} \tag{5.7}$$

where $\bar{X}$ is the averaged moisture content in the body, and $\bar{X}_{cr}$ is the corresponding critical-point value. For a hygroscopic body, it is more appropriate to recast this definition in terms of the moisture that can be freely expelled by drying into a humid gas. Thus the so-called free moisture content, the moisture content in excess of the equilibrium value, is used instead; that is

$$\Phi = (\bar{X} - X^*) / (\bar{X}_{cr} - X^*) \tag{5.8}$$

where X* is the equilibrium-moisture content.

The use of the mean moisture content as an index of the degree of drying contains an implicit assumption. The extent of drying at a mean moisture content will depend also on the relative extensiveness of the exposed surface per unit volume of material. Thus we should expect similar drying behaviour only for material unchanged in form.

The concept of the characteristic drying curve implies

$$f = f(\Phi) \tag{5.9}$$

Clearly, when moisture covers the surface, so that the material plays no role in the evaporative process, f = 1. At the critical point, the relative drying rate will take some lesser value f_{cr}. However, for fine-structured materials, the value of f_{cr} will be closely equal to unity.(8) Since the f function relates to the influence of the solid in hindering evaporation, and this restraint is already appearing during the first drying period, instead we choose a graphical critical point rather than the point of funicular transition as the "critical point". The critical point, then, is the first-appearing knee in the drying curve, as shown in Fig. 5.8.

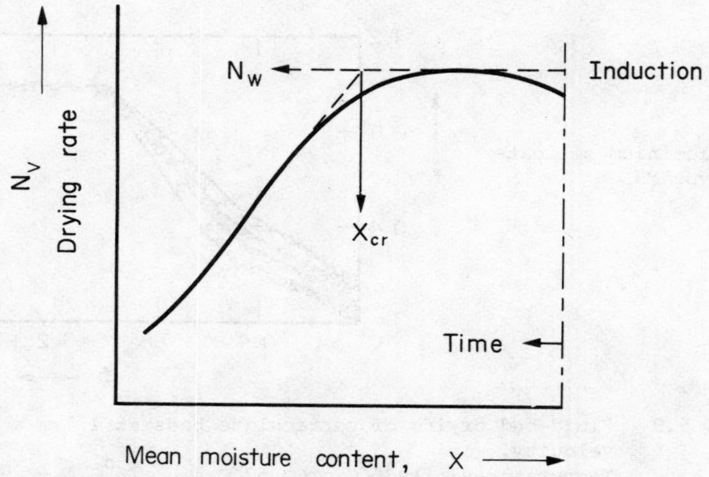

Fig. 5.8 Determination of graphical critical point from drying-rate data.

Initially, let us assume that this graphical critical point is independent of drying conditions and is specific to a given material. Later it will be shown that this assumption is not always valid, but rather fortuitously this invalidity is of minor consequence in a number of practical applications. With this assumption, we can set out formally the limits for the f-function, namely:

$$\Phi \geqslant 1 \quad , \quad f = 1$$

$$0 \leqslant \Phi \leqslant 1 \quad , \quad f_e \leqslant f \leqslant 1$$

For hygroscopic particles, f_e is zero.

By way of example, a characteristic drying curve is shown in Fig.5.9(a) for the fluid-bed drying of molecular-sieve particles.

(a) molecular sieve type 13 x 2.2 mm diameter

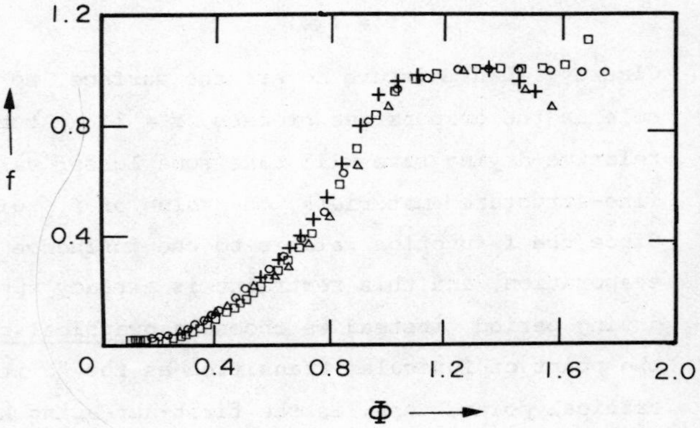

(b) aluminium silicate type M3

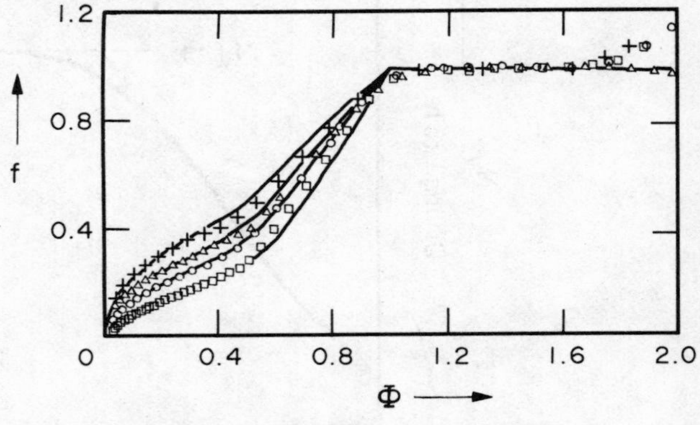

Fig. 5.9 Fluid-bed drying of particulate beds at 4.4 m s^{-1} air velocity.
Temperatures: □ 36°C, ○ 57°C, Δ 76°C, + 98°C .
After Schlünder (27).

A common curve is obtained for inlet-air temperatures between 36 and 98°C.[27] On the other hand, there is no common curve for the drying of aluminium silicate particles under the same conditions, Fig. 5.9(b). This difference is attributed to the lack of sharpness of the receding evaporative front in the latter particles.

Characteristic drying curves are also observed for slab-form material, as illustrated by data for the cross-circulation drying of samples of gypsum wallboard for a 250 per cent change in humidity potential.[5] These data are displayed in Fig. 5.10.

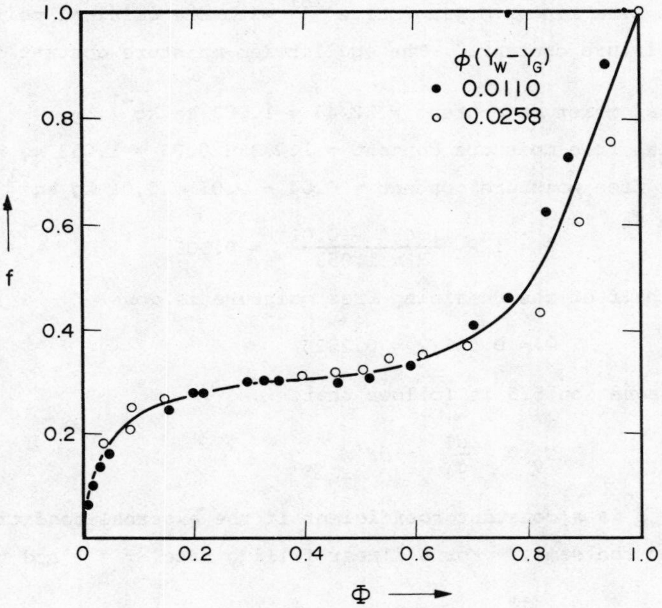

Fig. 5.10 Characteristic drying curve for 9.5 mm thick gypsum wallboard.

Most of the evidence for the concept of the characteristic drying curve, however, is indirect, being based on the corroboration of experimentally determined drying rates in laboratory dryers with those predicted from a single drying curve.[15, 36]

Footnote. Another dimensionless quantity for the drying rate which has been used to resolve drying curves into a single form is $N_Y a\tau/\rho_s(X_{cr} - X^*)$.[7, 24] For a slab of thickness b, this parameter is similar to the intensity of drying, which is defined on p. 178 as $N_W b/\rho_s X_o \mathcal{D}_a$, since the ratio $b^2/\mathcal{D}_a$ is equivalent to the time of drying τ.

There is no reason to believe that the f function should follow any particular graphical form, although the characteristic drying curve is often approximated in whole or part by linear segments for convenience. The following worked example illustrates the case when a single linear falling-rate curve is found.

Example 5.1. Skimmilk is to be spray-dried from a concentrate containing 48 per cent solids to a powder of 4 per cent moisture content (dry basis). If half the moisture is lost in 5 s, how long will it take to remove half of the remaining free moisture? Assume constant external conditions and that there is a linear drying curve(38) with the critical point at the initial moisture content. The equilibrium-moisture content is 0.03 kg kg^{-1}.

Initial moisture content = 52/48 = 1.083 kg kg^{-1}

Initial free moisture content = 1.083 - 0.03 = 1.053 kg kg^{-1}

Final free moisture content = 0.04 - 0.03 = 0.01 kg kg^{-1}

After 5s

$$\Phi = 1 - \frac{(1.053 - 0.01)}{2 \times 1.053} = 0.505$$

When half of the remaining free moisture is gone

$$\Phi = 0.505/2 = 0.2525$$

From equation 5.5 it follows that

$$N_V \alpha - \frac{d\Phi}{d\tau} = \beta f$$

where β is a constant coefficient if the external conditions remain the same. For a linear falling rate, $f = \Phi$ and thus

$$-\frac{d\Phi}{d\tau} = \beta\Phi$$

$$\therefore t = -\frac{1}{\beta}\int_1^{\Phi_t} \frac{d\Phi}{\Phi}$$

$$= -\frac{1}{\beta} \ln \Phi_t$$

On substitution to time t = 5s ,

$$\frac{1}{\beta} = \frac{5}{-\ln 0.505} = +\ 7.319$$

and the drying time to $\Phi = 0.2525$ becomes

$$t = -\ 7.319 \ln 0.2525 = 10.1 \text{ s} .$$

Thus it takes as much time to remove the further one-quarter as that for driving off the initial half of the moisture. This result is self-evidently the direct consequence of the slowing down of the drying process with time.

Extrapolation of data. Once a characteristic drying curve has been established, it would seem possible to estimate drying rates, and thus the time to dry between specified moisture levels, for any given set of process conditions. The maximum drying rate, when f = 1, can be determined by methods outlined in Chapter 6, while the relative drying rate f yields the fraction of this rate that is obtainable at a given moisture content. This procedure is only strictly valid if the relative resistance to moisture movement through the air and through the solid remain unchanged in the new conditions compared to those in the test from which the characteristic drying curve is derived. The sensitivity of this curve to changes of relative resistance has been explored.[9] Provided the mass-transfer Biot number, which is a measure of the relative resistance the solid affords, is high enough, no significant inaccuracies will result. Such conditions pertain, for example, with cross-circulated beds of porous materials, some 100 mm thick, or with rotary drying particles 10 mm in diameter.

When this extrapolation is possible, one may use values of the reciprocal rate ($d\tau/d\Phi$) to estimate drying times, since

$$t = \int_{\Phi_o}^{\Phi_t} \frac{d\tau}{d\Phi} \cdot d\Phi \qquad (5.10)$$

The value of $d\tau/d\Phi$ is found from the absolute rate N_V in the following way: now

$$N_V = -\frac{d}{d\tau}\left(\frac{\rho_s X}{a}\right) \qquad (5.11)$$

where a is the surface area of the drying goods per unit volume and ρ_s is the bulk density of the bone-dry solids. From the definition of the characteristic moisture content Φ, eqn.5.8, namely

$$\Phi = (X - X^*) \; / \; (X_{cr} - X^*) \qquad (5.8)$$

it follows that

$$\frac{d\Phi}{d\tau} = \frac{1}{(X_{cr} - X^*)} \cdot \frac{dX}{d\tau} \qquad (5.12)$$

so that from equations 5.11 and 5.12

$$\frac{d\tau}{d\Phi} = -\frac{\rho_s(X_{cr} - X^*)}{a\,N_V} \qquad (5.13)$$

Substitution of this expression for $d\tau / d\Phi$ into equation 5.10 yields the working relationship

$$t = -\frac{\rho_s(X_{cr} - X^*)}{a}\int_{\Phi_o}^{\Phi_t}\frac{d\Phi}{N_V} \qquad (5.14)$$

Equation 5.5 is used for N_V. The integrand may be evaluated numerically or graphically; the following worked example illustrates the numerical procedure.

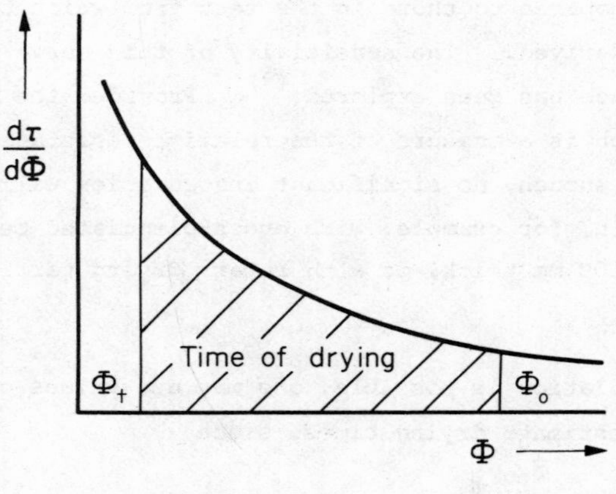

Fig. 5.11 Graphical determination of the time of drying.

Example 5.2. Gypsum wallboard of 10 mm thickness is being dried in a conveyor dryer in which the conditions are held constant at 110°C dry-bulb temperature and 54°C wet-bulb temperature. The board enters at a characteristic moisture content of 0.5 and leaves at $\Phi = 0.05$. Estimate the drying time.

Data: $K_o = 120$ g m^{-2} s^{-1} ; $(X_{cr} - X^*) = 0.76$ kg kg^{-1}

$\rho_s = 1650$ kg m^{-3}

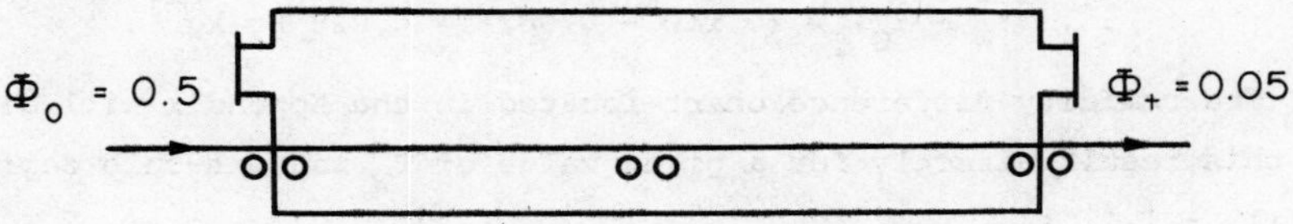

Fig.5.12 Board dryer

We have

$$t = -\frac{\rho_s(X_{cr} - X^*)}{a}\int_{\Phi_o}^{\Phi_t}\frac{d\Phi}{N_V}\ , \text{eq. 5.14,}$$

where $a = 2/b$, for drying from the two faces of the board with thickness b

and $N_V = f\ K_o\phi(Y_W - Y_G)$, eq. 5.5.

Since the external drying conditions are constant, only f varies with Φ. On combining these expressions, we find

$$t = -\frac{\rho_s b(X_{cr} - X^*)}{2\ K_o\phi(Y_W - Y_G)}\int_{\Phi_o}^{\Phi_t}\frac{d\Phi}{f} \qquad \text{(A)}$$

Values of the relative drying rate may be read off Fig. 5.10 :

Φ	f	1/f
0.5	0.315	3.17
0.45	0.31	3.23
0.4	0.305	3.28
0.35	0.30	3.33
0.3	0.293	3.42
0.25	0.285	3.51
0.2	0.275	3.64
0.15	0.26	3.85
0.1	0.237	4.22
0.05	0.19	5.26

$$\therefore \int = -0.05\left[\frac{3.17}{2} + 3.23 + 3.28 + \ldots\ldots + 4.22 + \frac{5.26}{2}\right]$$

$$= -1.635$$

Humidity potential. For $T_G = 110^oC$ and $T_W = 54^oC$, $Y_W = 0.110$ kg kg^{-1} and $Y_G = 0.082$ from psychrometric considerations.

$$\therefore (Y_W - Y_G) = (0.110 - 0.082) = 0.028 \text{ kg kg}^{-1} .$$

(The humidity-difference chart located in the Appendix will give this result directly for a given value of T_W and wet-bulb depression $(T_G - T_W)$).

Humidity-potential coefficient. As the humidity potential is small,

$$\phi = D/(D + Y_W)$$
$$= 0.622 / (0.622 + 0.110) = 0.850 .$$

Drying time. Equation A can now be evaluated.

$$t = \frac{-1650 \times 0.01 \times 0.76}{2 \times 0.12 \times 0.850 \times 0.028} \times [-1.635]$$
$$= 3590 \text{ s}$$

or about 1 h .

Note that the units of K_o must be expressed in basic SI units of kg m^{-2} s^{-1} .

5.4 Drying Theories

1. Simultaneous transport. The most rigorous methods of describing the drying process are derived from the concepts of irreversible thermodynamics in which the various fluxes are taken to be directly proportional to the appropriate "potential". Moisture may exist in either the frozen, liquid or vapour state. Mass balances for transfer within each state can be written in terms of moisture fluxes $\underline{J}_i$ that give rise to material accumulations, or depletions $\underline{I}_i$:

$$\frac{\partial(\rho_s X_i)}{\partial\tau} = -\text{div } \underline{J} + \underline{I}_i \tag{5.15}$$

Likewise, heat-energy balances can be set down :

$$\bar{C}_p \rho_s \frac{\partial T}{\partial \tau} = -\text{div } J_Q + \sum_4^3 H_i \underline{I}_i \tag{5.16}$$

where J_Q is the heat flux and H_i is the specific enthalpy of the moisture in the designated state. These considerations lead to complex partial differential equations for the moisture-content and temperature fields in the body. Further, these equations incorporate transport coefficients, which must be determined experimentally and are strong functions of moisture content. Such analyses are beyond the scope of an introductory book, and the interested reader is referred to Luikov's monograph (16) for such detail.

2. Flow through porous media. A less abstract, but closely similar approach is to describe the drying in terms of vapour diffusion and liquid movement as a result of gravitational and capillary forces in a porous structure. This method was developed by Krischer (12) in Germany in the late 1930's, and independently by soil physicists (3, 26) some twenty years later, an example of the slow diffusion of knowledge across linguistic and professional barriers. A good discussion of modern developments in this approach is given by Novak and Coulman. (21)

Soil physicists, however, are principally concerned with the moisture and temperature fields during the slow drying out of porous media, whereas process engineers are more concerned with the drying kinetics. In this regard, the calculations of Berger and Pei, (1) although restricted unrealistically to constant transport coefficients, nevertheless do provide some useful predictions for the dynamic drying behaviour of hygroscopic porous solids. The physical picture underlying the mathematical analysis depicts the moisture reaching the surface by the capillary flow of liquid due to a gradient of liquid-moisture content and the diffusion of vapour through the liquid-free pores due to a gradient of moisture-vapour pressure. Local thermodynamic equilibrium is assumed, and the sorptional isotherm is used to relate local moisture-vapour levels to the moisture content at that "place". Some of the computed rate-of-drying curves are shown in Fig. 5.13.

The predicted curves take the often observed experimental form: an initial adjustment in rate leading to a period of uniform rates, and a period of dimishing rates thereafter. The duration of the constant-rate period, as shown in Fig. 5.13(a), depends upon the magnitude of Lu_L, and thus the extent of internal liquid transfer. As long as this liquid movement can match the evaporation from the surface, the constant-rate period will endure. The initial adjustment may result in either an increase or decrease in rate, following a change in surface temperature, caused by an imbalance between the heat transferred to the body and the heat needed by evaporation at the surface.

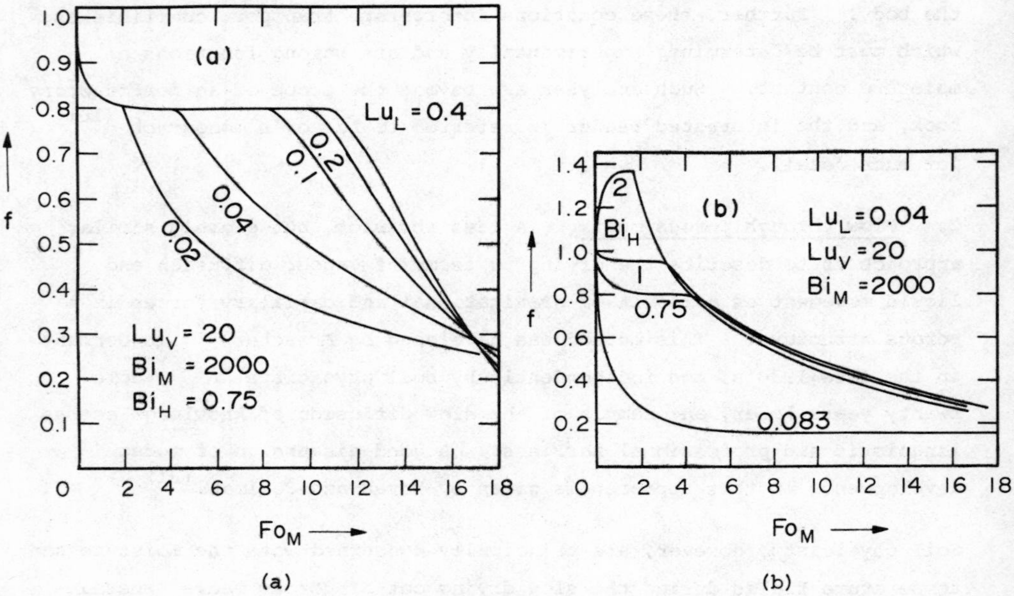

Fig. 5.13 Theoretical rate-of-drying curves for a uniform hygroscopic porous solid. After Berger and Pei.[1]

Bi_H	Biot number (heat transfer)	hb/λ
Bi_M	Biot number (mass transfer)	Kb/κ
Fo_M	Fourier number	$\kappa\tau/b^2$
Lu_L	Luikov number (liquid transport)	$\mathcal{D}_L/\kappa$
Lu_V	Luikov number (vapour transport)	$\mathcal{D}_V/\kappa$

3. Simplified models. All the foregoing methods involve substantial computational effort and have yet to be fully shaped into tools for process-engineering analysis. For practical calculations we must rely on cruder concepts, and two such models have been advanced: the wetted-surface and the receding-plane models.

Wetted-surface Model

This model[23] assumes that the drying rates lessen due to the smaller fraction of the surface that is wetted. It thus follows, with this view, that the relative drying rate f is equal to the fractional wetted area. To a first approximation, the area available for evaporation varies as the square of an equivalent wet length, whereas the moisture concentration would vary as this length cubed. Whence,

$$f = \Phi^{2/3} \qquad (5.17)$$

If account is taken that the evaporative interface may recede into the body, then the exponent on the characteristic moisture content will be somewhat less. Equation 5.17 should thus be replaced by the more general expression

$$f = \Phi^n \quad , \quad 0 < n < 2/3 \qquad (5.18)$$

where the exponent n is determined empirically. A value of $n = 0.6$ has been found in the drying of balsawood slats up to 9.5 mm in thickness.[23] These expressions are satisfactory as long as the material being dried is thin, and the averaged moisture content, which is used to determine Φ, is approximately equal to the surface moisture content. For thick materials, equation 5.17 should be corrected by a "thickness function", which depends upon the ratio of moisture content at the surface X_o to the mean moisture content $\bar{X}$; that is

$$f = [(X_o - X^*)/(\bar{X} - X^*)]^{2/3} \, \Phi^{2/3} \qquad (5.19)$$

The parameter $[(X_o - X^*)/(\bar{X} - X^*)]^{2/3}$ is found from a fit of the experimental data.

Although this approach has been used to calculate successfully from laboratory data the moisture-content profiles in the rotary drying of gypsum particles and the time to dry pharmaceutical tablets in a fluid bed,[11] the need to incorporate an empirical thickness function effectively reduces the method to one involving an empirical f function itself. This could be done directly from the experimental data without invoking any particular "model" of drying behaviour. Further, the model is based on a physically unrealistic picture of the way porous solids dry out, except perhaps when very thin. Although dry patches have been seen and photographed on the surface of moist granular beds as they dry out,[22] finely porous material when exposed to gentle air currents can have a significant fraction of its exposed surface dry before the evaporation from the whole surface is effected.[33]

Receding-plane Model

A study[6] of the temperature fields in drying brick-like material shows that, after an initial period has elapsed, two distinctive thermal regions appear. From the exposed surface inwards, the temperature fall from about the dry-bulb to about the wet-bulb value, which persists thereafter

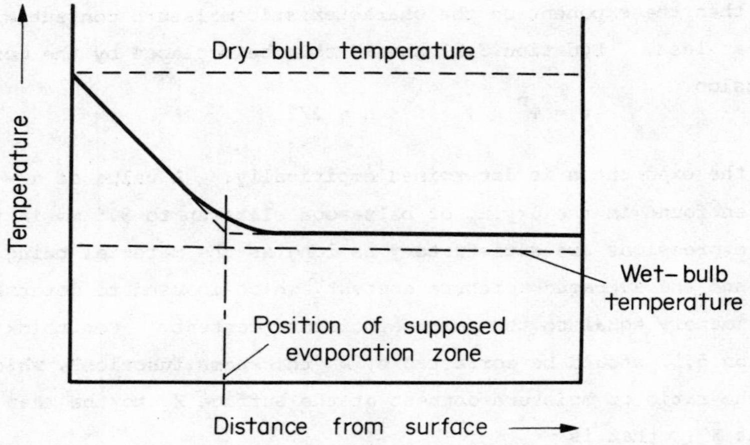

Fig. 5.14 Temperature profile in a brick during drying.[6]

throughout the remainder of the body (Fig. 5.14). It seems reasonable to assume that this discontinuity in temperature gradient represents the position of an evaporative zone within the body. Because the transition in slope is relatively sharp, the extent of the zone itself is probably small compared with the thickness of the body. Further evidence of the existence of a narrow evaporative zone within a porous body is provided by considering relative humidity levels in porous material.[17] Other corrobatory evidence of an evaporative front separating a dry region and one still wet is found from tests[20] on drying a Terylene bobbin made up from 3.2 mm thick pads of cloth which could be separately analysed for moisture content.

These considerations provide some substance for the receding-plane model of drying.[10] In this theory, it is assumed that moisture evaporates from a sharp front, which progressively withdraws into the body as it dries out, and then diffuses as moisture vapour to the exposed surface. This diffusion is taken to be the rate-determining step.

Consider an infinitely extensive porous slab in which an evaporative plane resides at a distance ζ below the exposed surface, as shown in Fig. 5.15.

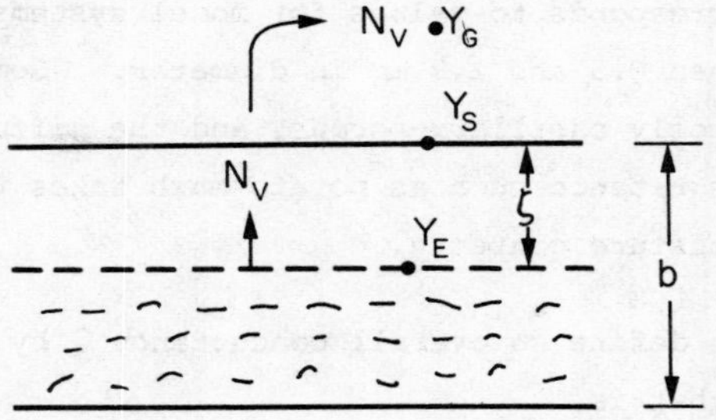

Fig. 5.15 Subsurface evaporation in a porous slab of thickness b.

The moisture-vapour rate through the air film above the solid surface is given by the expression

$$N_V = K_o \, \phi_S (Y_S - Y_G) \tag{5.20}$$

A similar expression* holds for the diffusion of moisture vapour through the dried-out zone below the surface

$$N_V = K_S \, \phi_E (Y_E - Y_S) \tag{5.21}$$

where K_S is the corresponding mass-transfer coefficient and the humidity-potential coefficient ϕ_E is approximately $D/(D + Y_E)$. The mass-transfer coefficient itself is a lumped parameter embodying a diffusion coefficient for transport of moisture vapour through a porous structure over a path length proportional to the distance ζ if the material is uniformly porous. Reference to the definition of mass-transfer coefficients in Section 4.4 leads to the identity:

$$K_S = F_S M_G = \frac{C\mathcal{D}_{WG}}{\zeta} \cdot \left(\frac{\Psi}{\xi}\right) \cdot M_G \tag{5.22}$$

where Ψ is the porosity (the fractional free space available for vapour flow) and ξ is the tortuosity. This latter quantity accounts for the crookedness of the network of pores and for the possible inclusion of small orifices and restrictions in an actual porous solid.

The ratio (ξ/Ψ) is Krischer's "diffusion-resistance coefficient" μ_D which varies from about 3 for some food powders to up to 30 for concrete blocks.[13]

Footnote* Strictly one should allow for the inflow of air into the porous body. However, at 50°C, $-N_G/N_V = \rho_G M_G/\rho_W M_W = 0.0007$ and is negligible.

The lower limit corresponds to values for model systems such as beds of glass spheres between 0.5 and 1.9 mm in diameter. Some materials, of course, are scarcely capillary-porous, and the diffusion-resistance coefficient for a substance such as potato mash takes values of several thousands at low moisture contents.

It is convenient to define an overall conductance $\mathcal{G}$ by analogy with these equations; that is,

$$N_V = \mathcal{G}\,\phi_E(Y_E - Y_G) \tag{5.23}$$

This new conductance can be related to the individual conductances K_o and K_S by considering the associated humidity potentials. Since

$$(Y_E - Y_G) \equiv (Y_E - Y_S) + (Y_S - Y_G) \tag{5.24}$$

it follows from equations 5.20, 5.21 and 5.23 that

$$\frac{N_V}{\mathcal{G}\phi_E} = \frac{N_V}{K_S\phi_E} + \frac{N_V}{K_o\phi_S} \tag{5.25}$$

and therefore

$$\frac{1}{\mathcal{G}} = \frac{1}{K_S} + \frac{1}{K_o}\left(\frac{\phi_E}{\phi_S}\right) \tag{5.26}$$

Equation 5.26 is simply a statement of the reciprocal law for summing conductances in series. At the beginning of the drying process, when $\zeta = 0$, $\phi_E = \phi_S = D/(D+Y_W)$. As the material becomes drier, ϕ_S increases in value towards 1 while ϕ_E becomes less. On the other hand, the numerical influence of the second term becomes increasingly minor as the resistance of the solid progressively dominates the drying process. Therefore, with little error, we may write

$$\frac{1}{\mathcal{G}} \simeq \frac{1}{K_S} + \frac{1}{K_o} \tag{5.27}$$

or

$$\mathcal{G} = K_o/[1 + K_o/K_S] \tag{5.28}$$

The ratio of the conductances, K_o/K_S, is called the <u>mass-transfer Biot number</u> Bi_M ; thus

$$\mathcal{G} = K_o/[1 + Bi_M] \tag{5.29}$$

A similar expression can be set down for the overall heat-transfer coefficient U in terms of the external heat-transfer coefficient and the internal thermal conductivity λ:

$$U = h/[1 + Bi_H] \tag{5.30}$$

where Bi_H is the heat-transfer Biot number, $h\,\zeta/\lambda$.

The heat received by the surface is employed in evaporating moisture in sensible heating of the slab. For unit surface, this heat balance yields

$$U(T_G - T_E) = \mathcal{G}\phi_E(Y_E - Y_G)\ \Delta H_{VE} + [\bar{C}_S\rho_S b]\ \frac{dT_E}{d\tau} \tag{5.31}$$

However, the rate of sensible heating is very small, so that almost all the heat input is used in vaporization of moisture. Equation 5.31 then reduces to the expression

$$m_{EG} = \frac{Y_E - Y_G}{T_E - T_G} = -\ \frac{U}{\mathcal{G}\phi_E\Delta H_{VE}} \tag{5.32}$$

The gradient m_{EG} represents the slope of the line connecting the state point for the bulk-air conditions (T_G, Y_G) and the state point for the conditions at the evaporative interface (T_E, Y_E). Equation 5.32 can be rewritten in terms of the external transfer coefficients h and K_o through equations 5.29 and 5.30:

$$m_{EG} = -\ \frac{h}{K_o}\ \frac{(1+Bi_M)}{(1+Bi_H)} \cdot \frac{1}{\phi_E\Delta H_{VE}} \tag{5.33}$$

When the evaporative interface rests at the exposed surface, $Bi_M = Bi_H = 0$, and $T_E = T_W$, so that equation 5.33 becomes

$$m_{WG} = \frac{Y_W - Y_G}{T_W - T_G} = -\ \frac{h}{K_o} \cdot \frac{1}{\phi_W\Delta H_{VW}} \tag{5.34}$$

Equation 5.34 represents the wet-bulb depression line. The relative magnitude of the hygrothermal gradients m_{EG} and m_{WG} depend only on the Biot numbers, that is the extent of the internal resistance to drying. We find by comparing equations 5.33 and 5.34 with each other that

$$\frac{m_{EG}}{m_{WG}} = \frac{(1+Bi_M)}{(1+Bi_H)} \tag{5.35}$$

When the internal resistance is relatively large, as always is the case at the end of drying, equation 5.35 reduces to

$$\frac{m_{EG}}{m_{WG}} \rightarrow \frac{Bi_M}{Bi_H} = \frac{K_o}{h} \cdot \frac{\lambda}{c\mathcal{D}_{WG}M_G} \cdot \frac{\xi}{\Psi} \tag{5.36}$$

on substituting the appropriate transport coefficients for the Biot numbers and equation 5.22 for K_S. The external transfer coefficients are related by the Chilton-Colburn expression, namely

$$K_o = \frac{\beta h}{\bar{C}_p} \cdot Lu^{2/3} \tag{4.47}$$

On substituting this expression for K_o into equation 5.36, and noting that $Lu = (cM_G)\mathcal{D}_{WG}\bar{C}_p/\lambda$, it follows that

$$\frac{m_{EG}}{m_{WG}} \rightarrow \beta\, Lu^{-1/3} \cdot \left(\frac{\xi}{\Psi}\right) \tag{5.37}$$

Now, $\beta\, Lu^{-1/3}$ is of order unity (for water vapour at 25°C, it is 0.997 when $\beta = 1.05$). Thus the ratio m_{EG}/m_{WG} depends only on the structure of the porous solid and, specifically, on the magnitude of the diffusion-resistance coefficient (ξ/Ψ) which normally varies between 3 and 30 for porous materials.

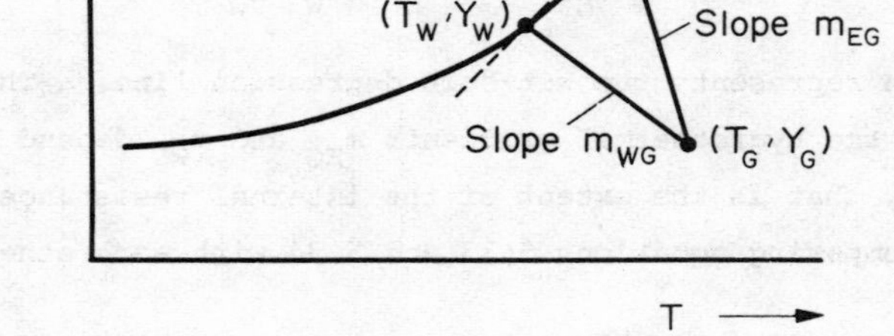

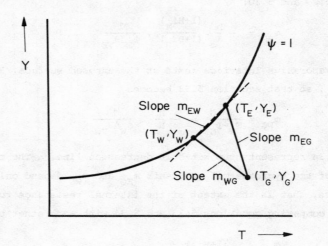

Fig. 5.16 Hygrothermal gradients.

Apparent wet-bulb temperature. From the geometry of Fig. 5.16, we have

$$(Y_E - Y_G) = (Y_E - Y_W) + (Y_W - Y_G) \tag{5.38}$$

or

$$m_{EG}(T_E - T_G) = m_{EW}[(T_E - T_G) - (T_W - T_G)] + m_{WG}(T_W - T_G) \tag{5.39}$$

whence

$$\frac{(T_G - T_E)}{(T_G - T_W)} = \frac{m_{EW} - m_{WG}}{m_{EW} - m_{EG}} = \frac{1 - m_{WG}/m_{EW}}{1 - m_{EG}/m_{EW}} \tag{5.40}$$

On noting that, with little error, $m_{EG}/m_{WG} = \xi/\Psi = \mu_D$ (eq.5.37), we have

$$\frac{(T_G - T_E)}{(T_W - T_G)} = \frac{1 - m_{WG}/m_{EW}}{1 - \mu_D m_{WG}/m_{EW}} \tag{5.41}$$

The quantity m_{WG}/m_{EW} is the ratio of the gradient of the wet-bulb depression line to the average slope of the moisture-saturation curve between humidities Y_E and Y_W. This ratio depends only on the wet-bulb temperature itself for a given vapour-gas system, and values for water vapour in air are shown plotted in Fig. 5.17.

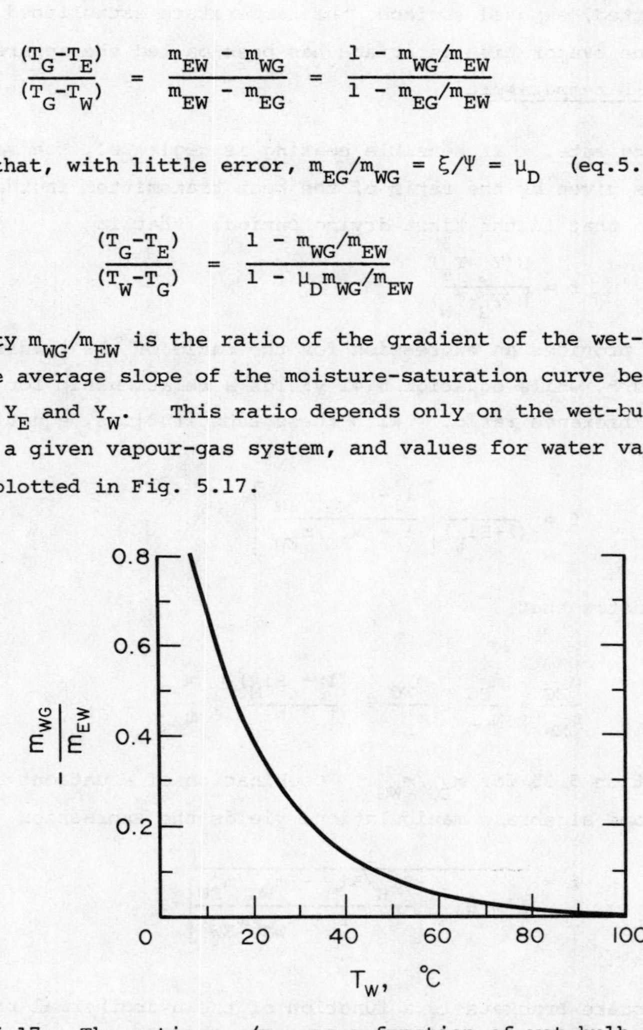

Fig. 5.17 The ratio m_{WG}/m_{EW} as a function of wet-bulb temperature. Water vapour in air. After Keey and Suzuki.[10]

Thus, for specified values of T_G and T_W, the temperature T_E at the evaporative interface approaches a limiting temperature when the Biot numbers are large, that is for sufficiently thick materials. The appearance of a steady evaporative temperature below the exposed surface of a drying body has been observed experimentally in the drying of bobbins wound with wool to a thickness of 25 mm over the spool.[20] Because of the likeness between this temperature equilibrium and that at a fully wetted, exposed surface, the temperature established within the body at the evaporative interface has been called the apparent or psuedo wet-bulb temperature.

Relative drying rate. If sensible heating is neglected, the relative drying rate is given by the ratio of the heat transmitted in the falling-rate period to that in the first drying period; that is

$$f = \frac{U(T_G - T_E)}{h(T_G - T_W)} \tag{5.42}$$

Equation 5.30 provides an expression for the ratio of the transfer coefficients U/h, while equation 5.41 yields a relationship for the temperature-difference ratio. With these substitutions, equation 5.42 becomes

$$f = \frac{1}{(1+Bi_H)}\left[\frac{1 - m_{WG}/m_{EW}}{1 - m_{EG}/m_{EW}}\right] \tag{5.43}$$

Further, one notes that

$$\frac{m_{EG}}{m_{EW}} = \frac{m_{EG}}{m_{WG}} \cdot \frac{m_{WG}}{m_{EW}} = \frac{(1 - Bi_M)}{(1 - Bi_H)} \cdot \frac{m_{WG}}{m_{EW}} \tag{5.44}$$

on using equation 5.35 for m_{EG}/m_{WG}. Combination of equations 5.43 and 5.44, after some algebraic manipulation, yields the expression

$$f = \frac{1}{1 + Bi_M\left[\dfrac{Bi_H/Bi_M - m_{WG}/m_{EW}}{1 - m_{WG}/m_{EW}}\right]} \tag{5.45}$$

The term in square brackets is a function of the hydrothermal properties of the material (Bi_H/Bi_M) and the wet-bulb temperature through the ratio m_{WG}/m_{EW}. This parameter has been called the evaporative-resistance coefficient γ.[10] For porous solids wet with water and drying in air, the coefficient varies from about 0.3 at 25°C to asymptotic values at the boiling point, as shown in Fig. 5.18. At temperatures near the boiling point, the ratio m_{WG}/m_{EW} becomes very small, due to the rapid rise of saturation humidity with temperature close to boiling, and thus the evaporative resistance coefficient γ approaches the limit of Bi_H/Bi_M. This ratio is less than unity since mass transfer is a slower process than heat transfer within porous materials. For beds of non-hygroscopic particles, values between 0.06 and 0.1 have been recorded for Bi_H/Bi_M.[35]

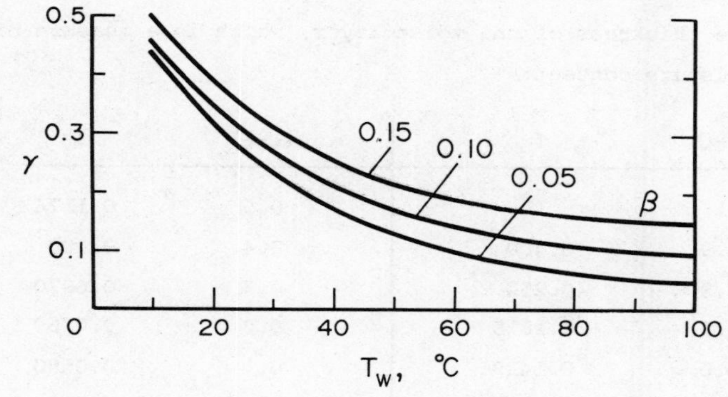

Fig. 5.18 The evaporative-resistance coefficient γ for water evaporating in air. The coefficient β is the ratio Bi_M/Bi_H. After Keey and Suzuki.[10]

The Biot number is directly proportional to the depth ζ at which the evaporative interface rests. If a Biot number Bi is defined in terms of the total body thickness b, then

$$Bi = Bi_M \left(\frac{b}{\zeta} \right) \tag{5.46}$$

and thus

$$Bi_M = Bi \left(\frac{\zeta}{b} \right) = Bi\ \xi \tag{5.47}$$

where ξ is the fractional depth of recession of the evaporative interface. Equation 5.45 can now be written in more convenient shorthand form,

$$f = \frac{1}{1 + \gamma Bi \xi} \tag{5.48}$$

Note that equation 5.48 forecasts a finite drying rate at the end of drying, when $\xi = 1$, a feature of drying porous, non-hygroscopic solids.

Example 5.3. Derive the characteristic drying curve for a bed of solids, for which $\gamma = 0.1$ and $Bi = 150$ (corresponding to a 30 mm bed of glass spheres[8]), by means of the receding-plane model of drying.

Equation 5.48 applies. We have $\gamma Bi = 0.1 \times 150 = 15$, so that equation 5.48 becomes

$$f = 1/(1+15\xi)$$

Values of f can be readily calculated as a function of $(1-\xi)$, the relative thickness of the moist layer, which is a measure of the mean moisture content:-

$(1-\xi)$	f	$(1-\xi)$	f
1	1	0.5	0.1176
0.9	0.400	0.4	0.100
0.8	0.250	0.3	0.0870
0.7	0.1818	0.2	0.0769
0.6	0.1429	0.1	0.0690
		0	0.0625

The data are plotted in Fig. 5.19 and show an upwards-concave curve, characteristic of the falling-rate period for porous solids. One should compare the form of this drying curve with that from a more rigorous calculation displayed in Fig. 5.6.

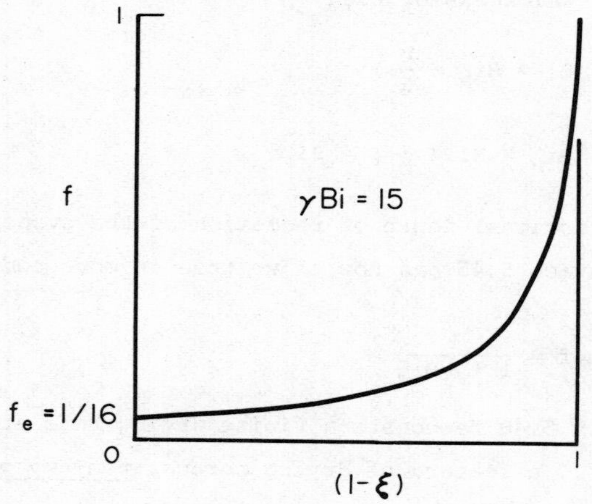

Fig. 5.19 Relative evaporation from bed.

The actual relationship between Φ and $(1-\xi)$ depends upon the moisture-content profile in the wet zone, but in the limit when this zone is uniformly moist

$$\Phi = 1 - \xi \tag{5.49}$$

Such limiting behaviour has been found in the drying of aluminium silicate particles in a through-circulating airstream in the early stages of the falling-rate period (to $\Phi = 0.5$).[12] More general behaviour is explored by Keey and Suzuki[10,34] who present inexplicit expressions for f in terms of the characteristic moisture content Φ.

To summarize: the receding-plane theory provides a simple, yet physically realistic, model of the way porous solids dry out. It can be used to look at aspects of drying not easily amenable to more formal analysis, such as drying-rate profiles in progressive dryers,[9] and the influence of drying conditions on the appearance of the critical point and the kinds of moisture-content profile that may arise. Such considerations now follow.

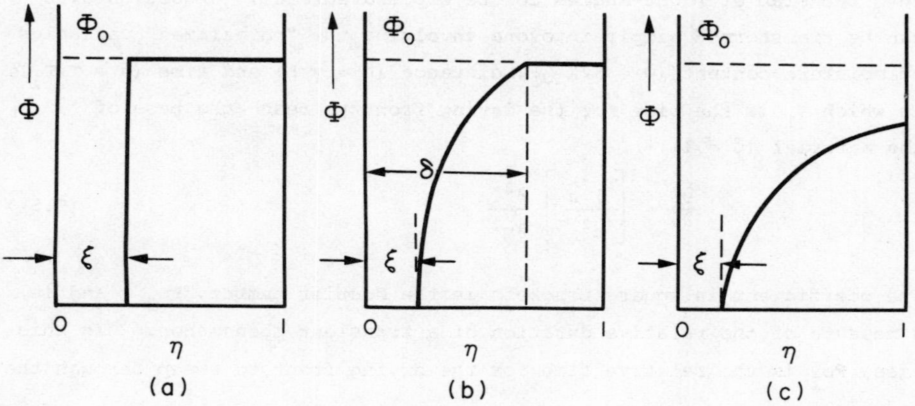

Fig. 5.20 Moisture-content profiles in an ideal, slab-form, porous body.

5.5 Intensity of Drying

Possible moisture-content profiles for the wet zone are illustrated in Fig. 5.20 for a porous solid being dried. The solid is assumed to be non-hygroscopic and slab-form, and the moisture held in liquid-free pores to be negligibly small.

Whenever the drying is very intensive, the variations in moisture content will be confined to a narrow zone and, in the limit, the profile takes the form of a step function, as shown in Fig.5.20(a). As the drying becomes less intense, so a definite moisture-content profile begins to appear in the wet zone. This situation is depicted in Fig. 5.20(b). The extent of this profile enlarges as the intensity of drying diminishes still further until, as shown in Fig. 5.20(c), the zone of reduced moisture content stretches throughout the whole material.

Let us suppose, in the first instance, that the moisture movement within this zone of reduced moisture content can be described by Fick's second law of diffusion[31] with a constant "moisture diffusivity", that is

$$\frac{\partial(\rho_s X)}{\partial\tau} = \mathcal{D}_a \frac{\partial^2(\rho_s X)}{\partial y^2} \tag{5.50}$$

in which $\mathcal{D}_a$ is the apparent diffusion coefficient and y is a distance into the slab at right-angles to the exposed surface. Equation 5.50 can be transformed simply into one involving the "normalized" variables of moisture content ($\Phi = X/X_{cr}$), distance ($\eta = y/b$) and time ($\theta = \tau/\tau_1$), in which τ_1 is the time for the drying front to reach the base of the material ($\delta = 1$):-

$$\frac{\partial\Phi}{\partial\theta} = \left[\frac{\tau_1\mathcal{D}_a}{b^2}\right]\frac{\partial^2\Phi}{\partial\eta^2} \tag{5.51}$$

The coefficient in square brackets is the Fourier number, Fo_1, and is a measure of the relative duration of a transient phenomenon. In this case, Fo_1 is the relative time for the drying front to sweep through the material.

Further, let us consider the drying process to the point in time when the constant-rate period has just ceased and the zone of reduced moisture content has penetrated a relative distance $\delta(\delta < 1)$ into the material. Equation 5.51, on integration over this zone from $\eta = 0$ to $\eta = \delta$, yields the differential equation

$$\frac{\partial}{\partial\theta}\left\{\int_o^\delta \Phi d\eta - \Phi_o\delta\right\} = Fo_1\left\{\left.\frac{\partial\Phi}{\partial\eta}\right|_{\eta=\delta} - \left.\frac{\partial\Phi}{\partial\eta}\right|_{\eta=0}\right\} \tag{5.52}$$

We can evaluate the surface moisture content from equation 5.55 by noting that $\eta = 0$, that is

$$\Phi|_{\eta=0} = \Phi_o - \frac{N_W b\delta}{2\rho_s X_{cr}\mathcal{D}_a} \tag{5.59}$$

If $\mathcal{D}_a$ is independent of moisture level, then a parabolic equation for Φ in η may be fitted to solve equation 5.52. Therefore, we assume

$$\Phi = a_o + a_1\eta + a_2\eta^2 \tag{5.53}$$

for which the unknown coefficients a_o, a_1 and a_2 can be found from the boundary conditions

$$\Phi|_{\eta=\delta} = \Phi_o \tag{5.54a}$$

$$\left.\frac{\partial\Phi}{\partial\eta}\right|_{\eta=\delta} = 0 \tag{5.54b}$$

$$\left.\frac{\partial\Phi}{\partial\eta}\right|_{\eta=0} = \frac{N_W b}{\rho_s X_{cr}\mathcal{D}_a} \tag{5.54c}$$

(The last boundary condition follows from equating the evaporation loss from a sodden surface N_W to the moisture-content gradient at the surface in terms of Fick's first law of diffusion.) Substitution of these foregoing boundary conditions, eq.5.54(a-c), into the polynomial expression for Φ, eq. 5.53, yields the profile

$$\Phi = \Phi_o - \frac{N_W b}{2\rho_s X_{cr}\mathcal{D}_a\delta}[\delta - \eta]^2 \tag{5.55}$$

Equation 5.52 may now be solved with the aid of these expressions for Φ and $\partial\Phi/\partial\eta$. The initial substitution yields

$$\frac{\partial}{\partial\theta}\left\{-\frac{N_W b}{2\rho_s X_{cr}\mathcal{D}_a\delta}\int_0^\delta (\delta-\eta)^2\, d\eta\right\} = Fo_1\left[0 - \frac{N_W b}{\rho_s X_{cr}\mathcal{D}_a}\right] \tag{5.56}$$

which reduces to

$$\frac{d}{d\theta}\left[\frac{\delta^2}{6}\right] = Fo_1 \tag{5.57}$$

from which we get the final solution

$$\delta = \sqrt{6Fo_1\theta} \tag{5.58}$$

for the penetration of the drying front.

When the surface moisture content has been reduced to zero, and the critical point attained, the depth of penetration is found from equation 5.59 to be given by

$$\delta = 2\Phi_o \rho_s X_{cr} \mathcal{D}_a / N_W b \tag{5.60}$$

Should the zone of reduced moisture content still not extend throughout the whole material, then $\delta < 1$ and, on noting $X_o = \Phi_o X_{cr}$,

$$\frac{N_W b}{\rho_s X_o \mathcal{D}_a} > 2 \tag{5.61}$$

Inequality 5.61 thus provides the criterion for the appearance of the profiles of the type depicted in Fig. 5.20(b). The dimensionless parameter $N_W b/\rho_s X_o \mathcal{D}_a$ has been called the <u>intensity of drying</u>[10] $\mathcal{N}$, since high values of $\mathcal{N}$ represent a narrow zone consequent on severe drying and small values of $\mathcal{N}$ a wide zone characteristic of gentle drying conditions.† Whenever, in fact, $\mathcal{N}$ is less than 2, the variations in moisture content are relatively small and, in the falling-rate period, the moisture contents everywhere are less than their initial values, as shown in Fig. 5.20(c). The drying zone then can be perceived as extending beyond the physical edges of the body. In the limit, when the drying is very gentle or the stuff very thin, the intensity of drying $\mathcal{N}$ becomes vanishingly small and the drying zone stretches out to infinity. At each stage of the drying process, then, the moisture contents are the same throughout the material, and the wetted-surface model of drying becomes appropriate.

Although the criterion for the appearance of a drying front within the moist material (eq. 5.61) is developed here for the case of a constant moisture diffusivity $\mathcal{D}_a$, a similar parameter is found for the more usual case of a diffusion coefficient which depends upon the moisture concentration, provided an appropriate 'mean' value for $\mathcal{D}_a$ is chosen. For instance, should $\mathcal{D}_a$ vary in an exponential fashion with moisture

†<u>Footnote</u>. The parameter $\mathcal{N}$ may be regarded as the ratio of the unhindered-drying flux N_W to an internal moisture flux which has dimensions of $\rho_S D_a X_o/b$. In the Russian literature[16], a criterion similar to $\mathcal{N}$ is called the Kirpichev number, for which the initial <u>free</u> moisture content is taken for X_o.

content, then this appropriate average is the logarithmic mean of the initial and critical-point values of the apparent diffusion coefficent.[34]

Example 5.4 Quartz sand of 0.06 mm diameter at an initial moisture content of 0.15 kg kg^{-1} is being dried in a 25 mm thick bed. What is the minimum drying rate per unit surface required in the initial drying period to ensure that a drying front will be found when the critical point is reached at a moisture content of 0.10 kg kg^{-1} ?

Data $\rho_s = 1550 \text{ kg m}^{-3}$.

X / kg kg^{-1}	$\mathcal{D}_a / 10^6 \text{ m}^2\text{s}^{-1}$
0.15	0.58
0.10	0.33

Data source: Tables 7.3 and A3 of reference 8.

The mean diffusivity is given by

$$10^6 \mathcal{D}_a = \frac{0.58 - 0.33}{\ln 0.58/0.33} = 0.443 \text{ m}^2\text{s}^{-1}$$

From equation 5.61, the maximum drying rate is given by

$$N_W = 2\rho_s X_o \mathcal{D}_a / b$$

$$= \frac{2 \times 1550 \times 0.15 \times 0.443 \times 10^{-6}}{0.025}$$

$$= 8.2 \times 10^{-3} \text{ kg m}^{-2} \text{ s}^{-1}$$

A drying rate of 8 g m^{-2} s^{-1} is scarcely attainable when materials are cross-circulated, as in tray and solid band dryers, and thus we should expect that the moisture contents throughout the material being dried are less than the initial value when the critical point is reached.

Example 5.5 How long would it take to reach the critical point for the case described in example 5.4 when the maximum drying rate for unit surface (N_W) is held at 8.2 g m^{-2} s^{-1} ?

The surface moisture content is found from equation 5.59 :

$$\Phi|_{\eta=0} = \Phi_o - \frac{N_W b\delta}{2\rho_s X_{cr} \mathcal{D}_a}$$

in which the penetration depth δ is given by equation 5.58, namely

$$\delta = \sqrt{6Fo_1 \theta}$$

$$= \sqrt{6 . \frac{\tau_1 \mathcal{D}_a}{b^2} . \frac{\tau}{\tau_1}}$$

$$= \sqrt{6 \mathcal{D}_a \tau / b^2}$$

Substitution of this expression into equation 5.59 yields

$$\Phi|_{\eta=0} = \Phi_o - \sqrt{\frac{3N_W^2 \tau}{2\rho_s^2 X_{cr}^2 \mathcal{D}_a}}$$

so that when $\Phi|_{\eta=0} = 0$, we find

$$\tau = \frac{2}{3} \mathcal{D}_a [\rho_s X_o / N_W]^2$$

On substituting numerical values, we get

$$\tau = \frac{2}{3} \times 0.443 \times 10^{-6} \times [\frac{1550 \times 0.15}{8.2}]^2 \times \frac{1}{10^{-6}}$$

$$= 237.5 \text{ s}$$

or about 4 min.

The critical point is reached very quickly under these conditions, and might not be observed experimentally.

5.6 Critical-point Curve

The onset of the critical moisture content, signalling the start of hindered drying, is an important event in the drying process. To devise process strategies by predicting drying behaviour, we need to know the magnitude of the critical moisture content. Values of the critical moisture content for certain materials have been collated[29], and reproduced in reference works such as the Chemical Engineers' Handbook[25] and elsewhere,[8] but the original data show clearly that the critical points are not specific properties of a given material. This variability follows directly from the definition of the critical moisture content as an average value, whereas the onset of hindered drying follows from changes in surface moistness. We thus expect the critical moisture content to depend upon the initial drying rate, the body thickness, as well as factors such as the temperature and internal structure of the material, which both influence the rate at which moisture can reach the surface.

To explore the way in which the critical point may vary, let us consider the drying of non-hygroscopic, slab-form material in terms of the receding-plane model of drying. The nature of the moisture-content profile at the critical point depends upon whether a drying front is present or not, as Fig. 5.21 demonstrates, and thus the conditions for low- and high-intensity drying will be examined separately. It will be assumed that the profiles are parabolic, corresponding to a constant moisture diffusivity $\mathcal{D}_a$.

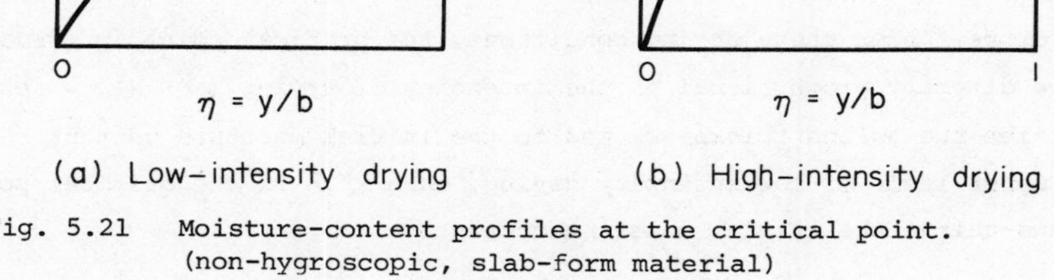

(a) Low-intensity drying (b) High-intensity drying

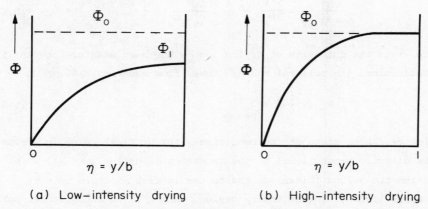

Fig. 5.21 Moisture-content profiles at the critical point. (non-hygroscopic, slab-form material)

Low-intensity drying. (Fig. 5.21 a) In this case, there exists a parabolic moisture-content profile

$$\Phi = \Phi_1[1 - (1-\eta)^2] , \quad 0 < \eta < 1 \tag{5.62}$$

where Φ_1 is the moisture content at the base of the material when the critical point is reached. Differentiation of equation 5.62 yields the moisture-content gradient at the surface:

$$\left.\frac{\partial\Phi}{\partial\eta}\right|_{\eta=0} = 2\Phi_1 \tag{5.63}$$

which can also be found by re-arranging Fick's first law of diffusion (eq. 5.54c); that is

$$\left.\frac{\partial\Phi}{\partial\eta}\right|_{\eta=0} = \frac{N_W b}{\rho_s X_{cr} \mathcal{D}_a} \tag{5.54c}$$

whence

$$\Phi_1 = \frac{N_W b}{2\rho_s X_{cr} \mathcal{D}_a} \tag{5.64}$$

The mean moisture content over the whole material is given by

$$\left.\frac{\bar{X}}{X_o}\right|_{cr} = \frac{\bar{\Phi}}{\Phi_o} = \frac{\int_o^1 \Phi d\eta}{\Phi_o} \tag{5.65}$$

which, from equation 5.62, yields

$$\left.\frac{\bar{X}}{X_o}\right|_{cr} = \frac{2\Phi_1}{3\Phi_o} = \frac{1}{3} \cdot \frac{N_W b}{\rho_s X_o \mathcal{D}_a} = \frac{1}{3}\mathcal{N} \tag{5.66}$$

where $\mathcal{N}$ is the intensity of drying. Now the mean moisture content is, by definition, the critical value; thus, from equation 5.66

$$X_{cr} = \frac{1}{3}\mathcal{N} X_o \tag{5.67}$$

Therefore, under these drying conditions, the critical point is predicted to be directly proportional to the intensity of drying ($\mathcal{N} < 2$), which contains the solids thickness, and to the initial moisture content. At the upper limit of low-intensity drying, when $\mathcal{N} = 2$, the critical point is two-thirds the initial moisture content.

High-intensity drying. (Fig. 5.21 b) In this case, the parabolic moisture-content profile,

$$\Phi = \Phi_o[1 - (1-\eta/\delta)^2], \quad 0 < \eta < \delta < 1 , \tag{5.68}$$

is confined to a region extending to a depth δ below the surface. There is also an undried region, $\delta < \eta < 1$, over which the moisture content has not changed, that is

$$\Phi = \Phi_o \quad . \tag{5.69}$$

The mean moisture content throughout the slab at the critical point is now

$$\left.\frac{\bar{X}}{X_o}\right|_{cr} = \frac{\bar{\Phi}}{\Phi_o} = \frac{\int_0^\delta \Phi d\eta + \int_\delta^1 \Phi_o d\eta}{\Phi_o} \tag{5.70}$$

which, on substituting Φ from equation 5.68, results in the expression

$$\left.\frac{\bar{X}}{X_o}\right|_{cr} = \frac{\Phi_o(0 + 2\delta/3) + \Phi_o(1-\delta)}{\Phi_o} = 1 - \delta/3 \tag{5.71}$$

The depth of the dried zone can be found by noting that the moisture-content gradient at the surface is given by differentiating equation 5.68:

$$\left.\frac{\partial\Phi}{\partial\eta}\right|_{\eta=0} = \frac{2\Phi_o}{\delta} \tag{5.72}$$

but also

$$\left.\frac{\partial\Phi}{\partial\eta}\right|_{\eta=0} = \frac{N_W b}{\rho_s X_{cr} \mathscr{D}_a} \tag{5.54c}$$

whence

$$\delta = \frac{2\rho_s X_o \mathscr{D}_a}{N_W b} = \frac{2}{\mathscr{N}} \tag{5.73}$$

(Note that, when $\mathscr{N} = 2$, $\delta = 1$, as we would expect for the transition between high- and low-intensity drying.) Combination of equations 5.71 and 5.73 yields the required expression for the critical point

$$X_{cr} = [1 - 2/3\mathscr{N}]X_o \tag{5.74}$$

Under high-intensity drying conditions then, the critical moisture content is still directly proportional to the initial moisture content, but becomes independent of rate as the intensity $\mathcal{N}$ itself becomes very large.

The predicted variation of the critical point with moisture content is shown in Fig. 5.22. The resultant envelope of critical points is called the critical-point curve.

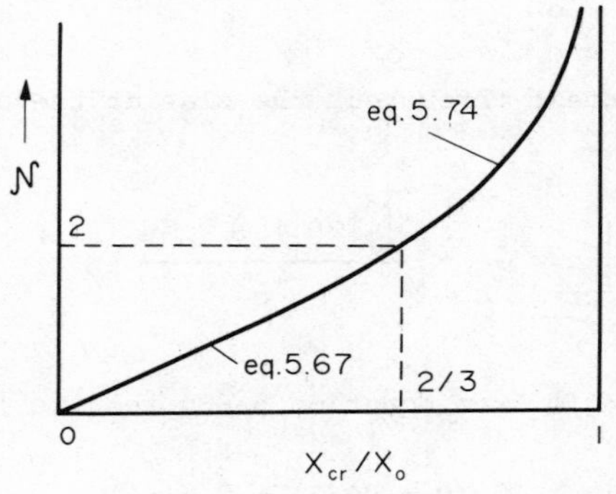

Fig. 5.22 Theoretical critical-point curve for non-hygroscopic slab-form material with a constant moisture diffusivity.

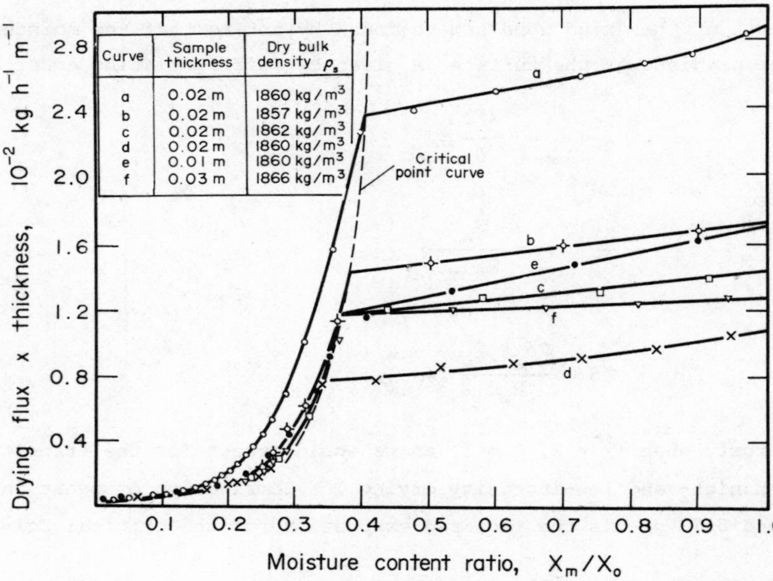

Fig. 5.23 Rate-of-drying and critical-point curves for samples of cross-circulated porous plates. After Krischer (14).

Some experimental drying-rate data for porous plates are reproduced in Fig. 5.23. While a critical-point curve of roughly similar form to the predicted curve of Fig. 5.22 can be drawn through the data, the measured rates correspond to a drying intensity $\mathcal{N}$ of only 0.008 for $X_{cr}/X_o = 0.4$, whereas equation 5.67 would suggest a much higher value of 1.2 for the drying intensity at this moisture-content ratio. Other data[4] for sea sand show a similar result, namely that the actual value of $\mathcal{N}$ is much smaller than the theoretical value for the corresponding critical moisture ratio X_{cr}/X_o.

One suspects that the principal source for this discrepancy lies in the moisture-content dependence of the diffusion coefficient, which does not remain invariant even for a bed composed of comparatively regular granules such as sand. However, it is still possible to retain the form of equations 5.67 and 5.74 to describe the critical-point curve provided a "more appropriate" definition of $\mathcal{N}$ is used to take account of the variableness of the diffusion coefficient. Endo and co-authors[4] suggest a logarithmic interpolation formula for $\mathcal{N}$, namely

$$\ln\mathcal{N} = \ln\mathcal{N}_{cr} - (\ln\mathcal{N}_{cr} - \ln\mathcal{N}_M)\,\Phi_{cr}/\Phi_o \qquad (5.75)$$

where $\mathcal{N}_{cr}$ is the drying intensity based on $\mathcal{D}_a$ at the surface moisture content at the critical point,

and $\mathcal{N}_M$ is the drying intensity for the logarithmic-mean value of $\mathcal{D}_a$ between the initial and surface critical-point values.

This method of defining $\mathcal{N}$ gives a satisfactory correlation of critical-point data for beds of granular material such as sand, alumina and powdered brick, even when the diffusion coefficient varies by several orders of magnitude.

The theoretical form of the critical-point curve suggests that the lumped rate $N_W b$ is a better basis for the normalization of bench-scale data than the absolute rate itself in drawing up the characteristic drying curve; that is the relative drying rate is defined as

$$f = \frac{N_V b}{N_W b} \qquad (5.76)$$

This artifice would permit modest variations in solids thickness to be encompassed. Physically, the product $N_W b$ represents the ratio of the drying rate per unit exposed area to the area presented per unit volume of drying goods, and thus may be regarded as the dimensional drying intensity having units of kg m^{-1} s^{-1}.

REFERENCES

1. Berger, D. and D.C.J. Pei, Drying of hygroscopic capillary-porous solids - a theoretical approach, Int.J.Heat Mass Transfer, 16, 293-302 (1973).
2. Cealgske, N.H. and O.H. Hougen, The Drying of Granular Solids, Trans.AIChE, 33, 283-314 (1937).
3. De Vries, D.A., Simultaneous Transfer of Heat and Moisture in Porous Media, Trans.Am.Geophys.Union, 39, 909-915 (1958).
4. Endo, A., J. Shishido, M. Suzuki and S. Ohtani, Estimation of Critical Moisture Content, AIChE Symposium Series, 73 (163), 57-62 (1977).
5. Fowler, L.G., Evaluation of a process design method for continuous dryers, B.E. Report, Univ. Canterbury N.Z. (1971).
6. Heertjes, P.M. and H.E. Tuider, Some observations on heat and water transport in a drying consolidated mass, De Ingenieur, 74(28)0 53-9 (1962).
7. Kachan, G. C., G. A. de Silva and W. Borzani, Drying of Penicillium Chrysogenum Majcelium, Lat.Am.J.Chem.Eng.Appl.Chem., 5, 151-6 (1975).
8. Keey, R.B., "Drying Principles and Practice", p.183 ff, Pergamon, Oxford (1972).
9. Keey, R.B., The Drying of Ideal Moist Solids, Paper MM-14, 2nd Australasian Heat Mass Transfer Conf., Sydney (1977).
10. Keey, R.B. and M. Suzuki, On the characteristic drying curve, Internat.J.Heat Mass Transfer, 17, 1455-1464 (1974).
11. Kisakurek B., R.D. Peck and J. Çakaloz, Generalized Drying Curves for Porous Solids, Can.J.Chem.Eng., 53, 53-9 (1975).
12. Krischer, O., Grundgesetze der Feuchtigkeitsbewegung in Trocknungsgutern, Kapillarwasserbewegung und Dampfdiffusion, Zeit.VDI, 82, 373-8 (1938).
13. Krischer, O., "Die wissenschaftlichen Grundlagen der Trocknungstechnik, 2/e, p.184-5, Springer, Berlin Göttingen Heidelberg (1963).
14. Krischer, O., ibid. p.296.
15. Krischer, O. and L.Jaeschke, Trocknungsverlauf in durchströmten Haufwerken bei geordneter und ungeordneter Verteilung, Chem.Ing.Techn., 33, 592-8 (1961).
16. Luikov, A.V., "Heat and Mass Transfer in Capillary-Porous Bodies", Pergamon, Oxford (1966).
17. Luikov, A.V., "Teoriya Sushki", 2/e p.155, Energiya, Moskva (1968).
18. Morgan, R.P. and S. Yerazunis, Heat and Mass Transfer between Evaporative Interface in Porous Medium and External Gas Stream, AIChEJ, 13, 132-140 (1967).

19. Morgan, R.P. and S. Yerazunis, Heat and Mass Transfer during Liquid Evaporation from Porous Materials, Chem.Eng.Prog.Symposium Ser., 63(79), 1-13 (1967).
20. Nissan, A.H., W.G. Kaye and J.R. Bell, Mechanism of Drying Thick Porous Bodies during the Falling Rate Period. I. The pseudo wet-bulb temperature. AIChEJ, 5, 103-110 (1959).
21. Novak, L.J. and G.A. Coulman Mathematical Model for the Drying of Rigid Porous Materials, Can.J.Chem.Eng., 53, 60-7 (1975).
22. Oliver, D.R. and D.L. Clarke, Some Experiments in Packed-bed Drying, Proc.Inst.Mech.Engrs., 187, 515-521 (1973).
23. Peck, R.E. and J.Y. Kauh, Evaluation of Drying Schedules, AIChEJ., 15, 85-8 (1969).
24. Peck, R.E., D.A. Max and M.S. Ahluwaha, Predicting drying times for thin materials, Chem.Eng.Sci., 26, 389-403 (1971)
25. Perry, H.E. ed., "Chemical Engineers' Handbook", 5/e, p.20-13, 20.30, McGraw Hill, New York (1973).
26. Phillip, J.R. and D.A. de Vries, Moisture Movement in Porous Materials under Temperature Gradients, Trans.Am.Geophys.Union, 38, 222-232, 594 (1957).
27. Schlünder, E.U., Fortschritte und Entwicklungstendenzen bei der Auslegung von Trocknern für vorgeformte Trocknungsgüter, Chem.-Ing.-Techn., 48, 190-8 (1976).
28. Sherwood, T.K., The Drying of Solids II, Ind.Eng.Chem., 21, 976-980, (1929).
29. Sherwood, T.K., The Air Drying of Solids, Trans.AIChE, 32, 150-168, (1936).
30. Sherwood, T.K. and E.W. Comings, The Mechanism of Drying of Clays, Trans.AIChE, 27, 118-133 (1932).
31. Sherwood, T.K., P.L. Pigford and C.R. Wilke, "Mass Transfer", p.67, McGraw Hill, New York (1975).
32. Suzuki, M., On Mass Transfer from Porous Materials, Paper A4, Annual Meeting SCE Japan, Akita (1974).
33. Suzuki, M., A. Endo, S. Ohtani and S. Maeda, Mass transfer from a discontinuous source, Proc. PACHEC '72, Paper 17-4, III, 267-276, Kyoto (1972).
34. Suzuki, M., R.B. Keey and S. Maeda, On the characteristic drying curve, AIChE Symposium Series, 73 (163), 47-56 (1977)
35. Suzuki, M. and S. Maeda, Mechanism of drying non-hygroscopic materials, Proc.III Japan. Heat Transfer Symposium, p.161, Sendai (1966).
36. Toei, R., S. Hayashi, M. Huaoka and T. Yamamoto, Calculation method of drying rate for through flow drying, Kāgaku Kōgaku, 30, 329-334 (1966).
37. Toei, R. and M. Okazaki, Drying Mechanism of Capillary-Porous Solid, Inzh.Fiz.Zh., 4, 464-475 (1970).
38. Trommelen, A.M. and E.J. Crosby, Evaporation and Drying of Drops in Superheated Vapours, AIChEJ., 16, 857-867 (1970).
39. Van Meel, D.A., Adiabatic convection batch drying with recirculation of air, Chem.Eng.Sci., 9, 36-44 (1958).

PLATE 5. A slurry-paste dryer. A pasty slurry is sprayed into a circulating bed of inert particles, and rapid drying results. The dried material is entrained in the offgas stream and recovered in a suitable collector. [William Boulton Ltd., Calmic Division, with permission.]

Chapter 6

PERFORMANCE OF DRYERS

6.1 Convective Dryers

In most dryers, the stuff being dried is warmed chiefly by convection from the hot surrounding air which takes away the evaporated moisture. Convective dryers are marketted in a diverse range of equipment types, of which a few are illustrated in Fig.6.1. The simplest kind is the stove fitted with shelves for the drying goods (Fig.6.1a). These shelves are often placed in sets in racks, and the system may be made continuous by moving the racks on trucks through a drying tunnel. If the material is piece-form, the higher drying rates can be achieved by through-circulating the material through perforated shelves (Fig.6.1b). Continuous operation is also obtained by drying on endless bands, either solid or perforated (Fig.6.1c). Particulate, free-flowing material can be dried in other ways in which the particles are partially or totally airborne. Such particles may be tumbled through a rotary dryer (Fig.6.1d), or fluidized in beds (Fig.6.1e), or blown through drying ducts (Fig.6.1f).

Rarely is convection the sole mode of heat transfer in such dryers; there may be radiation received from hot areas visible to the drying goods and conduction from warm surfaces supporting them. However, pure convection will be considered first, and the analysis of multiple modes of heating will be treated later.

The maximum drying rate, when the solid plays no role in hindering the loss of moisture, can be found from the convective heat-transfer coefficient h_c directly, since

$$N_W = h_c \, (T_W - T_G) \, / \, \tilde{Q}_V \qquad (6.1)$$

where $\tilde{Q}_V$ is the heat required to drive off unit mass of moisture and bring it to the bulk-gas temperature T_G (Fig.6.2). This quantity is given by the enthalpy needed to vaporize unit mass of moisture at T_W, since the heat-transfer coefficient is obtained from the nett convection reaching the surface.

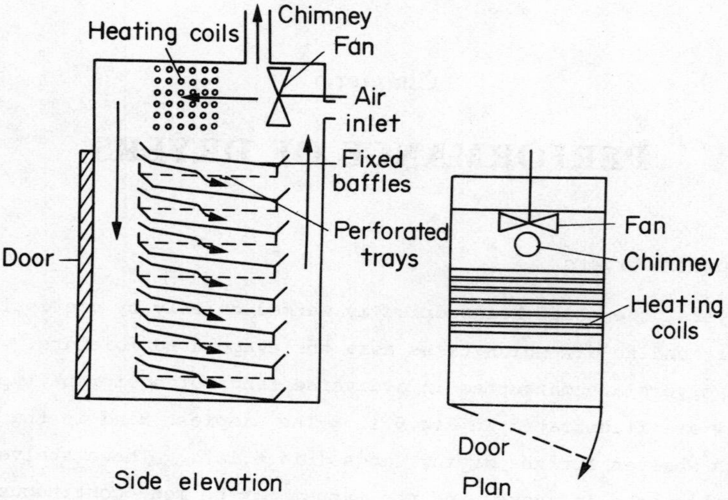

(a)

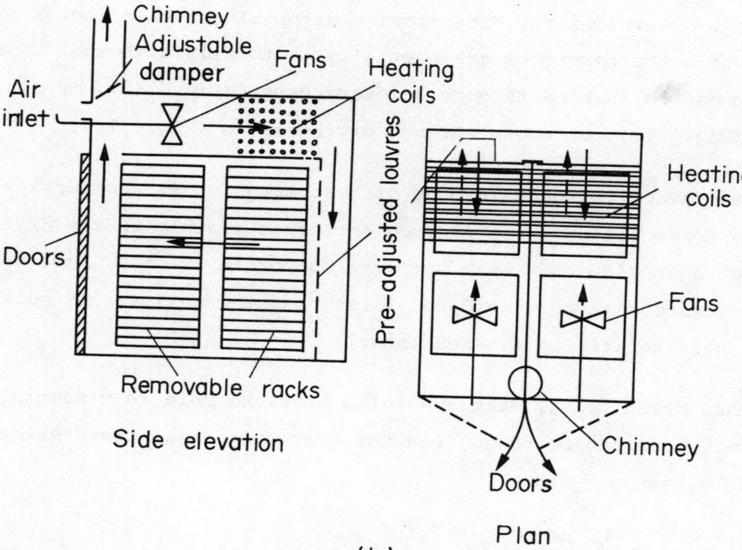

(b)

Fig.6.1 Convective dryers: (a) cross-circulation tray dryer;
(b) through-circulation tray dryer;

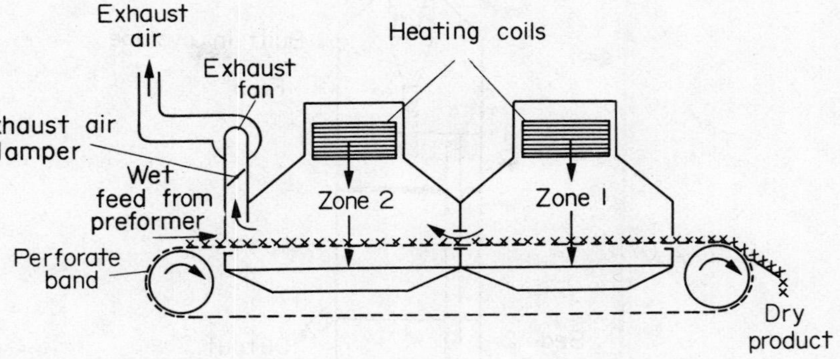

(c)

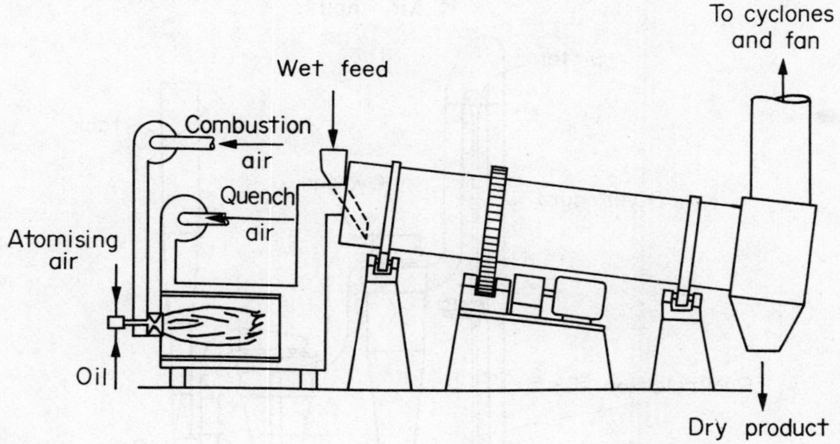

(d)

Fig.6.1 (cont.) Convective dryers: (c) through-circulation band dryer;
(d) rotary dryer;

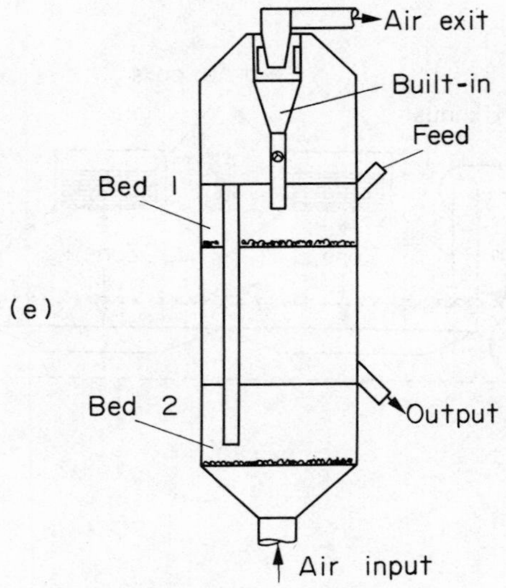

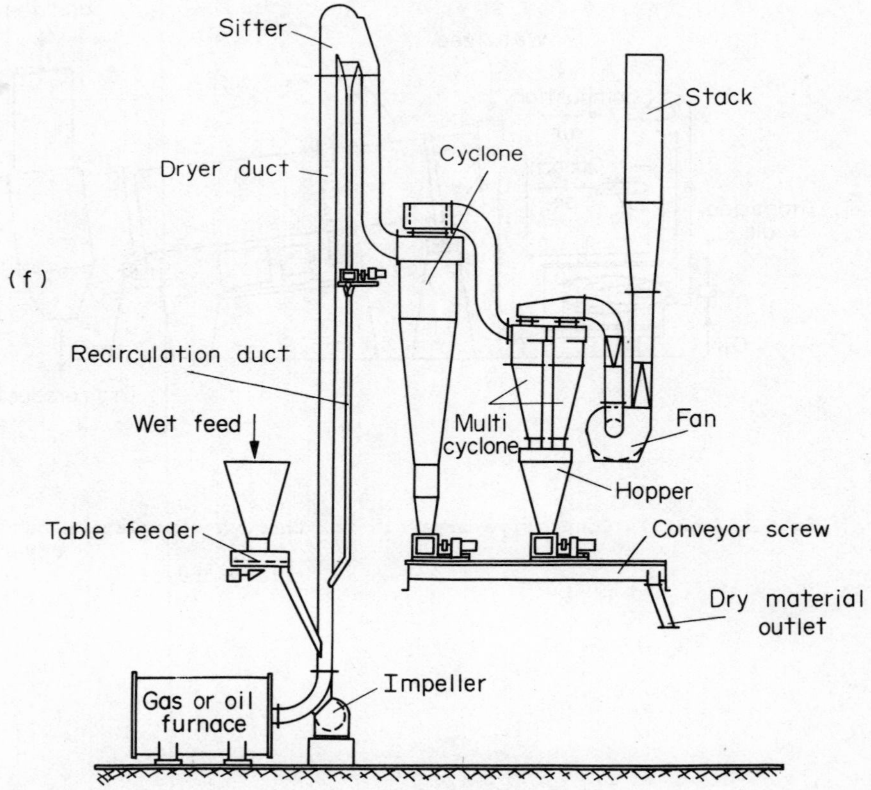

Fig.6.1 (cont.) Convective dryers: (e) two-stage fluid-bed dryer; (f) airlift dryer.

Thus

$$\tilde{Q}_V = \Delta H_{VW} \tag{6.2}$$

Whence, from equations 6.1 and 6.2,

$$N_W = h_c(T_G - T_W) / \Delta H_{VW} \tag{6.3}$$

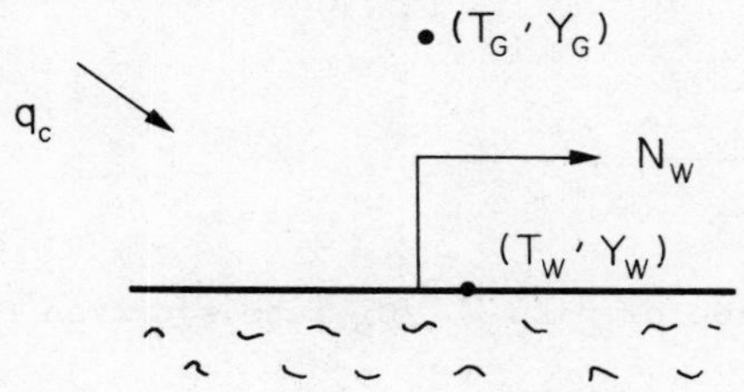

Fig. 6.2 Unhindered convective drying.

On the other hand, the unhindered drying rate can also be found from the humidity difference between the surface and the bulk air:

$$N_W = K_o\phi(Y_W - Y_G) \tag{6.4}$$

Mass-transfer coefficients, however, are usually less readily found than heat-transfer coefficients, and thus it is convenient to invoke the analogy between the transport phenomena.[4] Direct evidence from drying conditions is scarce, apart from those tests already mentioned in Chapter 4 on evaporating liquids from small porous objects. Slessor and Clelland[37] describe experiments on the drying of tows of synthetic fibre. ('Ardil') impregnated successively with six volatile liquids. A number of 8 mm fibre ropes were drawn side by side over a rotating drying cage, having a hollow shaft through which air was blown at a velocity of 0.5 m s^{-1} over a total width of 457 mm. Their results could be expressed by

$$j_M = 1.05\ j_H \tag{6.5}$$

which may be compared with the Chilton-Colburn expression[4] (eq.4.45)

$$j_M = \frac{K_o}{G_T} Sc^{2/3} = \beta j_H = \beta . \frac{h_c}{\bar{C}_P G_T} Pr^{2/3} \tag{6.6}$$

Re-arrangement of equation 6.6 gives (cf. eq. 4.47)

$$K_o = \beta h_c Lu^{2/3} / \bar{C}_P \tag{6.7}$$

From the definition of the psychrometric coefficent σ (cf. equations 4.49 to 4.51), one has

$$\sigma = \beta Lu^{2/3} \phi \bar{C}_{PY}/\bar{C}_P \tag{6.8}$$

so that

$$\frac{\beta Lu^{2/3}}{\bar{C}_P} = \frac{\sigma}{\phi \bar{C}_{PY}} \tag{6.9}$$

On inserting this value of $\beta Lu^{2/3}/\bar{C}_P$ into equation 6.7, one gets

$$\phi K_o = h_c \sigma / \bar{C}_{PY} \tag{6.10}$$

so that equation 6.4 now becomes

$$N_W = h_c (Y_W - Y_G) \sigma / \bar{C}_{PY} \tag{6.11}$$

When water evaporates into air, the psychrometric coefficient is put equal to 1. For other vapour-gas pairs, this coefficient can be evaluated by (cf. eq. 4.55)

$$\sigma = [\frac{Lu}{Lu_{AW}}]^{2/3} \tag{6.12}$$

where Lu_{AW} is the limiting Luikov number for the air-water system and taken to be 1.170. Some values of Lu for liquids evaporating into air are given in Table 4.3.

Nonhebel and Moss[30] give an approximate relationship for equation 6.10 when σ = 1. Their expression, transformed into base SI units, is

$$\phi K_o = h_c/1090 \tag{6.13}$$

and corresponds to the "average" drying situation for which the humid heat is 1090 J $kg^{-1}K^{-1}$.

These various methods of computing the unhindered drying rate are illustrated in the following worked examples.

Example 6.1 Estimate the maximum drying rate from a water-wet surface being convectively heated by air such that $h_c = 30\ W\ m^{-2}K^{-1}$. The dry-bulb and wet-bulb temperatures are respectively 100^oC and 50^oC.

For $T_W = 50^oC$, $\Delta H_{VW} = \tilde{Q}_V = 2382\ kJ\ kg^{-1}$ or $2.382 \times 10^6\ J\ kg^{-1}$

Therefore, from eq. 6.1

$$N_W = 30 \times (100-50) / 2.382 \times 10^6$$

$$= 6.30 \times 10^{-4}\ kg\ m^{-2}s^{-1}$$

$$\text{or } 0.630\ g\ m^{-2}s^{-1}$$

Note, that if the units of $\tilde{Q}_V$ are retained as <u>kJ kg^{-1}</u>, then the value of N_W is calculated in units of <u>g m^{-2}s^{-1}</u>, which have a more convenient size than the base units themselves in expressing drying rates.

Example 6.2 Repeat example 6.1, using the Chilton-Colburn analogy.

From the appended psychrometric chart for $T_G = 100$, $T_W = 50^oC$,

$$Y_G = 0.0637,\ Y_W = 0.0875$$

Thus

$$\bar{C}_{PY} = \bar{C}_{PG} + \bar{C}_{PW}Y_G$$

$$= 1.008 + (1.88 \times 0.0637)$$

$$= 1.128\ kJ\ kg^{-1}\ K^{-1}$$

From equation 6.11 for $\sigma = 1$,

$$N_W = 30 \times (0.0875 - 0.0637) / 1.128$$

$$= 0.633\ g\ m^{-2}s^{-1}$$

The small difference between this and the former estimate is well within the uncertainty of the transfer coefficient.

Example 6.3 Repeat Example 6.1, employing the approximate Nonhebel-Moss relationship (eq.6.13).

From eq. 6.13, $\phi K_o = 30/1.09 = 27.5\ g\ m^{-2}s^{-1}$

Now,

$$N_W = \phi K_o (Y_W - Y_G)$$
$$= 27.5 \times (0.0875 - 0.0637)$$
$$= 0.655 \text{ g m}^{-2}\text{s}^{-1}$$

In this case, the estimate is close to the others as the true humid heat is only slightly larger than 1.09 kJ $kg^{-1}K^{-1}$.

Humidity-difference chart. The accuracy of drying rate calculations is improved by reference to the humidity-difference chart in the Appendix. The humidity difference is smaller, sometimes an order smaller than the absolute humidities themselves. Thus the reading of standard humidity charts for differences can often beget significant errors. On the other hand, with the humidity-difference chart, one can read off on the ordinate scale the difference directly from the wet-bulb temperature along the horizontal axis and the wet-bulb depression as a parameter. For the example considered above, for instance, the wet-bulb temperature (T_W) is 50°C and the depression ($T_G - T_W$) is (100-50) = 50°C; from the vertical axis one reads a value $\Pi = (Y_W - Y_G)$ of 0.0238 kg kg^{-1}.

Influence of process conditions. The influence of various process conditions on the drying rate is shown in Fig.6.3 for a convective heat-transfer coefficient of 30 W $m^{-2}K^{-1}$. Contours of constant relative humidity are plotted across those for constant wet-bulb temperature with the dry-bulb temperature plotted along the abscissae. For a given drying rate, the relative humidity of the drying atmosphere can be increased only if the dry-bulb temperature is also raised. Suppose, for example, drying is taking place at a dry-bulb temperature of 80°C and a relative humidity of 20 per cent. Should this relative humidity be raised to 30 per cent, then the dry-bulb temperature must be increased by 50°C to 110°C to maintain the same drying rate. At the same time, however, the severity of the drying process has increased markedly, the wet-bulb temperature rising from about 47°C to 80°C. On the other hand, the more rigorous drying conditions need less heat for their sustenance, as reference to Fig.3.19 shows for the heat demand of an ideal dryer, this quantity falling from 3.7 MJ kg^{-1} to 2.85 MJ kg^{-1} for the conditions cited. Thus, the need to limit drying temperatures, because of the thermal sensitivity of the stuff to be dried, or for any other reason, can bring heavy economic penalties.

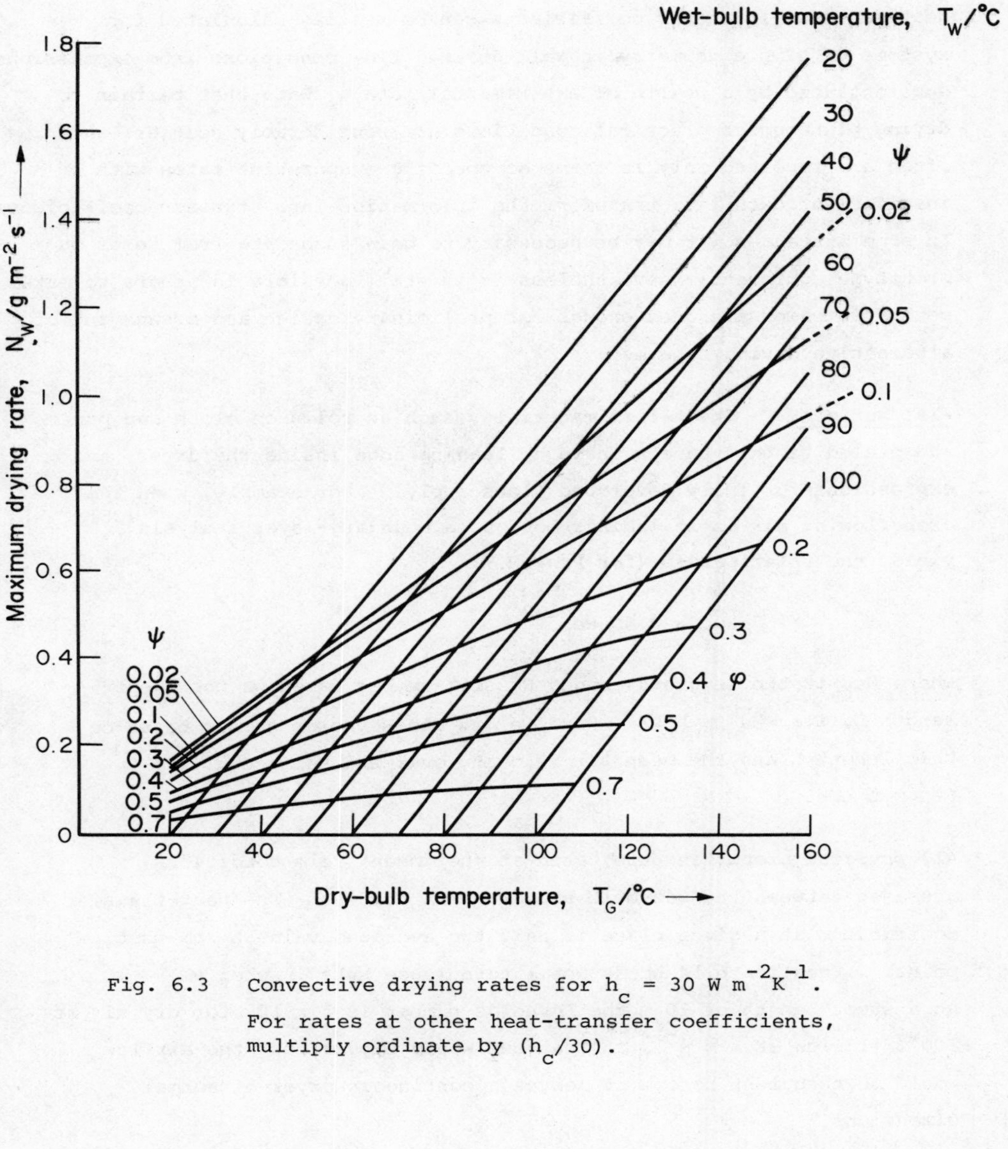

Fig. 6.3 Convective drying rates for $h_c = 30$ W m^{-2}K^{-1}. For rates at other heat-transfer coefficients, multiply ordinate by (h_c/30).

HEAT-TRANSFER COEFFICIENTS

Heat-(or mass-)transfer coefficients can be readily calculated for systems of simple geometry in well-defined flow conditions from expressions substantiated by a wealth of experimental data. Data that pertain to drying plant under practical conditions are considerably scarcer, and often are reported only in terms of specific evaporation rates with insufficient detail to transform the information into transfer coefficients. In some instances, it may be necessary to gain such data from tests on prototype equipment; nevertheless it is still possible in others to make estimates which are good enough for preliminary design and assessment of alternative drying schemes.

Flat surfaces. Sheet-form materials, such as rolls of cloth and paper and peeled timber veneers, have no leading edge inside the dryer, and expressions for fully developed flows apply. For example, when the crossflowing air is in laminar motion, a boundary-layer analysis[25] yields the relationship (for $Pr \sim 0.7$)

$$\overline{Nu_L} = 0.60\ Re_L^{1/2} \tag{6.14}$$

where $\overline{Nu_L}$ is the length-averaged Nusselt number based on the swept length L, i.e. $\overline{Nu_L} = \bar{h}_c L/\lambda$, and Re_L is the Reynolds number based on this length L and the mean air velocity over the surface $\hat{u}$, i.e. $Re_L = \hat{u}\ L/\nu$.

All physical properties are taken at the "mean" film conditions, averaged between the wet-bulb and dry-bulb states. The heat-transfer coefficient at a given place is half the averaged value $\bar{h}_c$ to that point. Equation 6.14 holds until turbulence sets in when $Re_L \sim 10^5$. At a swept length of 10 m the Reynolds number is 5×10^5 for dry air at $100^\circ C$ flowing at 1 m s^{-1}, so that one would expect that the airflow would be turbulent before it leaves a continuous dryer of normal dimensions.

For fully developed turbulent flow, right from the leading edge, it can be shown that

$$\overline{Nu_L} = 0.032\ Re_L^{0.8} \tag{6.15}$$

for $Pr \sim 0.7$ over the Reynolds number range between 5×10^5 to 10^7.

The following worked example indicates the magnitude of the transfer rate forecast for representative drying conditions.

Example 6.4. A unit of effective length 20 m for drying sheets is to be worked with a dry-bulb temperature of 80°C and a wet-bulb temperature of 50°C. Estimate (a) the convective heat-transfer coefficient and (b) the maximum drying rate for air flowing at 1.5 m s^{-1} over the drying sheet.

From tabulated data for moist air we obtain the following values:

State	$\lambda/Wm^{-1}K^{-1}$	$10^6\nu/m^2s^{-1}$
dry-bulb	0.0303	20.2
wet-bulb	0.0280	18.9
mean	0.0292	19.6

$$\therefore\ Re_L = \hat{u}\ L/\nu = 1.5 \times 20/\ 19.6 \times 10^{-6}$$

$$= 1.53 \times 10^6$$

The flow is turbulent as $Re_L > 10^5$ and will be assumed to be fully developed from the point of intake of the sheet.

(a) From equation 6.15

$$\bar{h}_c = [\ \overline{Nu}_L]\ \frac{\lambda}{L} = [\ 0.032\ Re_L^{\ 0.8}\ ] \cdot \frac{\lambda}{L}$$

$$= 0.032 \times [1.53 \times 10^6]^{0.8} \times 0.0292/20$$

$$= 4.14\ W\ m^{-2}K^{-1}$$

This heat-transfer coefficient is an order less than that observed in tray dryers for which the length of the swept surface is much less. For this reason, the drying rate in long dryers for sheet-form material is normally enhanced, for example by directing air-jets onto the surface when drying timber veneers or, with open fabrics such as textile sheets, by blowing air through the stuff.

(b) The specific heat demand $\tilde{Q}_V$ is given by $\Delta H_{VW} = 2382\ kJ\ kg^{-1}$. The maximum drying rate then becomes (eq.6.1)

$$N_W = h_c(T_G - T_W) / \tilde{Q}_V$$
$$= 4.14 \times (80 - 50) / 2382$$
$$= 0.052 \text{ g m}^{-2}\text{s}^{-1}$$

Although equations 6.14 and 6.15 are well attested for heat transfer to and from dry surfaces, their application to transfer under drying conditions is less certain, particularly if the swept length is short. Tests[1] on the drying of linen stretched over a frame show that the transfer coefficient depends upon the square root of the airspeed, but experiments[34] drying very wet, small slabs, 50 mm x 52 mm in extent, of pressed sulphite-paper pulp yield the dependence

$$h_c \propto \hat{u}^{0.6} \tag{6.16}$$

for cross-circulating air velocities between 2 and 12 m s^{-1}. The ends of the slabs were sealed so that evaporation occurred only normally to the airstream. Furthermore, equation 6.15 considerably underestimates the heat-transfer data[33] found on drying sand spread over trays 305 mm square in depths between 13 and 51 mm. One might presume that in these tests the material surface is scoured by the airflow due to the influence of the projecting edges of the tray. Moreover, in an actual dryer, the air is deflected by baffles and internal heating coils, while the motion, especially in short dryers , is influenced by the way the air enters and leaves. For these reasons, Treybal[40] suggests a "practical" relationship, which corresponds to the expression

$$\overline{Nu}_L = 0.055\, Re_L^{0.8} \tag{6.17}$$

to estimate heat-transfer to material laid on trays.

<u>Jet-impinged surfaces</u>. Since 1950 it has become common to improve the drying of sheetform materials by exposing the strip of material to high-velocity air-jets directed onto the surface. Such jets, which may cover from 2 to 6 percent of the wet surface, induce very high heat-transfer rates under the axis of each jet, whose influence spreads to about 20 to 50 jet diameters on either side.[9] The results of an extensive experimental study[19] with both round and rectangular nozzles are summarized in Fig.6.4. In these experiments, a sealed porous plate was impinged

with air-jets at velocities up to 200 m s^{-1} at various spacings between the nozzles and the surface. Once the ratio of this gap to the characteristic dimension of the nozzle (hole diameter or slot width) exceeds 5, then the effect of the air-jet begins to fall off.

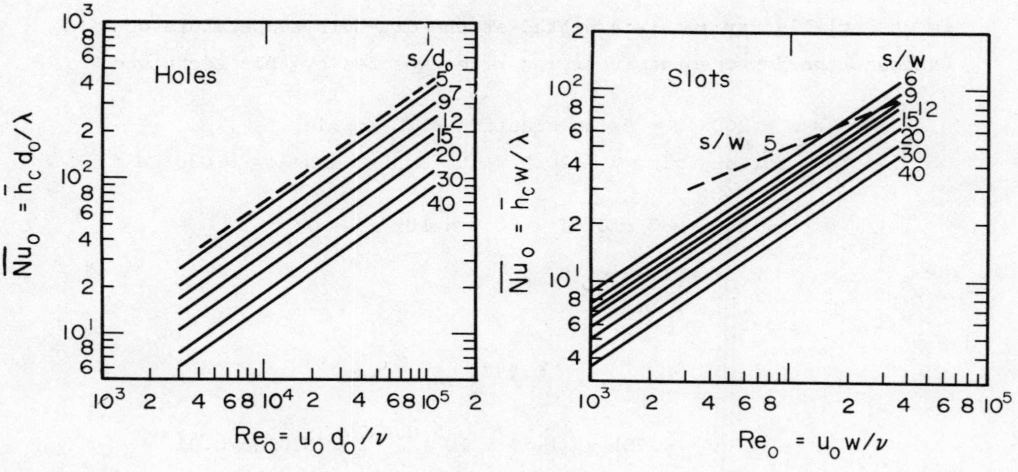

Fig.6.4 Average heat-transfer coefficients for jet-impinged surfaces. Data of Kerscher and colleagues.(19)

Circular orifices

$$s/d_o \leqslant 5 , \quad \overline{Nu}_o = 0.075\, Re_o^{0.745} \tag{6.18}$$

$$s/d_o > 5 , \quad \overline{Nu}_o = 0.320\, Re_o^{0.745}\, (s/d_o)^{-0.828} \tag{6.19}$$

Rectangular orifices

$$s/w \leqslant 5 , \quad \overline{Nu}_o = 0.901\, Re_o^{0.437} \tag{6.20}$$

$$s/w > 5 , \quad \overline{Nu}_o = 0.135\, Re_o^{0.697}\, (s/w)^{-0.351} \tag{6.21}$$

In the foregoing expressions, $\overline{Nu}_o$ is the surface-averaged Nusselt number based on the characteristic dimension of the orifice. Thus, for round holes of diameter d_o , $\overline{Nu}_o = \bar{h}_c d_o/\lambda$; for slots of width w, $\overline{Nu}_o = \bar{h}_c w/\lambda$. Likewise, Re_o is the Reynolds number based on the orifice velocity u_o and the appropriate characteristic dimension, either d_o or w. The parameter s is the distance between the nozzle opening and the surface.

The following worked example illustrates the benefit of jet impingement.

Example 6.5. The dryer of example 6.4 is to be fitted internally with a ventilating hood to direct air onto the drying sheet. The openings in the hood are rectangular, of width 40 mm and are set at a distance of 200 mm above the surface. The mean slot velocity is 80 m s^{-1}. The air is essentially dry and is admitted at the dry-bulb temperature of 80°C. Estimate the improvement in drying rate expected by this technique.

Now, s/w = 200/40 = 5, so equation 6.20 applies.
At 80°C for dry air, $\nu = 20.9 \times 10^{-6}$ m^2s^{-1} and $\lambda = 0.0303$ W $m^{-1}K^{-1}$

$$\therefore Re_o = 80 \times 0.04 / 20.9 \times 10^{-6}$$
$$= 1.53 \times 10^5$$
$$\text{and } \bar{h}_c = [\overline{Nu}_o] \frac{\lambda}{w}$$
$$= [0.901\ Re_o^{0.437}] \frac{\lambda}{w}$$
$$= 0.901 \times (1.53 \times 10^5)^{0.437} \times 0.0303/0.04$$
$$= 126 \text{ W m}^{-2}\text{K}^{-1}$$

This is a thirtyfold increase in the heat-transfer coefficient compared with the case when the surface is not impinged. Almost certainly the installation of the ventilating hood would be worthwhile with an improvement of this magnitude. Moreover, the provision of air-jets provides a means of "tuning" the drying operation.

Single objects. If the surface swept by the air is short and stubby, then an eddy may form just downstream of the leading edge. Smoke tests about a single slat, 75 mm x 6.3 mm show that fast-moving air scours the upstream surface and curls around the sharp fore-edge,[7] as illustrated in Fig. 6.5. Such eddying begins when the Reynolds number based on the body thickness b, $Re_b = \hat{u}b/\nu$, exceeds 245.[39]

By measuring the sublimation from naphthalene-coated slabs, 300 mm x 250 mm of varying thickness, Sørenson[39] obtained the correlation

$$j_H = h_c/\bar{C}_p \hat{u}\ \rho_G = 0.152\ Re_b^{0.179}\ [Re_L - 49.8_b^{0.61}]^{-0.5} \quad (6.21)$$

The term in the square brackets represents the Reynolds number for the laminar boundary layer from the point of re-attachment (point A of Fig. 6.5). For convenience of calculation, equation 6.22 is represented in graphical form in Fig. 6.6.

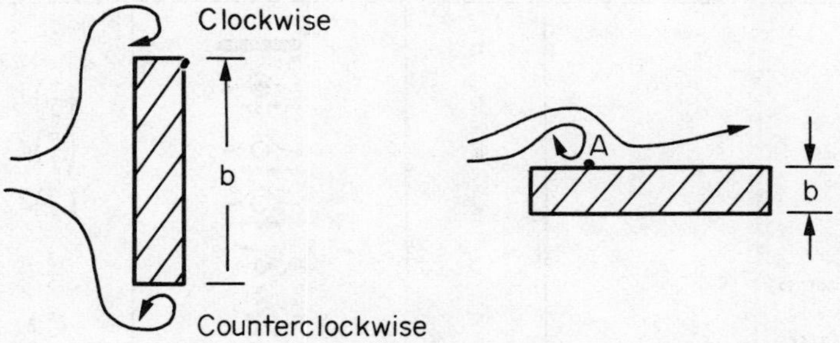

Fig. 6.5 Flow patterns about blunt objects

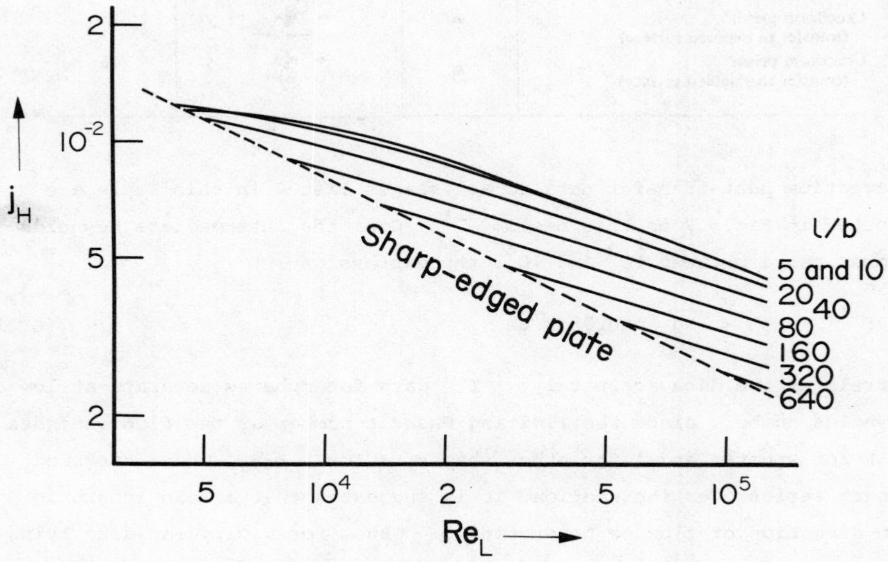

Fig. 6.6 Heat-transfer coefficients for blunt bodies.[39]

Krischer[21] shows that a common correlation for any body shape can be found by choosing the appropriate streamed length in defining the characteristic dimension in the Nusselt and Reynolds number. A list of the characteristic streamed length ℓ of objects of various shapes is given in Table 6.1.

TABLE 6.1 STREAMED LENGTHS FOR BLUNT BODIES IN CROSSFLOWING AIR

Shape	Symbol	Dimensions	Streamed length
Flat plate	□	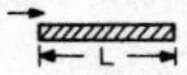	L
Cylinder	○		$\left(\frac{\pi}{2}\right)d$
Sphere	●		$\left(\frac{\pi}{2}\right)d$
Disc	⊗		$\left(\frac{\pi}{4}\right)d$
Prism (flow onto corner)	△		$\frac{3}{2}b$
Prism (flow onto face)	▽		$\frac{3}{2}b$
Angular prism (flow onto corner)	∧		$2b$
Angular prism (flow behind corner)	∨	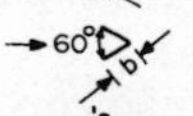	$2b$
Cruciform prism (transfer to exposed surface)	+		$2b$
Cruciform prism (transfer to shielded surface)	+		$2b$

Convective heat-transfer data to all shapes listed in this Table are plotted in Fig.6.7 on this basis.[21] Over the intermediate Reynolds number range between 10^2 and 10^4, the expression

$$\overline{Nu}_\ell = 0.75 \sqrt{Re_\ell} \qquad (6.23)$$

correlates the data adequately. The data for spheres separate at low Reynolds number, since the limiting Nusselt number as the flow vanishes is π for spheres and 1 for other shapes. For a body, whose streamed length varies over its section, it is suggested that a mean length in the direction of flow be taken for ℓ. Thus, for a circular disc lying endwise into the airstream, $\ell = \pi d/4$ and for an elliptical pellet in the same orientation $\ell = \pi a/2$, where a is the half-axis parallel to the airflow direction.

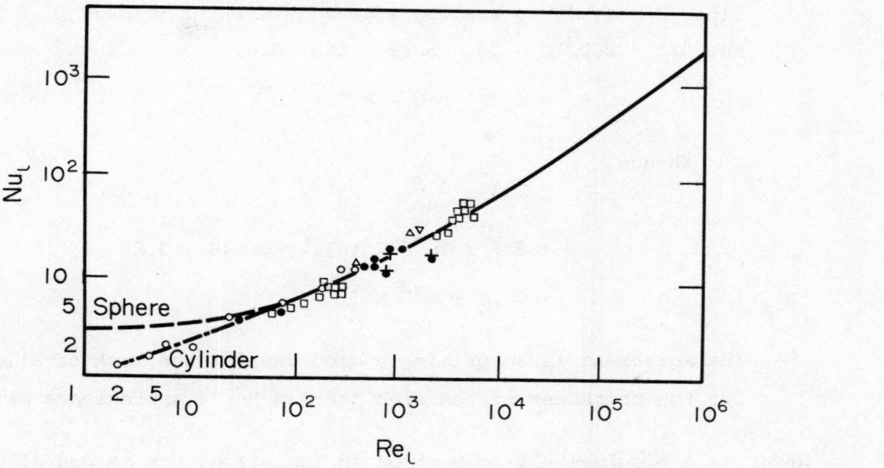

Fig. 6.7 Convective heat transfer to single bodies in crossflow.[21] Symbols for data points are given in Table 6.1.

The following worked example illustrates to what extent Sørensen's and Kirscher's methods agree in estimating heat transfer to a short slab.

Example 6.6 A moist ceramic slab 200 mm long and 10 mm thick is being dried in a crossflowing airstream at 2.5 m s^{-1}. The air may be assumed to be perfectly dry at 100°C.

From tables of thermophysical data, $\lambda = 0.0317$ W $m^{-1}K^{-1}$ and $\nu = 23 \times 10^{-6}$ $m^2 s^{-1}$.

a) Krischer's method: The streamed length ℓ is 200 mm, thus

$$Re_\ell = \hat{u}\,\ell/\nu = 2.5 \times 0.2 / 23 \times 10^{-6}$$
$$= 2.17 \times 10^4$$

$$h_c = [Nu_\ell] \cdot \frac{\lambda}{\ell}$$

$$= [0.75\, Re_\ell^{\frac{1}{2}}] \cdot \frac{\lambda}{\ell}$$

$$= 0.75 \times (2.17 \times 10^4)^{\frac{1}{2}} \times 0.0317/0.2$$

$$= \underline{17.5 \text{ W m}^{-2} \text{ K}^{-1}}$$

b) Sørensen's method:

$$Re_b = \hat{u}\, b/\nu = 2.5 \times 0.01/23 \times 10^{-6}$$
$$= 1.087 \times 10^3$$

Since $Re_b > 245$, a leading-edge eddy will be formed.
Now $\ell/b = 200/10 = 20$, so from Fig. 6.6,

$$j_H = h_c/\bar{C}_p \, \rho \, \hat{u} = 8.7 \times 10^{-3}$$

Whence,

$$h_c = j_H(\bar{C}_p \, \rho \, \hat{u})$$

$$= 8.7 \times 10^{-3} \times 1011 \times 0.946 \times 2.5$$

$$= 20.8 \text{ W m}^{-2} \text{ K}^{-1}$$

The agreement is surprisingly good despite the lack of allowance for the thickness of the slab in Krischer's approximate method.

Never is a single small body dried in isolation, but as one of an array of items placed on a shelf or moving band. Clearly, if these items are close enough together, then the motion about one body will interact with the motion about its neighbours. With an in-line arrangement of 32 mm x 8 mm thick slabs coated with naphthalene, Miller[28] finds that no apparent interaction occurs when the slabs are set at a pitch equal to the swept length of the object, but the assembly behaves like a continuous surface subject to turbulent airflow at shorter spacings. Thus, the flow within the spaces of close-stacked timber, for instance, will correspond to motion through rectangular ducts, rather than flow about a series of bluff bodies. Under these conditions, one might expect Krischer's expression to apply, with the width of the stack as the appropriate swept length ℓ, as equation 6.23 was obtained in the first instance[22] from data on heat transfer within ducts of rectangular cross-section.

<u>Through-circulated beds</u>. To improve drying rates, the heated air can be blown through a thin bed of the moist material which must be particulate. The material is usually moved through the dryer on some form of perforate band. Very wet feedstocks have to be preformed into discrete ribbons to yield appropriate "particles". Alternatively, hot air can ventilate a stack of trays over which the material tumbles in helter-skelter fashion. Typically, the bed thickness would be about 50 mm, and the through-circulating air velocity from 1 to 1.5 m s^{-1}. For 5 mm particles, these velocities correspond to a particle Reynolds number ($Re_p = u \, d_p/\nu$) between 220 and 330 for perfectly dry air at 100°C.

Wilke and Hougen[45], from a study of evaporation from water-soaked, porous Cellite pellets in a bed of 0.093 m^2 in extent and 23.5 mm deep, obtain the correlation

$$j_H = j_M = 1.82\ Re_p^{-0.51} \qquad (6.22)$$

for Reynolds numbers below 350. Above 350, the flow though the bed becomes turbulent and the alternative correlation

$$j_H = j_M = 0.989\ Re_p^{-0.41} \qquad (6.23)$$

applies. Although laboratory data[10] on the through-circulation of extruded rods of a filter cake with a chalky consistency do not uphold this expression, Sherwood and co-authors[35] find that more general mass-transfer data from nine separate investigations, covering a particle-diameter range between 2 and 19 mm and Reynolds numbers between 10 and 2500, support equation 6.23 with a preferred coefficient of 1.17 and an exponent of (- 0.415). The Wilke-Hougen correlation is also supported in the through-flow drying of beds of clay particles and sawdust in superheated steam.[41]

With distended beds, when the through-circulating air moves fast enough to move the particles to a greater or lesser degree as in fluid-bed and spouting-bed systems, it seems that the appropriate correlating parameters are εj_H and $Re_p/(1-\varepsilon)$, where ε is the voidage of the bed.

The results[35] of a large number of experiments, including studies of the evaporation from liquid-soaked porous spheres made from filter-paper fibre in eight packing geometries, suggest that

$$\varepsilon j_H = 1.0\ [Re_p/(1-\varepsilon)]^{-\frac{1}{2}} \qquad (6.24)$$

over the ranges $0.25 < \varepsilon < 0.5$

$$40 < Re_p/(1-\varepsilon) < 4000$$

The voidage ε of fluidized beds can be correlated to the excess velocity over the incipient velocity for fluidization for particle sizes between 0.66 and 7.55 mm with a mean deviation of ± 7 per cent, as illustrated in Fig. 6.8.

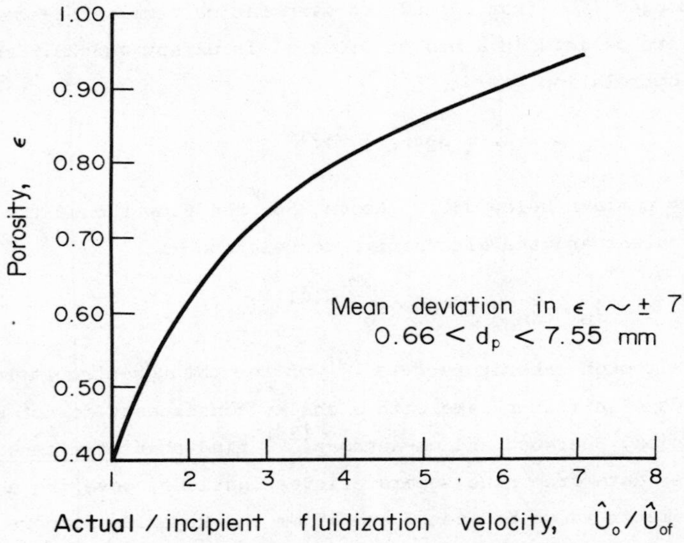

Fig. 6.8 Voidage fractions of fluidized beds. After Mosberger.[29]

An alternative approach is developed by Krischer and colleagues.[21] He shows that the ratio of the characteristic dimension of a body to its swept length may be related to the equivalent ratio for an array to yield the same Nusselt number. On this basis, it is possible to extend equation 6.23 to apply to fixed and fluidized beds besides single particles. The method is explained in detail by Keey.[14]

The foregoing expressions (eq. 6.22 - 6.24) should be regarded only as a <u>rough guide</u> to heat-transfer coefficients in the <u>absence</u> of field data, particularly for distended-bed conditions. As a recent (1975) review[20] demonstrates for the prediction of heat transfer in spouted-bed dryers, considerable variation in estimates can occur from expressions derived from laboratory data. Partly, this spread stems from the small-scale nature of most of the original experiments themselves; but partly, because even randomly packed beds may contain certain orderly clusters of particles and, in certain moisture-content ranges, the particles may become cohesive and bind into aggregates. Further, it is always more difficult to ensure uniformity of bed loading with the larger-scale equipment used commercially compared with that obtainable under laboratory conditions.

Example 6.7. Suppose particulate material (5 mm diameter) is being fed to a perforated-band dryer and is spread into a layer 50 mm thick and through-circulated with dry air at 80°C at a velocity of 1.25 m s^{-1}. what is the mean heat-transfer coefficient under these conditions ?

Now, $Re_p = ud_p/\nu = 1.25 \times 0.005/20.9 \times 10^{-6} = 299.$

Since $Re_p < 350$, we use eq.6.22 for the heat-transfer coefficient.

$$\text{i.e. } j_H = 1.82\, Re_p^{-0.51}$$

$$= 1.82[299]^{-0.51} = 0.099$$

$$h_c = j_H[\bar{C}_p \rho u]$$

$$= 0.099 \times 1011 \times 0.946 \times 1.25$$

$$= 118 \text{ W m}^{-2}\text{K}^{-1}$$

This value is about the same as that found when the cross-circulated surface is scoured by air-jets (Example 6.5) and is considerably greater than that for the equivalent solid-band dryer.

Influence of Prandtl number. The foregoing expressions (eq. 6.14 ff) assume that Pr ~ 0.7, which is correct only for perfectly dry air. In other cases, the computed value of h_c should be corrected by a factor $(Pr/0.7)^{1/3}$, which accounts for the change in the relative magnitude of the thermal and momentum boundary layers. However, often the correction will be negligible if the moisture content of the gas is small, since the Prandtl number of all diatomic gases is essentially the same. The presence of moisture influences the value of the Prandtl number slightly, and the correction becomes significant for very humid gases. The Prandtl number of steam at 100 kPa and 100°C is 1.06, to give a boundary layer correction factor of 1.148 to the computed value of the heat-transfer coefficient.

6.2 Multiple Heating Modes

Multiple shelves. As noted in the previous section, the drying solid in a "convective" dryer rarely gets all its warmth from convected heat. Consider the case of particulate solids put on trays which are stacked one above each other so that the air-gap between the trays is fairly close. The air warms up by convection both the exposed surface of the solids and the underside of the supporting tray above. In turn, this tray radiates heat to the solids below, while heat is also conducted through the body of the solids. This situation is sketched in Fig. 6.9.

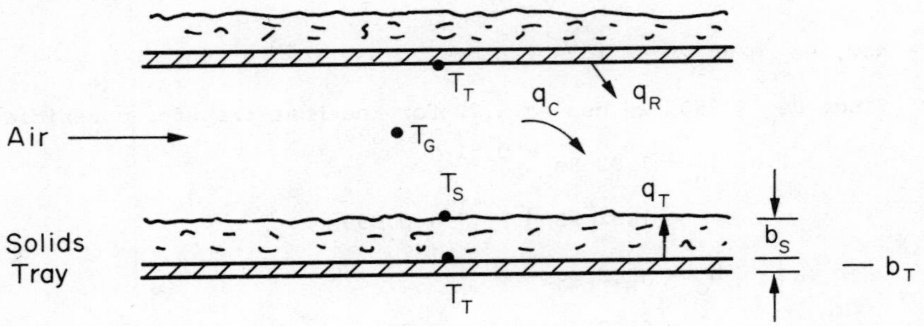

Fig. 6.9 Heating modes for solids on stacked shelves.

From equations 6.1 and 6.4, the maximum drying rate for unit exposed surface is given by

$$N_W = K_o\phi(Y_S - Y_G) = (q_C + q_R + q_T) / \tilde{Q}_V \tag{6.25}$$

where q_C is the convective, q_R the radiative and q_T the conductive heat fluxes respectively; $\tilde{Q}_V$ is the specific heat demand (eq. 6.2.). For convenience, we evaluate each of these fluxes as the product of a coefficient and the appropriate temperature difference; thus

$$q_C = h_C(T_G - T_S) \tag{6.26}$$

$$q_R = h_R(T_T - T_S) \tag{6.27}$$

$$q_T = h_T(T_T - T_S) \tag{6.28}$$

The estimation of the convective coefficient has already been examined in the previous section.

The radiative coefficient h_R is found from the Stefan-Boltzmann fourth-power law of radiation as follows:[15]

$$h_R = \sigma\left[\frac{1}{1/\varepsilon_S + 1/\varepsilon_T - 1}\right](T_S + T_T)(T_S^2 + T_T^2) \tag{6.29}$$

where σ is the Stefan-Boltzmann constant, which takes a value 5.6697×10^{-8} W m^{-2}K^{-4}, ε_S is the emissivity of the damp surface and ε_T

is the emissivity of the supporting tray. At 50°C the emissivity of materials such as oak, paper and plaster and even water itself ranges between 0.9 and 0.95 so that ε_S will probably be only slightly less than 1.[38] The emissivity of the shelf is likely to be less: Spiers[38] gives a value of 0.56 for rolled sheet steel at 50°C. All temperatures are evaluated on the absolute (Kelvin) scale.

If the surface of the solids acts as a black body and $\varepsilon_T \sim 0.5$, then equation 6.29 becomes approximately

$$h_R \simeq 2\sigma(T_S + T_T)(T_S^2 + T_T^2) \qquad (6.30)$$

Unlike h_C, the radiative coefficient is highly temperature-dependent. For $T_S = 60^\circ C$ and $T_T = 90^\circ C$, the coefficient takes a value of 19 W m^{-2}K^{-1}. Thus radiation plays a significant role in "convective" dryers such as tray and multiple-deck dryers.

The conductive coefficient h_T is found from the Ohmic law of aggregating thermal resistances, that is

$$\frac{1}{h_T} = \frac{b_S}{\lambda_S} + \frac{b_T}{\lambda_T} \qquad (6.31)$$

Often the thermal resistance of the tray (b_T/λ_T) is negligible. The thermal conductivity of utterly dry porous solids may be correlated with porosity for solids of similar structure because of the wide disparity between the conductivity of the solid skeleton and that of the included air. Jakob[13] notes that the thermal conductivity of bone-dry building materials can be correlated to within ± 15 per cent with their bulk density which reflects the porosity of the substances concerned. These values are set out in Table 6.2, together with other values collated by Cammerer[3a] for loose materials of various kinds.

The thermal conductivity of air-dried timber and other building materials is about 15 per cent greater than the corresponding value for perfectly dry stuff.

TABLE 6.2 THERMAL CONDUCTIVITY OF BONE-DRY POROUS MATERIALS

(Adapted from data of Cammerer[3a] and Jakob[13])
At temperatures between 0 and 24°C

Bulk density/ $kg\ m^{-3}$	Thermal conductivity / $W\ m^{-2}K^{-1}$ loose materials granular	powdery	fibrous	building materials	timber
100	-	-	0.059	-	-
200	0.093	0.046	0.052	0.053	0.056
300	0.105	0.056	0.058	0.059	0.074
400	0.116	0.066	0.070	0.070	0.092
500	0.13	0.077		0.083	0.110
600	0.14	0.087		0.101	0.129
800	0.16	0.116		0.140	0.148
1000	0.19	0.151		0.186	0.166
1200	0.21	0.19		0.24	0.185
1400	0.24	0.22		0.30	0.204
1600	0.29			0.38	
1800	0.36			0.50	
2000	0.43			0.64	
2200	0.52			0.83	
2400	-			1.14	

Jakob[13] also presents a correction factor f to estimate the thermal conductivity of building materials when moist from values in Table 6.2 for these when bone-dry. Values of this factor f are given in Table 6.3 below.

TABLE 6.3 MOISTURE-CORRECTION FACTOR f TO ADJUST VALUES OF λ IN TABLE 6.2 FOR VOLUMETRIC MOISTURE CONTENT Ψ_W

Ψ_W	0.01	0.025	0.05	0.10	0.15	0.20	0.25
f	1.3	1.55	1.75	2.1	2.35	2.55	2.75

These factors must be regarded as being <u>very rough</u>, and only applicable at room temperature, as the thermal conductivity of a moist porous solid can exceed the conductivity of the moisture itself at intermediate moisture contents above certain temperatures when the moisture may move through a vaporization-condensation process towards the exposed surface. This effect is illustrated in Fig. 6.10.

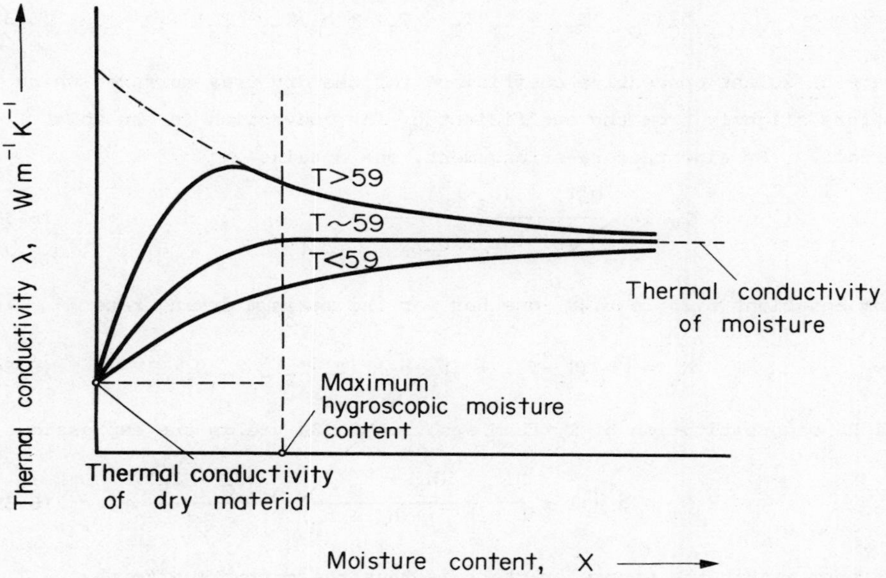

Fig. 6.10 Variation of thermal conductivity of moist porous solids with temperature and moisture content.[21]

Such variations in thermal conductivity have been observed[12] in the drying out of beds of fibreglass sheets at temperatures around 70°C, when maxima in the thermal conductivity appear when the pores are about half-full of moisture.

For a 50 mm thick bed of dry granular solids of bulk density 1600 kg m^{-3} on a thin metal tray, the conductive coefficient h_T becomes 5.8 W $m^{-2}K^{-1}$ from the values presented in Table 6.2, and thus the effect of conduction is likely to be significant also in multiple-shelf dryers. Under unhindered-drying conditions, when the solids are fully soaked with moisture, the conducted heat will be much greater as the thermal conductivity will approach that of the moisture itself. For instance, the thermal conductivity of pinewood across the grain is 0.51 W $m^{-1}K^{-1}$ at a moisture content of 0.44 kg kg^{-1}, which is above the fibre-saturation point.[17] This value compares with one of 0.60 W $m^{-1}K^{-1}$ for water at 20°C.[26]

The temperature of the underside of the supporting shelf or deck can be found by equating the convective gain in heat to the heat dissipated by conduction and radiation; namely

$$h_C'(T_G - T_T) = h_T(T_T - T_S) + h_R(T_T - T_S) \qquad (6.32)$$

where h_C' is the convective coefficient for the <u>dry</u> tray surface, which differs slightly from the coefficient h_C for convection to the moist surface.   By algebraic re-arrangement, one obtains

$$T_T = \frac{h_C'T_G + (h_T+h_R) \, T_S}{(h'_C + h_T + h_R)} \qquad (6.33)$$

From equations 6.25 to 6.28, one has for the maximum drying rate

$$N_W = \left[h_C(T_G-T_S) + (h_R+h_T) \, (T_T-T_S)\right] \, / \, \tilde{Q}_V \qquad (6.34)$$

which, on substitution of T_T from equation 6.32, yields the expression

$$N_W = h_C\left[1 + \frac{h_C'}{h_C} \, \frac{(h_R + h_T)}{(h'_C + h_T + h_R)} \, \right] \, \frac{(T_G - T_S)}{\tilde{Q}_V} \qquad (6.35)$$

The term within the square brackets is thus the correction to the convective heat-transfer coefficient due to the influence of conduction and radiation.   Because of this <u>overheating factor</u>, the temperature T_S is above the wet-bulb temperature T_W for the hygrothermal state of the air.   The actual surface temperature can be found by noting that the humidity adjacent to the surface (Y_S) can be expressed in terms of the temperature there (T_S) through the moisture-saturation relationship. For example, we have from equation 4.10

$$Y_S = \left(\frac{M_W}{M_G}\right) \, \frac{1}{\left[P \, \exp(a/T_S - b) - 1\right]} \qquad (4.10)$$

where a and b are best-fit constants from the Antoine equation (eq.4.9) and P is the total pressure.   On substituting equation 4.10 into equation 6.25, and noting $\phi \, K_o = h_C\sigma/\bar{C}_{PY}$ (eq. 6.10) one gets

$$N_W = \frac{h_C\sigma}{\bar{C}_{PY}} \left[\frac{M_W}{M_G} \cdot \frac{1}{\left[P \, \exp(a/T_S - b) - 1\right]} - Y_G\right] \qquad (6.36)$$

Whence, finally, on eliminating N_W from equations 6.35 and 6.36, there results the following implicit expression for the unknown surface temperature T_S :

$$\left[1 + \frac{h'_C}{h_C} \frac{(h_R + h_T)}{(h'_C + h_T + h_R)}\right] \frac{(T_G - T_S)}{\tilde{Q}_V} - \frac{\sigma}{\bar{C}_{PY}} \left[\frac{M_W}{M_G} \cdot \frac{1}{[P \exp(a/T_S - b) - 1]} - Y_G\right]$$

$$= 0 \qquad (6.37)$$

Since T_S is only slightly above the wet-bulb temperature T_W , the unknown temperature can be found quickly by the iterative method of false position with T_W as the starting value.

Example 6.8. Pellets prepared from an organic paste are dried in thin layers on shelves in a multiple-tray dryer. Estimate by how much convection is boosted by conduction and radiation and the elevation of the surface temperature over the wet-bulb temperature caused thereby.

Data:				
	Dry-bulb temperature	(T_G)	=	$70^\circ C$
	Wet-bulb temperature	(T_W)	=	$30^\circ C$
	Shelf temperature	(T_T)	=	$60^\circ C$
	Air velocity	$(\hat{u})$	=	$1.5\ m\ s^{-1}$
	Bed thickness	(b_S)	=	40 mm
	Tray length	(L)	=	1 m
	Tray thickness	(b_T)	=	5 mm

Convective coefficient, h_C

From the appended tables of thermophysical data, we find:

State	Pr	$\lambda / W\ m^{-1} K^{-1}$	$10^6 \nu / m^2 s^{-1}$
dry-bulb	0.712	0.0294	19.9
wet-bulb	0.747	0.0261	16.2
mean	0.730	0.0278	18.1

$$Re_L = \hat{u}\, L/\nu = 1.5 \times 1/18.1 \times 10^{-6}$$

$$= 8.29 \times 10^4$$

The flow will be almost turbulent. Use the "practical" relationship (eq. 6.17) for h_C :

i.e. $$\bar{Nu}_L = 0.053\ Re_L^{0.8}$$

$$= 0.053 \times (8.29 \times 10^4)^{0.8}$$

$$= 456.$$

and $$h_C = \bar{Nu}_L\ \lambda/L$$

$$= 456 \times 0.0275\ /\ 1$$

$$= 12.6\ W\ m^{-2}K^{-1}$$

The Prandtl number correction is $(0.730/0.7)^{1/3} = 1.014$ which can be neglected.

Radiation coefficient, h_R

Suppose the overheating is small, and in the first instance $T_S = 35^oC$. Further, since the material is a wet paste, put $\varepsilon_S = 0.95$ and let $\varepsilon_T = 0.5$, so from equation 6.29

$$h_R = 5.6697 \times 10^{-8} \times \left[\frac{1}{(1/0.95 + 1/0.5 - 1)} \right] \times (308 + 333)\ (308^2 + 333^2)$$

$$= 3.65\ W\ m^{-2}K^{-1}.$$

Conductive coefficient, h_T

Since the paste is very wet, assume $\lambda_S \to \lambda_W \simeq 0.5\ W\ m^{-1}K^{-1}$. The thermal conductivity of the shelf, assumed sheet steel, is $40\ W\ m^{-1}K^{-1}$. Whence from equation 6.31

$$\frac{1}{h_T} = \frac{0.040}{0.5} + \frac{0.005}{40}$$

$$= 0.080$$

(Note, the thermal resistance of the tray is negligible).

$$\therefore\ h_T = 1/0.080 = 12.5\ W\ m^{-2}K^{-1}$$

Overheating coefficient, F_o

We shall assume that $h_C \simeq h_C'$, and thus the factor F_o is found from equation 6.35 to be

$$F_o = \left[1 + \frac{(h_R + h_T)}{(h_C + h_T + h_R)} \right]$$

$$= \left[1 + \frac{(3.65 + 12.5)}{(12.6 + 12.5 + 3.65)} \right]$$

$$= 1.56$$

The effect of conduction, and to a lesser extent radiation, is to boost the heat transfer to the drying surface by 56 per cent. The apparent convective transfer coefficient ($F_o h_c$) is 19.7 W m^{-2} K^{-1}.

<u>Surface temperature</u>, T_S

We wish to find the appropriate root of the equation $f(T_S) = 0$, where from equation 6.37

$$f = 1.56\ (T_G - T_S)/Q - \frac{1}{\bar{C}_{PY}} \left[\frac{0.622}{(P \exp(a/T_S - b) - 1)} - Y_G \right]$$

For $T_G = 70^\circ C$ and $T_W = 30^\circ C$, $Y_G = 0.01066$ for the humidity chart in the Appendix.

$$\bar{C}_{PY} = 1.006 + (1.87 \times 0.01066)$$

$$= 1.026 \text{ kJ kg}^{-1} K^{-1}$$

$$\tilde{Q}_V = 2501 - 2.44(T_S - 273)$$

$$= 3167 - 2.44\ T_S$$

where the temperature T_S is evaluated in Kelvins.

With P = 1 bar (10^5Pa), the Antoine constants are given by a = 4288 and b = 10.93 for the vapour pressure in bar.

Thus

$$f = \underbrace{\frac{1.56(343 - T_S)}{3167 - 2.44\ T_S}}_{\text{I}} - \underbrace{\frac{1}{1.026} \left[\frac{0.622}{\exp(\frac{4288}{T_S} - 10.93) - 1} - 0.01066 \right]}_{\text{II}}$$

T_S/K	I	II	f
308	0.02260	0.02164	+ 0.00096
309	0.02198	0.02320	- 0.00122
308.4	0.02235	0.02225	- 0.00010 close

The estimated surface temperature is 308.4 K or 35.2°C , which is 5.2°C higher than the wet-bulb temperature. The difference in surface temperature (0.2°C) over that assumed for the calculation of the radiative coefficient is extremely slight, and thus the calculation need not be repeated.

Gas radiation. In other dryers, in which the drying solids are exposed to a relatively large volume of gas, as in a rotary dryer for instance, the solids may receive additional heat by way of gas radiation. While some gases, principally monatomic and diatomic gases with structural symmetry, do not radiate, others are fairly good radiators of heat. Among these, vaporized moisture is always present in the air space above drying solids and carbon dioxide in combustion gases used in high-temperature dryers. The emission of radiation takes place only over a few narrow wavebands and is thus temperature-dependent. The quantity of radiation received by the drying surface, in turn, depends upon the self-absorption of radiant heat, and thus the mean radiation path or so-called beam length.

Specifically, let us consider in the axis of a rotary dryer a single particle, falling through the air space, after being thrown from the upper lifting flights as the chamber rotates (see Fig. 6.11). The mean beam length for this situation is that for an infinitely long space, half-cylindrical in section, filled with gas radiating to a spot on the centre of the flat side. Hsu[12a] gives this beam length L as 0.63 D, where D is the diameter of the chamber.

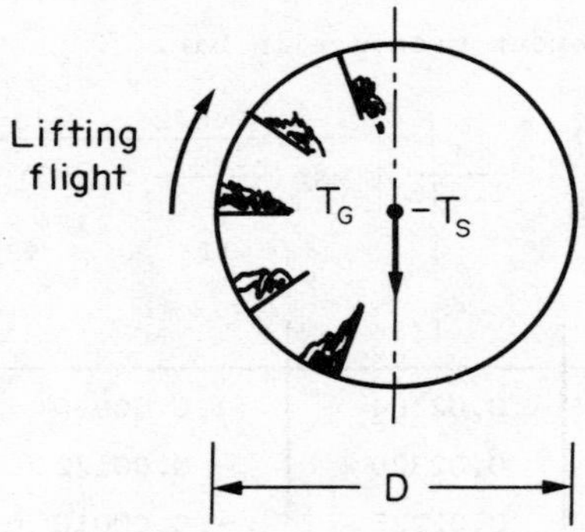

Fig. 6.11 Section through rotary dryer.

The gas emissivity ε_G is a function of pL , where p is the partial pressure of the radiating gas. Values of ε_G for water vapour in air are shown in Fig. 6.12. The emissivity ε_G increases with pL , as anticipated, but decreases with temperature.

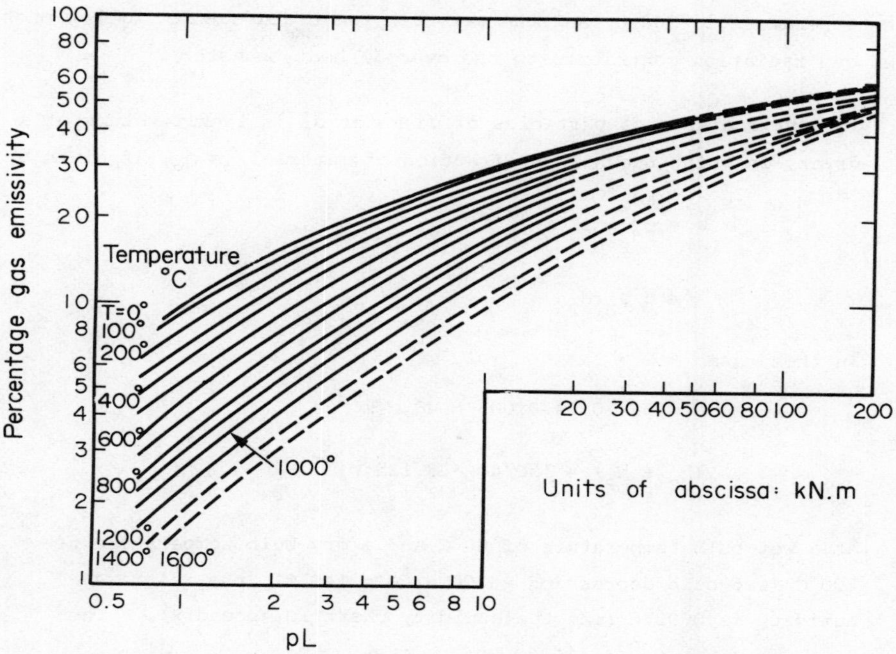

Fig. 6.12 Gas-emissivity chart for water-vapour in air. Adapted from Eckert.[8]

The nett radiated heat received by the particle's surface, if a black body, is given by

$$q_R = \sigma(\varepsilon_{GG}T_G^{\ 4} - \alpha_{GS}T_S^{\ 4}) \qquad (6.38)$$

where ε_{GG} is the gas emissivity at a temperature T_G and α_{GS} is the gas absorptivity at a temperature T_S, assumed to be equal to ε_{GS}. In practice, the surface is not a perfect absorber, and thus equation 6.38 becomes

$$q_R = \sigma\, \varepsilon_S'(\varepsilon_{GG}T_G^{\ 4} - \varepsilon_{GS}T_S^{\ 4}) \qquad (6.39)$$

where ε_S' is a "mean surface" emissivity, and is taken to be the average between the true value ε_S and 1. The following calculation illustrates the magnitude of such radiation.

Example 6.9. A rotary dryer of 3 m diameter is being used to dehydrate spherical pellets. When the volumetric hold-up is 4 per cent, the observed overall heat-transfer coefficient is 150 W $m^{-3}K^{-1}$ (based on the total chamber volume). The particles are 5 mm in diameter. If the gas temperature is 200°C, and the wet-bulb temperature 60°C, by how much does gas radiation contribute to the overall heat-transfer ?

The surface area of particles of diameter d_P is a unit volume of a dryer, in which the hold-up fraction of material is ϕ_H, is given by

$$a = \frac{\pi d_P^2}{\pi d_P^3/6} \times \phi_H$$

$$= 6\ \phi_H/d_P$$

In this case

$$a = 6 \times 0.04/0.005 = 48\ m^2 m^{-3}$$

$$\therefore\quad (h_C + h_R) = 150/48 = 3.125\ W\ m^{-2}K^{-1}$$

At a wet-bulb temperature of 60°C and a dry-bulb temperature of 200°C (wet-bulb depression = 200 - 60 = 140°C), the relative humidity is 0.0076 (see the humidity chart in Appendix). The vapour pressure p^o is 19.92 kPa. Then

$$p = \psi p^o = 0.0076 \times 19.92 = 0.151\ kPa$$

$$\text{and } pL = 0.151 \times (0.63 \times 3) = 0.285\ kN\ m$$

This value is offscale in Fig. 6.12. Very roughly one may assume that the plotted curves can be extrapolated to smaller values of pL. Such an extrapolation yields:

$$\varepsilon_{GG} = 0.04 \quad\text{and}\quad \varepsilon_{GS} = 0.03$$

$$\text{while}\quad \varepsilon'_S \simeq (0.9 + 1)/2 = 0.95$$

Thus, from equation 6.39,

$$q_R = 5.6697 \times 0.95\ (0.04 \times 4.73^4 - 0.03 \times 3.33^4)$$

$$= 88.0\ W\ m^{-2}$$

$$h_R = 88.0\ /\ (200 - 60)$$

$$= 0.629\ W\ m^{-2}\ K^{-1}$$

Therefore,

$$\frac{h_R}{(h_C+h_R)} = \frac{0.629}{3.125} = 0.2$$

Gas radiation contributes by 20 per cent to the total heat transfer.

Since the gases are confined within cramped spaces in most commercial dryers, gas radiation will play a minor, if not negligible role under normal circumstances.

6.3 Contact Drying

In contact drying, the moist material covers a hot surface which supplies the heat required for the drying process. The use of internally heated cylinders is a convenient way of drying very wet and fluid materials which can be spread over a hot, rotating drum. Thin webs of porous materials can be stretched over the hot surface, as in papermaking machines, when usually pressing felts are run over the web as well to eliminate as far as possible any air-gap between the web and the hot surface. These arrangements can be very elaborate, as Fig. 6.13 shows.

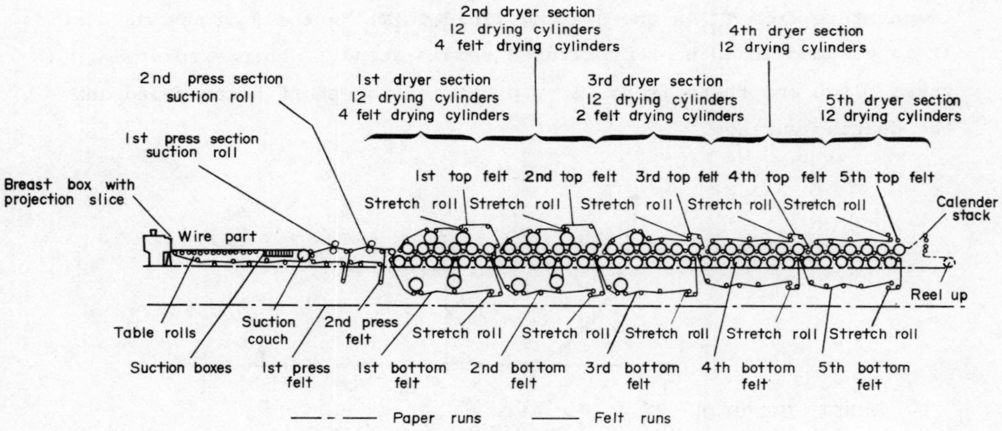

Fig.6.13 Drying section of a papermaking machine.

Other forms of contact dryers include vacuum dryers of various kinds, in which the heating medium is confined in chests or tubes within the drying chamber. One example of this sort of dryer is shown in Fig.6.14. The sketch illustrates a continuously worked, band dryer, in which the material to be dried is fed onto a number of narrow bands, made from fine wire mesh for flexibility, which move through the vacuum chamber. The upper part of each endless band is supported on a number of steam chests which heat up the moist material by conduction.

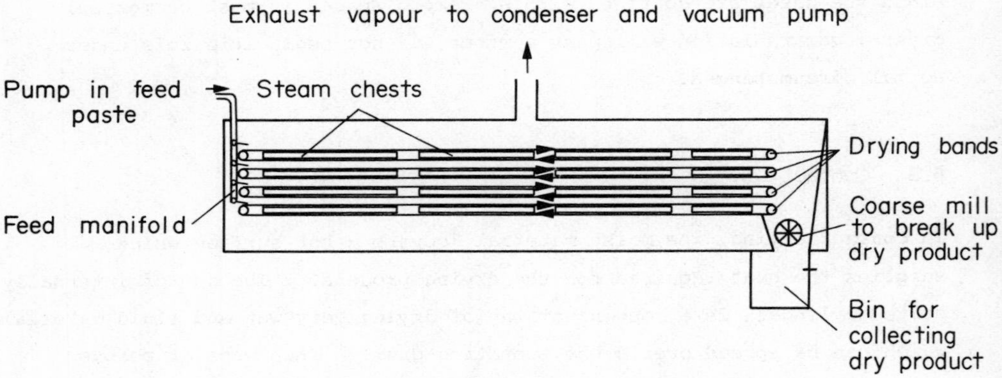

Fig. 6.14 A "solid-band" dryer for vacuum working.

Let us consider a moist material lying on a hot flat plate of infinite extent. Fig. 6.15 illustrates the temperature profile for the fall in temperature from T_H in the heating fluid to T_G in the surrounding air. It is assumed that the temperatures remain steady, unhindered drying takes place and there is no air-gap between the stuff being dried and the heating surface.

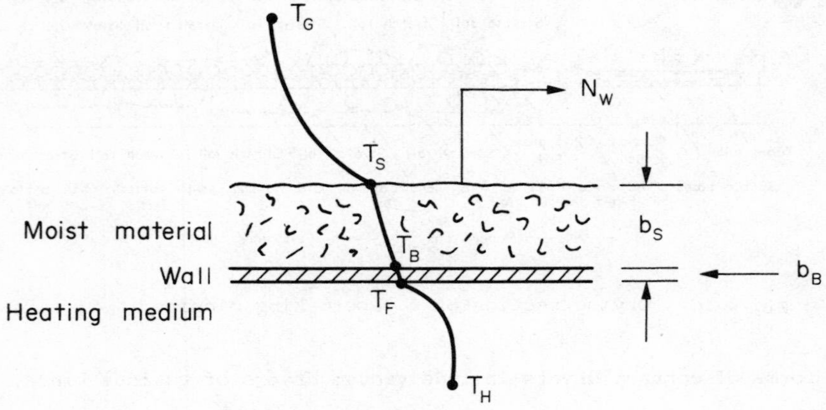

Fig.6.15 Temperature profile in conductive drying.

The heat conduced through the wall and material is dissipated by evaporation of moisture and convection from the moist surface to the surrounding air. A heat balance yields:

$$U(T_H - T_S) = N_W \Delta H_{VS} + h_C(T_S - T_G) \tag{6.40}$$

where U is the overall heat-transfer coefficient. This coefficient is found from the reciprocal law of summing resistances in series:

$$\frac{1}{U} = \frac{1}{h_H} + \frac{b_B}{\lambda_B} + \frac{b_S}{\lambda_S} \tag{6.41}$$

in which h_H is the heat-transfer coefficient for convection inside the heating fluid. If condensing steam is used, this coefficient is very large normally and the corresponding resistance $1/h_H$ negligible. Re-arrangement of equation 6.40 yields an expression for the maximum drying rate:

$$N_W = [U(T_H - T_S) - h_C(T_S - T_G)] / \Delta H_{VS} \tag{6.42}$$

Equation 6.42, as it stands, would give an overestimate of the maximum drying rate for the case of contact drying over heated rolls, when there are significant heat losses from the ends of the drum and only part of the drum's surface can be used for drying. In the roller-drying arrangements shown in Fig. 6.16, only a fraction a of the drum's periphery is available from the point of pick-up to the point where the solids are peeled off.

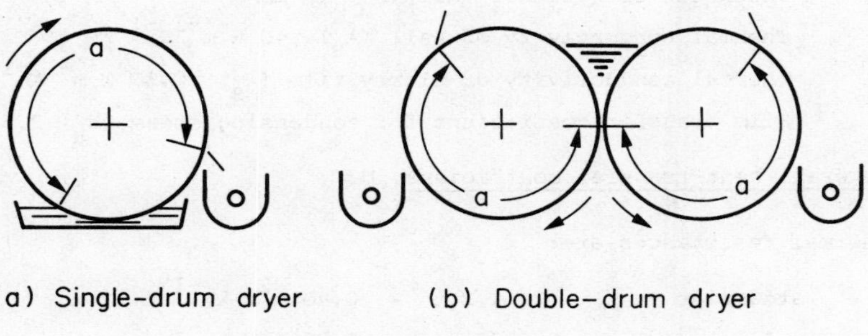

Fig.6.16 Roller-dryer arrangements.

Let q_E be the heat loss per unit area from the ends. The ratio of the end areas to cylindrical surface, from a drum of diameter D and length L, is $2(\frac{1}{4}\pi D^2) / \pi DL$ or $D/2L$. Equation 6.42 for the maximum drying rate under roller-drying conditions thus becomes

$$N_W = [a\ U(T_H - T_S) - h_C(T_S - T_G) - Dq_E/2L] / \Delta H_{VS} \quad (6.43)$$

The total evaporation from the drum is $N_W a(\pi DL)$. Equation 6.43 could be refined further, as it neglects the effect caused by the small portion of the drum's surface being covered by the slurry in the feed trough, as well as thermal conduction through the axial shaft to the bearing mounts. The use of equation 6.43 to estimate the maximum drying rate is illustrated in the following worked example.

Example 6.10 A single rotating drum of 1.250 m diameter and 3 m wide is internally heated by saturated steam at 0.27 MPa. As the drum rotates, a film of slurry 0.1 mm thick is picked up and dried. The dry product is removed by a knife, as shown in Fig. 6.16a. About three-quarters of the drum's surface is available for evaporating moisture. Estimate the maximum drying rate, when the outside-air temperature (T_G) is 15°C and the surface temperature 50°C, and compare the effectiveness of the unit with a dryer without end-effects and in which all the surface could be used for drying.

Data Heat-transfer coefficient (h_C) 50 W $m^{-2}K^{-1}$
Thickness of cylinder wall (h_B) 10 mm
Thermal conductivity of wall (λ_B) 40 W $m^{-1}K^{-1}$
Thermal conductivity of slurry film (λ_S) 0.10 W $m^{-1}K^{-1}$
Film transfer coefficient for condensing steam (h_H) 2.5 kW $m^{-2}K^{-1}$

Overall heat-transfer coefficient, U.

The thermal resistances are:

steamside	1/2.5	= 0.40 $m^2K\ kW^{-1}$
wall	0.01/0.04	= 0.25 $m^2K\ kW^{-1}$
filmside	$0.0001/0.1 \times 10^{-3}$	= 1.0 $m^2K\ kW^{-1}$

$\therefore$ overall resistance = 0.40 + 0.25 + 1.0 = 1.65 $m^2K\ kW^{-1}$

U = 1/1.65 = 0.606 kW $m^{-2}\ K^{-1}$

Wall temperature, T_B

At 0.27 MPa, the steam temperature is 130°C. If it is assumed that the temperature drops between the steam and the film surface are directly proportional to the respective thermal resistances, it follows that

$$\frac{(T_H - T_B)}{(T_H - T_S)} = \frac{(0.40 + 0.25)}{1.65} = 0.3939$$

$$\therefore T_B = T_H - 0.3939\ (T_H - T_S)$$

$$= 130 - 0.3939\ (130 - 50)$$

$$= 98.5^{\circ}C.$$

Heat losses from ends, q_E

By interpolation in Table 3.1 for an emissivity ~ 1 and an air temperature of $15^{\circ}C$ with a drum temperature of $98.5^{\circ}C$, one finds:

$$q_E = 1184\ W\ m^{-2}$$

Maximum drying rate, N_W

From equation 6.43

$$N_W = [aU(T_H - T_S) - h_C(T_S - T_G) - Dq_E/2L]/\Delta H_{VS}$$

$$= \frac{[0.75 \times 0.606 \times (130 - 50) - 0.05(50 - 15) - (1.25 \times 1.184)/6]}{2382}$$

$$= 0.0144\ kg\ m^{-2} s^{-1}$$

The ideal maximum rate is given by equation 6.42 for an endless surface:

$$N_N = [U(T_H - T_S) - h_c(T_S - T_G)]/\Delta H_{VS}$$

$$= [0.606\ (130 - 50) - 0.05\ (50 - 15)]/2382$$

$$= 0.0196\ kg\ m^{-2} s^{-1}$$

Therefore, the effectiveness of the dryer is 0.0144/0.0196 = 0.735.

The predicted thermal efficiency η is

$$\eta = 1 - \frac{h_c(T_S - T_G) + Dq_E/2L}{aU(T_H - T_S)}$$

$$= 1 - \frac{[0.05(50 - 15) + (1.25 \times 1.184)/6]}{0.75 \times 0.606 \times (130 - 50)}$$

$$= 0.945$$

These estimates may be compared with the range of values found in practice, quoted by Nonhebel and Moss:[31]

	this estimate	typical range
specific evaporation/$g m^{-2} s^{-1}$	14.4	7 - 11
thermal efficiency	0.945	0.4 - 0.7

The typical performance is somewhat less than the estimated maximum evaporative capacity, although values as high as 25 $g\ m^{-2} s^{-1}$ have been reported.[31] As the solids dry out, so the thermal resistance of the film increases and the evaporation falls off accordingly. Heat losses through the bearing of the drum shaft have been neglected, but the effect of radiation is accounted for in the value of h_c taken. In the case of drying organic pastes, the heat losses have been determined as 2.5 $kW\ m^{-2}$ over the whole surface, compared with 1.75 $kW\ m^{-2}$ estimated herein for the cylindrical surface. The inside surface of the drum has been assumed to be clean; any scale would reduce the heat transfer markedly.

The temperature of the drying material rises from a value T_S at the exposed surface to a higher one T_B for stuff at the wall. This higher temperature can be deduced as shown in the worked example above, or as follows. An alternative form of the heat balance (eq.6.40) yields the expression

$$\lambda_S(T_B - T_S)/b_S = N_W \Delta H_{VW} + h_C(T_S - T_G) \tag{6.44}$$

but also it follows from equation 6.36 that

$$N_W = \frac{h_C\sigma}{\bar{C}_{PY}}\left[\frac{M_W}{M_G}\cdot\frac{1}{[P\exp(a/T_S - b) - 1]} - Y_G\right] \tag{6.45}$$

whence

$$T_B = T_S + \frac{b_S}{\lambda_S}\left\{\frac{h_c\sigma\Delta H_W}{\bar{C}_{PY}}\left[\frac{M_W/M_G}{(P\exp(a/T_S - b) - 1)} - Y_G\right] + h_C(T_S - T_G)\right\} \tag{6.46}$$

which defines T_B explicitly.

For constant hygrothermal conditions, the base temperature T_B is directly proportional to the thickness of the material over the hot surface. When the wet-bulb temperature is high, and layer of material thick enough, the temperature T_B will reach the boiling point of the moisture. Under these conditions, a mixed vapour-air layer interposes between the material and the heating surface. This is known as the Leidenfrost effect, and the phenomenon causes a greatly increased thermal resistance to heat transfer to hinder drying.

Effect of ventilation. Often roller dryers are fitted with ventilating hoods to reduce the discharge of moist gases into the building housing the equipment. If outside air is used for this purpose, then the surface will receive further convective cooling and its temperature will fall. The consequential increase in conductive heat transfer is offset by the extra convective cooling, and the drying rate wanes as well, as Kröll's calculations[24] show. On the grounds of energy conservation, it is better to ventilate at higher temperatures, so reducing the ideal heat demand for reasons outlined in Section 3.5.

As the air passes over the cylinder, it can easily become foggy. Suppose outside air with state conditions (I_A, Y_A), point A in Fig.6.17, is drawn over the dryer. For unit mass of moisture so evaporated, the humid enthalpy of the air increases by ΔH_{VS}. As the amount of air is reduced, the exhaust stream becomes steamier and, should the air be restricted enough, the final hygrothermal state of the air will lie in the fog region, say at point B in Fig. 6.17. The use of preheated air will eliminate the risk of fogging conditions, even if the amount of air is not increased.

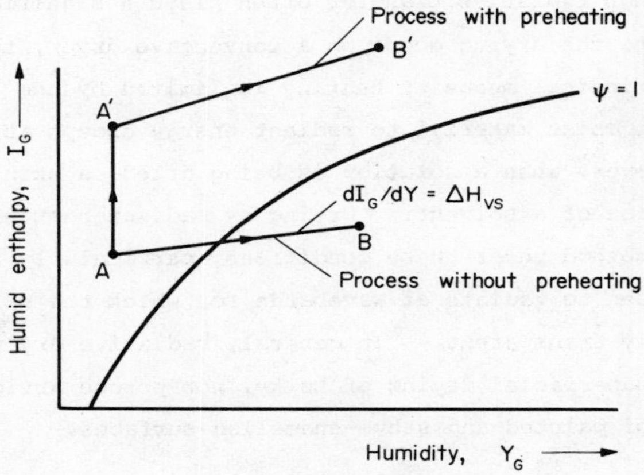

Fig. 6.17 Air conditions over a ventilated roller dryer.

Effect of radiation. Usually the influence of radiation in contact drying is small, but not always. One example of such an exception is to be observed in the operation of a vacuum drying chamber. In this case, the solids are dried in an evacuated environment by contact with a hot tray or jacket, but the solids will also receive radiant heat from the warm sides of the chamber itself. Thus equation 6.42 for the maximum drying rate becomes

$$N_W = [U(T_H - T_S) + h_R(T_R - T_S)] / \Delta H_{VS} \tag{6.44}$$

where T_R is the temperature of the radiating surface and h_R is the radiative coefficient, which may be estimated from equation 6.29. When $h_R \sim 5\ W\ m^{-2}K^{-1}$, $T_H = 130^oC$, $T_R = 100^oC$, $T_S = 50^oC$ and $U \sim 20\ W\ m^{-2}K^{-1}$, radiation provides about 14 per cent of the heating load. Towards the end of drying, however, the overall heat-transfer coefficient will fall as the solids dry out, while the radiative coefficient will remain little different. Under these conditions, the contribution of radiation is considerable.

6.4 Less-common Heating Modes

There are a number of specialized heating techniques which are useful in certain circumstances. Both radiative and dielectric heating methods fall into this group of methods.

Radiation. Although radiative transfer often plays a significant role in supplying heat to the drying goods on a convective dryer, the use of radiation as the principal means of heating is limited by the poor permeability of the moist material to radiant energy except at certain wavelengths. However, when a solution is being dried, a skin may form to hinder evaporation of a solvent. Drying by radiant heat may be the only satisfactory method under these conditions, particularly if an emitter can be chosen to radiate at wavebands for which the skin and solution are largely transparent. In general, radiative drying will be useful for the superficial drying of bulky, non-porous articles, and in the drying of painted and stove-enamelled surfaces.

Radiators include electric lamps, electric sheathed filaments and gas-heated units. Gas radiators enable the moist surface to be warmed up by infra-red radiation directly, as well as by convected heat from the gaseous products of combustion. A compact design of the dryer can result from the judicious disposition of the radiators and gas-baffle plates.[36]

In principle, equations 6.29 and 6.30 could be used to estimate the radiant transfer, but rarely are the properties of the surface known to a sufficient degree of exactness for such an analysis to be accurate enough for process design. Some emission data for common radiators are tabulated below.

TABLE 6.4 ENERGY EMISSION OF INFRA-RED RADIATORS. Adapted from data of Nonhebel and Moss.[32]

Type	Temperature/K	Radiating area/ $m^2 (kW\ input)^{-1}$	Emitted energy/ $kW\ m^{-2}$
electric lamp	2500	0.0092	103
sheathed filament	1020	0.55*	1.3
high-temperature gas	1070	0.41	0.83
low-temperature gas	610	0.60	0.56

* mean of range

<u>Dielectric heating</u>. In dielectric heating, the material being dried is sandwiched between two electrodes. The arrangement may be regarded as being a resistive condenser, for which the vector diagram and equivalent electrical circuit is shown in Fig. 6.18 for the situation where there is no air-gap between the material and the adjacent electrode.

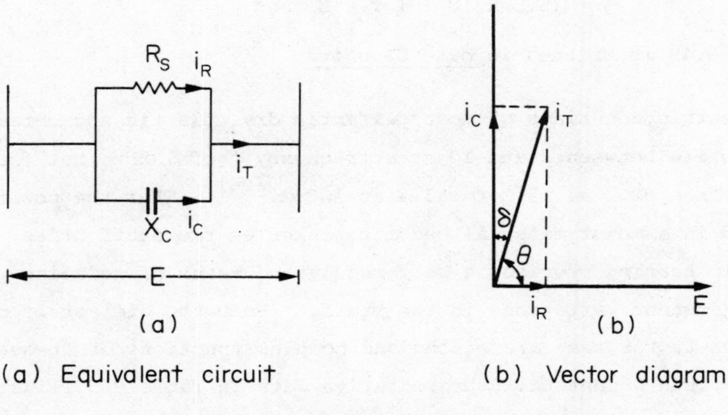

(a) Equivalent circuit (b) Vector diagram

Fig. 6.18 Electrical diagrams for a flat, dielectrically heated solid without air-gaps.

Only the in-phase current furnishes power which dissipates as heat. The angle δ in Fig. 6.18(b) is very small. Thus the power factor ($\cos\theta$)

is given approximately by the expression

$$\cos\theta \simeq \tan\delta = i_R/i_C \tag{6.45}$$

Since the resistive and capacitives branches can be considered to be parallel,

$$\frac{i_R}{i_C} = \frac{X_C}{R_S} = \frac{1}{2\pi f C R_S} \tag{6.46}$$

where C is the capacitance, R_S the resistance and X_C the impedance of the system. The resistance of the solids is found by combining equations 6.45 and 6.46:

$$R_S = 1/2\pi f\, C(i_R/i_C) = 1/2\pi f\, C \tan\delta \tag{6.47}$$

The power thus delivered becomes

$$P = V^2/R_S = 2\pi f\, CV \tan\delta \tag{6.48}$$

which expression may be used to estimate the heating effect when the electrical properties of the material to be dried are known. For this purpose, it is more convenient to re-write equation 6.48 in terms of the voltage gradient E (in V m^{-1}) and the dielectric constant ε to yield the heat dissipated per unit volume (W m^{-3}):

$$\tilde{q} = (5.5 \times 10^{-11})\, f\, \varepsilon\, E \tan\delta \tag{6.49}$$

Equation 6.49 is written in base SI units.

The dielectric constants of most perfectly dry, plastic and ceramic materials lie between 2 and 10 at a frequency f of 1 MHz, but for water it falls from 82.2 at 15°C to 55.3 at 100°C.[44] Thus the power generated in a moist material becomes weaker as the stuff dries, so dielectric heating provides a self-regulating means of reducing moisture-content variations in the stuff. Both the dielectric constant ε and loss-factor tanδ are strong and complex functions of frequency as well as moisture content. Representative data in Table 6.5 illustrate these variations for pinewood over a practical frequency range at two levels of moistness.

TABLE 6.5 DIELECTRIC DRYING OF PINEWOOD IN RADIAL SECTION
Adapted from Kröll's data[24] for $E = 100$ kV m^{-1}

	X = 0.09			X = 0.20		
Frequency/MHz	ε	$\tan\delta$	$\tilde{q}$/W m^{-3}	ε	$\tan\delta$	$\tilde{q}$/W m^{-3}
0.1	3.8	0.03	6.3	6.0	0.65	215
1.0	3.2	0.04	70.4	4.5	0.18	446
10.0	2.7	0.057	84.6	3.7	0.08	1630

The above table shows that the volumetric heating rate $\tilde{q}$ would be very small at frequencies below 0.1 MHz, and operational frequencies at least of order 1 MHz are preferred. The high-frequency equipment needed for commercial operations is very expensive and is costly to run. Dielectric drying would only be considered when other more conventional heating methods involve grave disadvantages.

Because of uncertainties in the value of $\varepsilon\tan\delta$ in a given case, some preliminary experimental data with small-scale or prototype equipment are normally essential for evaluating dielectric drying techniques.

6.5 Transfer Units

So far in this chapter we have considered the maximum drying rate possible for particular ways of heating the material to be dried. The actual rate in a given case will depend also upon the extent to which the material hinders the evaporative process, as noted in Chapter 5. While commonly the performance of dryers is still quoted in terms of the <u>specific evaporative capacity</u>, the mass evaporated over unit surface in unit time, particularly in trade literature, other criteria of performance are desirable which are less dependent on process conditions.

Let us consider the humidification of a gas which is picking up moisture from the stuff being dried. Let y be the fractional approach of the gas to saturation, as shown in Fig. 6.19, so that y = 0 at saturation.

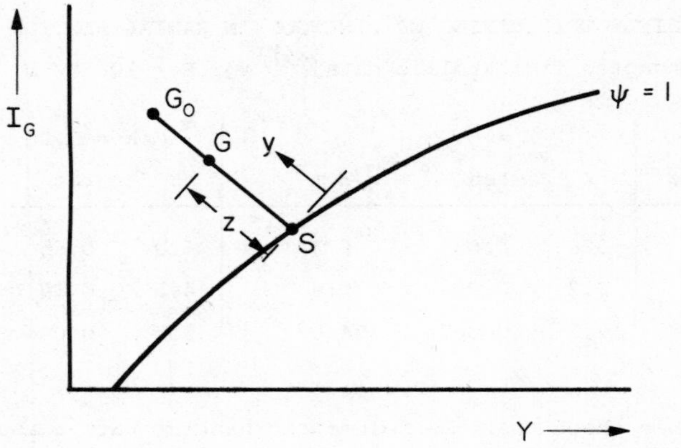

Fig. 6.19 Humidification path G_oS of drying gas

It follows that

$$-\frac{dy}{y} = \frac{dY_G}{(Y_S - Y_G)} \tag{6.50}$$

where Y_S is the humidity at the saturation-state point S and Y_G the humidity at point G. The negative sign appears because the approach y shrinks as the gas humidity Y_G increases towards Y_S.

Let Δs be the surface of material needed for the gas to progress a distance (1-z) between states G_o and G. Further suppose n_e is the number of unit surfaces s_e contained in this total surface Δs. Then,

$$-\int_1^z \frac{dy}{y} = \frac{\Delta s}{s_e} = n_e \tag{6.51}$$

whence

$$n_e = \ln(1/z) \tag{6.52}$$

or

$$z = e^{-n_e} \tag{6.53}$$

Thus the larger n_e becomes, the smaller is the approach z. For unit surface, $n_e = 1$, $z = e^{-1} = 0.368$; the gas is nearly two-thirds its way to saturation.

Inherent in the integration of equation 6.51 to give equation 6.52 is the assumption that the point of approach, or <u>pulling point</u>, remains in the same place. This situation is only true for the adiabatic removal of non-hygroscopic moisture. In all other instances, such as non-adiabatic drying or driving off bound moisture, the pulling point will wander. Such shifts are illustrated in Fig. 6.20. In both cases, the humidification path is now curved on the enthalpy-humidity diagram.

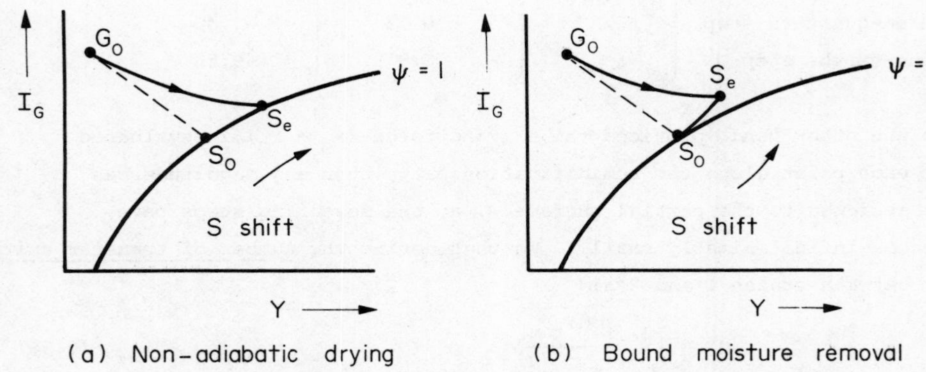

Fig. 6.20 Wandering of pulling point S during drying.

The accuracy of the calculation can, however, be arbitrarily increased to any desired degree of accuracy by choosing smaller approach distances z. For example, one could take the half-step value for which the state distance is decreased to z = 1/2. The corresponding half-step surface is thus

$$s_{\frac{1}{2}} = s_e \ln(1/0.5) = 0.693\ s_e \qquad (6.54)$$

The total surface, as before, is the multiple $n_{\frac{1}{2}}$ of the chosen partial surface $s_{\frac{1}{2}}$:

$$\Delta s = n_{\frac{1}{2}} s_{\frac{1}{2}} = n_e s_e \qquad (6.55)$$

The smaller the partial surface taken, the greater is the number of difference steps n needed, as set out in Table 6.6.

TABLE 6.6 DEFINITIONS OF PARTIAL SURFACES FOR HUMIDIFICATION

Measuring unit	s	z	n/n_e
unit surface	s_e	0.368	1
half-step surface	$s_{\frac{1}{2}}$	0.5	1.445
three-quarters step	$s_{3/4}$	0.75	3.48
nine-tenths step	$s_{0.9}$	0.9	9.53

On the other hand, provided the driving force $(Y_S - Y_G)$ is evaluated at each point along the humidification path, then any magnitude can be assigned to the partial surface s, as the measuring steps have become infinitesimally small. We thus define the number of transfer units N_t between states 1 and 2 as

$$N_t = \int_1^2 \frac{dY_G}{(Y_S - Y_G)} \tag{6.56}$$

If equation 6.56 is compared with equations 6.50 and 6.51, it is seen that N_t is identical to n_e, the number of unit surfaces s_e. Since most dryers have a uniform cross-section, N_t may also be regarded as the number of unit lengths (or heights). Therefore, we have

$$N_t = \frac{\text{length (or height) of dryer}}{\text{length (or height) of transfer unit}} \tag{6.57}$$

Equation 6.57, in this form, is identical to the transfer-unit concept first introduced by Chilton and Colburn[6] to describe gas-absorption processes in packed columns.

For humidification in a progressive dryer, it has been shown that

$$\int_{Y_{GO}}^{Y_{GZ}} \frac{dY_G}{(Y_W - Y_G)} = \int_o^Z \frac{K_o \phi a dz}{G} \tag{4.69}$$

(for f = 1), so that the length of a transfer unit (Z_t) is given by

$$Z_t = Z \Big/ \int_o^Z \frac{K_o \phi a dz}{G} \tag{6.58}$$

For the limiting case when K_o is independent of distance and $\phi \simeq 1$, equation 6.58 reduces to the expression

$$Z_t = G/K_o a \tag{6.59}$$

The following worked example illustrates the expected magnitude of Z_t in a simple case of convective drying.

Example 6.11. Estimate the length of a transfer unit for the case of cross-circulating slabs 10 mm thick at a velocity of 2.5 m s^{-1} at 100°C. Data: K_o = 20 g m^{-2}s^{-1}.

At 100°C ρ_G = 0.942 kg m^{-3} so that the specific gas rate G becomes

$$G = 2.5 \times 0.942 = 2.355 \text{ kg m}^{-2}\text{s}^{-1}$$

If a is the interfacial area per unit volume and b is the slab thickness, then a = 1/b . So,

$$a = 1/0.01 = 100 \text{ m}^{-1}$$

$$\therefore \quad Z_t = \frac{G}{K_o a} = \frac{2.355}{0.02 \times 100} = 1.18 \text{ m}$$

This is of the same order as the swept length of a tray dryer, so that in this case $N_t \simeq 1$.

Data for the number of transfer units (N_t) are not generally available for dryers. Values of N_t, however, normally range between 1 and 3 for continuously worked, progressive dryers. The next worked example shows how N_t may be estimated for such a dryer removing non-hygroscopic moisture.

Example 6.12. Estimate the number of transfer units in reducing the moisture content of non-hygroscopic solids from 1 kg kg^{-1} to 0.2 kg kg^{-1} in an adiabatic progressive dryer with co-current movement of solids and material.

Data:	Inlet-air humidity	= 0.008 kg kg^{-1}
	Air/solids ratio	= 20
	Wet-bulb temperature	= 50°C
	Dryer pressure	= 110 kPa

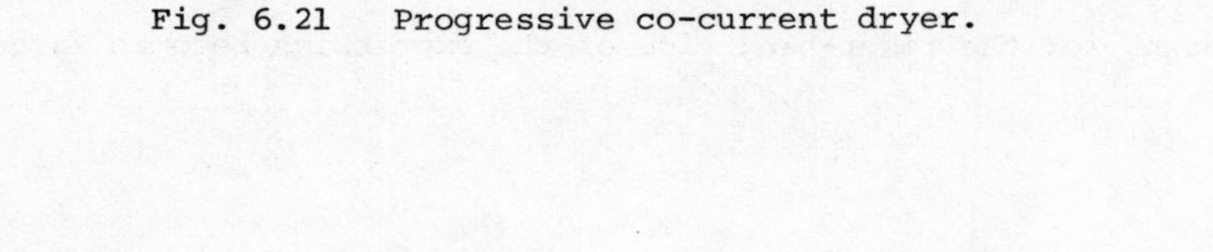

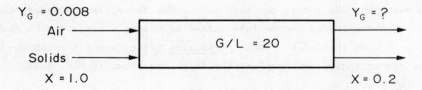

Fig. 6.21 Progressive co-current dryer.

At 50°C the vapour pressure of water is 12.33 kPa

$$\therefore Y_W = 0.622 \times 12.33/110$$

$$= 0.0697 \text{ kg kg}^{-1}$$

This remains constant throughout the dryer since the drying is adiabatic. Let us now evaluate the humidity driving force at moisture-content decrements ΔX of 0.1, since by mass balance

$$\Delta Y_G = L\ \Delta X/G = \Delta X/20$$

X	Y_G	$(Y_W - Y_G)$	$\Delta Y_G/(Y_W - Y_G)_m$
1.0	0.008	0.0617	
			0.084
0.9	0.013	0.0567	
			0.092
0.8	0.018	0.0517	
			0.102
0.7	0.023	0.0467	
			0.113
0.6	0.028	0.0417	
			0.128
0.5	0.033	0.0367	
			0.146
0.4	0.038	0.0317	
			0.171
0.3	0.043	0.0267	
			0.207
0.2	0.048	0.0217	
			Σ = 1.043

i.e. there is just over 1 transfer unit.

<u>Backmixing</u>. In those dryers wherein the solids being dried can tumble or be shifted by the drying-gas stream, the material does not progress uniformly through the plant, but becomes to some extent mixed up. Further, the piston-like flow of the drying-gas itself can be disrupted by eddying around baffles, heating coils and other obstacles, and at places where the gas enters and leaves. Under these circumstances, the humidity driving forces become less marked as damper air mixes with the drier air nearer the inlet. This <u>backmixing</u> brings about a moisture-content discontinuity or "jump" at the inlet, as shown in Fig. 6.22.

Clearly, when such backmixing occurs, the equality of equation 4.69 breaks down, for the right-hand side of the expression becomes larger.

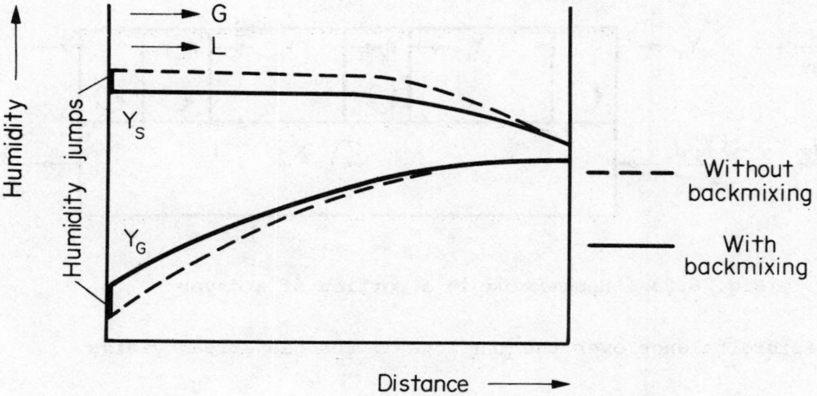

Fig. 6.22 Humidity profiles in co-current drying with and without backmixing.

Let the number of external transfer units (ETU) be defined by

$$N_{te} = \int_{Y_{G,in}}^{Y_{G,out}} \frac{dY_G}{(Y_S - Y_G)} \qquad (6.60)$$

where $Y_{G,in}$ and $Y_{G,out}$ are the "external" inlet-gas and outlet-gas humidities respectively: let the number of internal transfer units (ITU) be defined through the "internal" mass-transfer coefficient K_o by

$$N_{ti} = \int_o^Z \frac{K_o \phi a d_Z}{G} \qquad (6.61)$$

We expect $N_{te} \leq N_{ti}$. The ratio N_{te}/N_{ti} represents the approach of a dryer to its maximum effectiveness as a moisture exchanger when unmixed flows of both contacting phases prevail.

Should backmixing take place, then the movement can be depicted as a diffusive dispersion or as progression through a series of well-mixed sections. The choice is often a matter of taste, but the latter picture gives rise to a simpler analysis.[18,40] In this model, the greater the number of mixing sections, the closer the flow becomes a piston-like movement. Consider a portion of a counter-current dryer, in which the drying gas goes through in well-mixed zones of equal extent while the material to be dried is completely mixed up in this part of the dryer (Fig. 6.23).

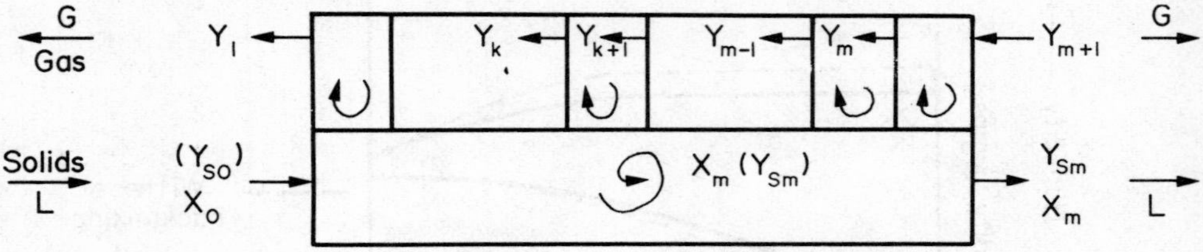

Fig. 6.23 Backmixing in a portion of a dryer

A moisture balance over the kth zone in the gas stream yields

$$-G(Y_{k+1} - Y_k) = K_o\phi a(Y_{Sm} - Y_k)\delta Z \tag{6.62}$$

in which the parameter G is understood to be the dry-gas flow over unit cross-section of the dryer and δZ is the length of the mixing zone (measured in the direction of the solids flow). The change in humidity across this zone follows directly from equation 6.62

$$Y_k - Y_{k+1} = \left[\frac{K_o\phi a\ \delta Z}{G}\right](Y_{Sm} - Y_k) \tag{6.63}$$

The term in square brackets is the number of internal transfer units for the kth stage, N_{ik} say. Re-arrangement of equation 6.63 thus yields

$$\frac{(Y_{Sm} - Y_k)}{(Y_{Sm} - Y_{k+1})} = \frac{1}{1 + N_{ik}} \tag{6.64}$$

Equation 6.64 may be used as a recurrence formula for each stage to get the overall change in humidity:

$$\frac{(Y_{Sm} - Y_1)}{(Y_{Sm} - Y_{m+1})} = \left[\frac{1}{1+N_{ik}}\right]^m = [1+N_i/m]^{-m} \tag{6.65}$$

where N_i is the total number of internal transfer units.

Now, the total number of <u>external transfer units</u> N_e is given by eq.6.60.

It is possible to gain an analytical solution to this equation under the following circumstances. Suppose the moisture isotherm can be linearized by an expression of the form

$$Y_S = \alpha X + \beta \tag{6.66}$$

where α and β are constant coefficients over the moisture-content range of interest. On noting that the moisture content X is also linearly related to the gas humidity Y_G, equation 6.60 then becomes

$$N_e = \frac{1}{(\Lambda-1)} \ln\left[\frac{Y_{S0} - Y_1}{Y_{Sm} - Y_{m+1}}\right] \tag{6.67}$$

where Λ is the moisture-sorption factor, $\alpha G/L$, which is analogous to the extraction factor of the general mass-transfer literature. The details of the solution are given in most introductory mass-transfer texts, e.g. Treybal.[41]

The moisture-sorption factor is composed of a product of concentration-difference ratios, namely:

$$\Lambda = \alpha \cdot \frac{G}{L} = \left[\frac{Y_{S0} - Y_{Sm}}{X_0 - X_m}\right] \cdot \left[\frac{X_0 - X_m}{Y_1 - Y_{m+1}}\right] \tag{6.68}$$

Thus,

$$\Lambda = (Y_{S0} - Y_{Sm}) \,/\, (Y_1 - Y_{m+1}) \tag{6.69}$$

(The first term of equation 6.68 follows from the linearization (eq.6.66) of the isotherm, while the second is the result of an overall moisture balance across the considered portion of the dryer). From equation 6.69 comes the identity:

$$\left[\frac{Y_{S0} - Y_1}{Y_{Sm} - Y_{m+1}}\right] \equiv 1 + [\Lambda - 1]\left[1 - \frac{(Y_{Sm} - Y_1)}{(Y_{Sm} - Y_{m+1})}\right] \tag{6.70}$$

which can be checked by substituting for Λ in the right-hand side of the expression.

Comparison of equations 6.65, 6.67 and 6.70 yields the required relationship between N_e and N_i

$$N_e = \frac{1}{(\Lambda-1)} \ln\,[1 + (\Lambda-1)\,\{1 - (1+N_i/m)^{-m}\}] \tag{6.71}$$

Suppose there are n sections, as shown in Fig. 6.23, in the whole dryer. There are, therefore, n mixing zones in the solids phase and mn mixing zones in the gas phase. Further, the total number of transfer units for the whole dryer are simple multiples of N_e and N_i respectively, that is:

(a) ETU $N_{te} = nN_e$ (6.72a)

(b) ITU $N_{ti} = nN_i$ (6.72b)

Equation 6.71 may thus be re-written in terms of N_{te} and N_{ti} by inspection:

$$N_{te} = \frac{n}{(\Lambda-1)} \ln \left[1 + (\Lambda-1) \{1 - (1+N_{ti}/mn)^{-m}\}\right] \quad (6.73)$$

A similar relationship holds for co-current movement of air and solids, and an "inverse" expression can be deduced for the case when there are fewer mixing zones in the gas phase than in the solid.[18]

Normally, the sorption factor Λ is closely equal to unity and, in the special case when $\Lambda = 1$, equation 6.73 simplifies to

$$N_{te} = n\ (1 - (1+N_{ti}/mn)^{-m}) \quad (6.74)$$

Further, in the limit when both streams are thoroughly mixed, as in fluid-bed drying, m=n=1 and equation 6.74 reduces to

$$N_{te}/N_{ti} = 1/(1+N_{ti}) \quad (6.75)$$

Thus, when the number of internal transfer units is large, the effect of backmixing in blurring the humidity potentials for drying is substantial. Nevertheless, well-mixed systems should not be discounted on this score alone, as the reduction in drying potential is often more than matched by the increase in the mass-transfer coefficient induced by the mixing process.

The influence of backmixing on the performance of dryers with 1 and 2 transfer units is displayed in Fig.6.24 for a range of values of m and n when $\Lambda = 1$. Note that various pairs of (m,n) values can be fitted to the same effectiveness ratio N_{te}/N_{ti}. One pair, however, may be the most "physically realistic" in modelling the motion of the material and the air.

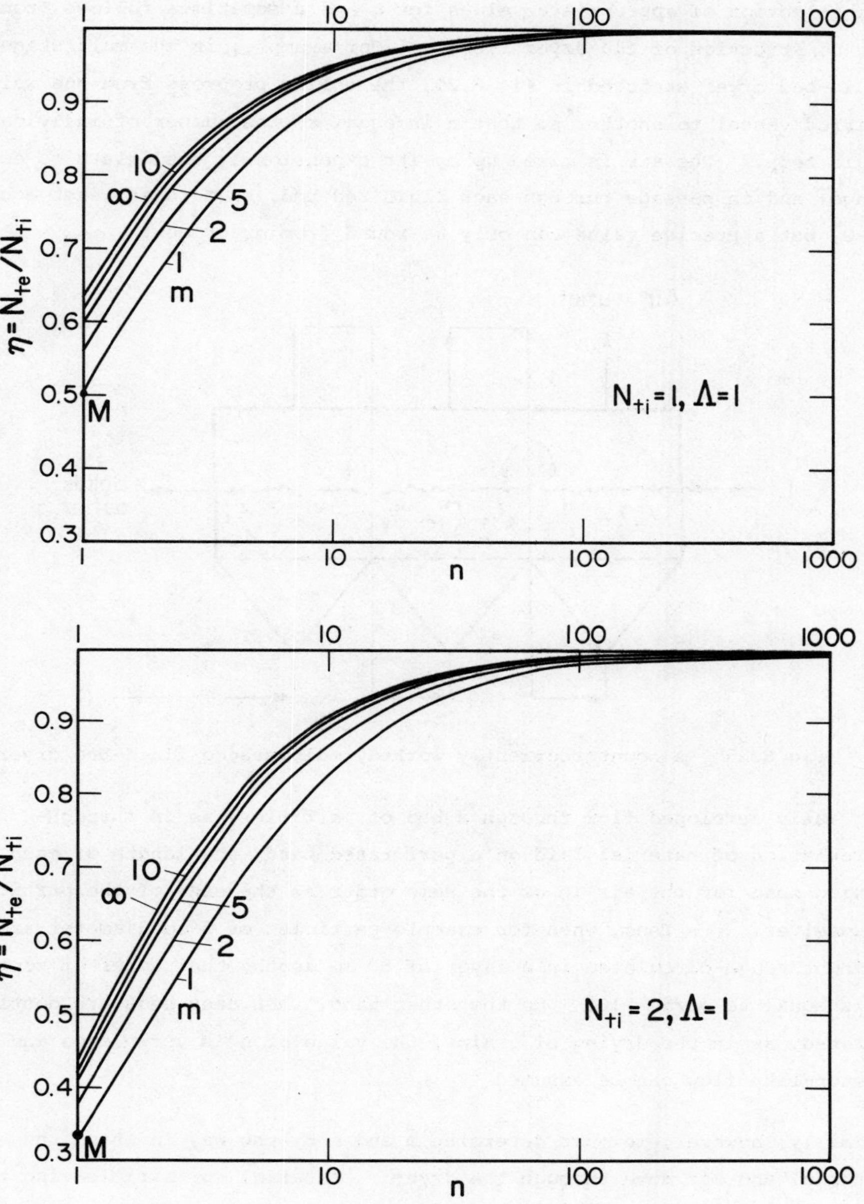

Fig. 6.24 Effectiveness ratio ($\eta = N_{te}/N_{ti}$) for dryers with mn mixing zones in the gas and n mixing zones in the material phase. Point M represents complete mixing.

The selection of appropriate values for m and n sometimes follows from the construction of the dryer itself. For example, in the multistaged fluid-bed dryer sketched in Fig.6.25, the solids progress from one well-stirred vessel to another so that n is equal to the number of individual fluid beds. The air is mixed up by its expansion at the inlets to each stage, and in passage through each fluidized bed, so m is at least equal to 2, but a precise value can only be found from experiment.

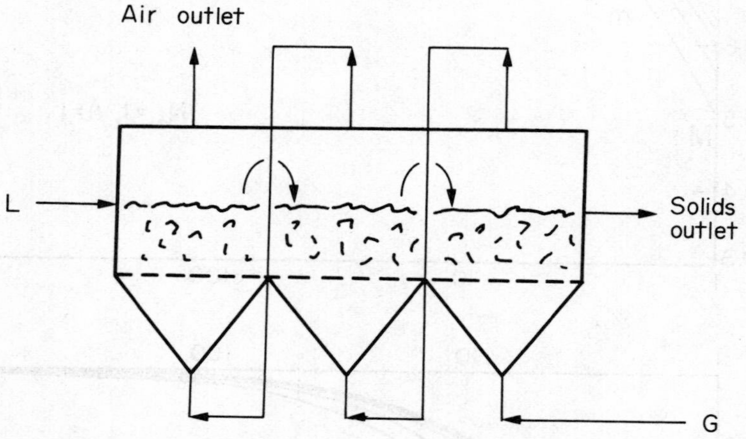

Fig.6.25 A countercurrently worked, multistaged fluid-bed dryer.

For fully developed flow through a bed of particles, as in through-circulation of material laid on a perforated band, the length of each mixing zone for the air is of the same order as the size of the particles themselves.[2] Thus, when for example particles of 5 mm diameter are being through-circulated in a layer of 50 mm depth, the number of zones n is equal to 50/5 = 10. On the other hand, when deep beds are gently aerated, as in the drying of grains, the value of n is very large and piston-like flow can be assumed.

Normally, however, we must determine m and n by the way in which the material and air move through the dryer. Material (or air) leaving the dryer will have spent an infinite range of periods in the dryer, some material bypassing straight to the outlet, other material lingering longer in quiescent parts. This range of dwell-times in the dryer can be represented by the <u>external-age distribution</u> E, the fraction of material emerging in the outlet stream with a residence time between θ and θ + dθ being given by Edθ. For convenience, we compute the duration of this process in terms of relative time, normalized with respect to the so-called space-time in the vessel, that is

$$\text{(space-time)} = \text{(volume of vessel)} / \text{(volumetric flow rate)} \qquad (6.76)$$

It thus follows that

$$\int_0^\infty E d\theta = 1 \qquad (6.77)$$

since all the material must emerge eventually. The E function is measured by injecting a tracer substance into the material stream close to the inlet for a very short time, and measuring its concentration at an appropriate downstream point for some time thereafter. From each concentration reading $\langle C \rangle$, the external-age distribution $E d\theta$ can be deduced, since

$$E(\theta) = \frac{\langle C \rangle}{\int_0^\infty C d\theta} = \frac{\langle C \rangle}{\Sigma \langle C \rangle \delta\theta} \qquad (6.78)$$

A typical outlet-response curve is shown in Fig. 6.26a, and transformed into the E-curve in Fig. 6.26b.

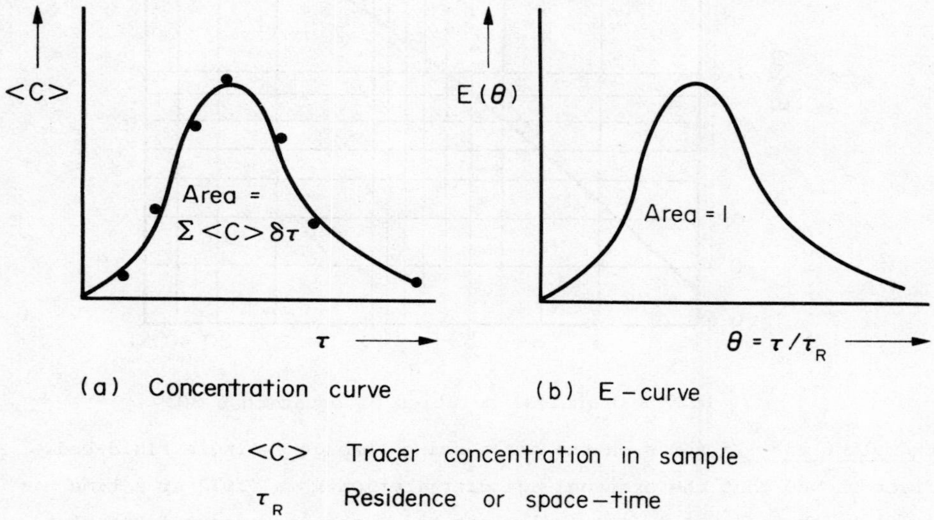

Fig. 6.26 Emergence of tracer in outlet from dryer.

For n well-mixed zones in a vessel, the corresponding external-age distribution can be shown to be[2]

$$E = \frac{n^n \theta^{n-1}}{(n-1)!} \cdot e^{-n\theta} \tag{6.79}$$

Large values of the factorial are tedious to calculate by hand, but many modern electronic calculators now have factorial functions hard-wired. Nevertheless, it is convenient (for $n > 5$) to replace the factorial $(n-1)! = m!$ by the exponential expression $m^m e^{-m} \sqrt{2\pi m}$, which permits non-integer values of n to be used to model the mixing process. To evaluate n, we estimate E at the residence time $\tau = \tau_R$ ($\theta=1$) for which

$$E(1) = \frac{n^n e^{-n}}{(n-1)!} \tag{6.80}$$

Equation 6.80 can be solved iteratively for n . The solution is graphically displayed in Fig. 6.27 .

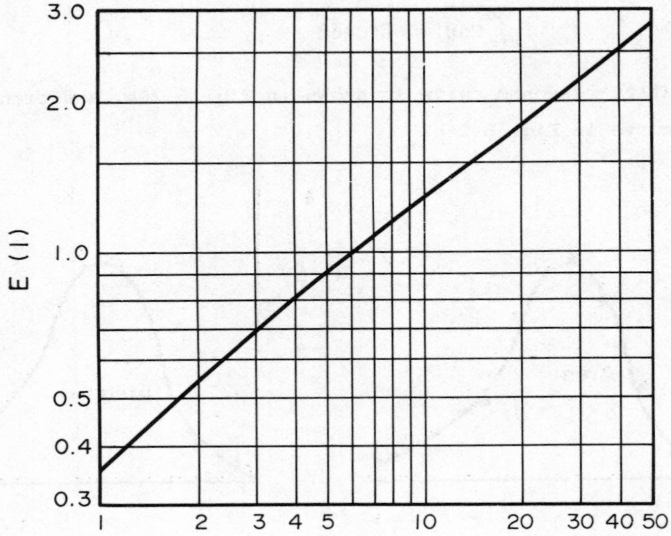

Fig.6.27 Graphical solution of equation 6.80.

<u>Example 6.13</u> If tests on the air passing through a single fluid-bed dryer showed that the external-age distribution E is 0.672 at a time equal to the space-time τ_R , calculate the number of <u>external</u> transfer units when the number of <u>internal</u> transfer units estimated from data on mass-transfer coefficients is 0.75. The gas/solids flow-rate is 20 (dry-matter basis) and the moisture isotherm is approximately given by $Y_S = 0.04X$.

Equation 6.73 applies:-

Now, $\Lambda = \alpha G/L = 0.04 \times 20 = 0.8$

Since there is a single fluid-bed, n = 1.

The number of mixing zones for the air motion is found through equation 6.80. Reference to Fig.6.27 suggests that, at E = 0.672, m ≃ 3, and this value can be checked by direct substitution,

$$\text{i.e.} \quad \frac{3^3 e^{-3}}{2 \times 1} = 0.672$$

Equation 6.73 now becomes

$$N_{te} = \frac{1}{(0.8 - 1)} \ln\left[1 + (0.8-1)\left\{1 - (1+0.75/3)^{-3}\right\}\right]$$

$$= 0.513$$

Thus $\eta = 0.513/0.75 = 0.684$. The number of transfer units, which would be estimated from the moisture-concentration changes, is about two-thirds that estimated from the mass-transfer coefficient.

6.6 Loading

A dryer with few transfer units may be a poor moisture exchanger due to the choice of process conditions or limitations inherent in the equipment or both. It is thus desirable to separate the effects of process and plant in assessing the kinetic performance of the dryer. For a given set of process conditions, a particular dryer will have a fixed capacity for abstracting moisture from the material to be dried. Keey (16) has defined the degree to which a dryer is "loaded" in terms of the ratio of the actual moisture removed to the maximum possible. The quantity is called the loading ratio ε. The magnitude of this ratio depends upon the manner whereby the contacting streams move relative to each other, whether in the same or opposite directions as sketched in Fig.6.28.

For convenience, we label all the streams from the solids-inlet end of the dryer. Thus Y_{GO} is the inlet-air humidity in co-current movement of air and solids, but Y_{GZ} is the same humidity in countercurrent movement as the air enters at the place where the solids are discharged. It is assumed that the solids enter at or above the critical point, and so the surface-air humidity at the inlet for the solids, Y_{SO}, is equal to the wet-bulb humidity there, Y_{WO}. In adiabatic drying, the wet-bulb humidity takes a constant value Y_W throughout the dryer.

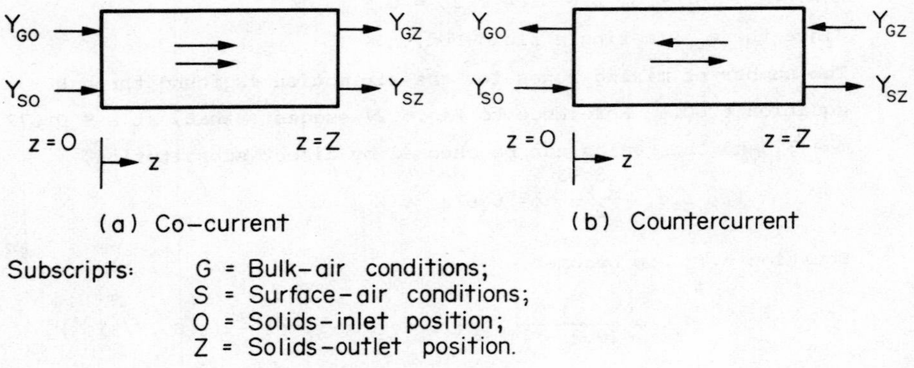

Fig. 6.28 Operational modes for continuous drying.

When the dryer is fully loaded, the humidity of leaving air is equal to the appropriate adiabatic-saturation humidity, which is equal to the surface-air humidity at the air outlet. This quantity is Y_{SZ} in co-current and Y_{SO} in countercurrent drying.

Therefore, the loading ratio for co-current drying is given by

$$\varepsilon = \frac{Y_{GZ} - Y_{GO}}{Y_{SZ} - Y_{GO}} \tag{6.81}$$

and for the countercurrent case

$$\varepsilon = \frac{Y_{GO} - Y_{GZ}}{Y_{SO} - Y_{GZ}} \tag{6.82}$$

To determine the relationship between loading and the number of transfer units, these two modes of operation are considered separately. The derivation is given elsewhere,[4, 16] but the published analysis for the countercurrent case is in error, and a corrected analysis is given below. In the derivation, we need to have an explicit relationship between the relative drying rate f and the distance into the dryer z, but such an expression is not available from theoretical grounds. However, for materials which exhibit a linear falling-rate period, the empirical rate expression

$$N_V/N_{VO} = f \exp \pm [P] \qquad (6.83)$$

where

$$P = [N_1 - (1-f)N_2/\ln f] \qquad (6.84)$$

holds closely.(4) In equations 6.83 and 6.84, N_V is the drying rate at a position in the dryer where the relative drying rate is f, N_{VO} is the rate at the solids inlet, N_1 is the number of internal transfer units in the first ("constant-rate") period and N_2 is the number in the second or falling-rate period of drying. The number of internal transfer units, being defined through equation 6.61, is directly proportional to the considered distance for constant $K_o\phi$. The negative sign in equation 6.83 is taken in co-current flow, and the positive sign in countercurrent flow.

Co-current loading. From the definition of the loading ratio ε (eq.6.81), one finds

$$\frac{1}{(1-\varepsilon)} = \frac{(Y_{SZ} - Y_{GO})}{(Y_{SZ} - Y_{GZ})} \qquad (6.85)$$

The rate expression (eq. 6.83) yields the equation

$$\frac{N_{VO}}{N_V} = \frac{1}{f\exp[-P]} = \frac{(Y_{SO} - Y_{GO})}{(Y_{SZ} - Y_{GZ})} \qquad (6.86)$$

by considering the humidity driving forces. Finally, one obtains the following identities from the definition of the relative drying rate f (eq. 5.6) :

$$f \equiv \frac{(Y_{SZ} - Y_{GZ})}{(Y_{WZ} - Y_{GZ})} \equiv \frac{(Y_{SZ} - Y_{GZ})}{(Y_{SO} - Y_{GZ})} \qquad (6.87)$$

on noting that the solids are assumed to enter at a moisture content at or above the critical point and assuming a uniform wet-bulb temperature throughout the dryer. Equation 6.87 is thus restricted to adiabatic drying conditions. Combination of equations 6.85, 6.86 and 6.87 leads to the expression

$$\frac{1}{(1-\varepsilon)} - \frac{1}{f\exp[-P]} = \frac{Y_{SZ} - Y_{SO}}{Y_{SZ} - Y_{GZ}} = \frac{(f-1)}{f} \qquad (6.88)$$

which after some manipulation produces an explicit equation for the loading ratio:

$$\varepsilon = 1 - \frac{f \exp[-P]}{\{1 - (1-f)\exp[-P]\}} \qquad (6.89)$$

where

$$[P] = [N_1 - (1-f)\, N_2/\ln f] \qquad (6.84)$$

Equation 6.89 notes the twin influences of the material (f value) and of the plant (N_1 and N_2 values) in loading the dryer. If the solids remain wholly within the first period of drying in the equipment, f = 1 and $N_2 = 0$, so equation 6.89 reduces to

$$\varepsilon = 1 - f \exp[-N_1] \tag{6.90}$$

This equation is of the same form as that (eq. 4.72) deduced previously for the humidification of a gas in the presence of a thoroughly sodden material.

Countercurrent loading. The expressions, equivalent to equations 6.85 to 6.87, for this mode of operation, are respectively:

$$(1-\varepsilon) = \frac{(Y_{SO} - Y_{GO})}{(Y_{SO} - Y_{GZ})} \tag{6.91}$$

$$\frac{N_V}{N_{VO}} = f \exp[P] = \frac{(Y_{SZ} - Y_{GZ})}{(Y_{SO} - Y_{GO})} \tag{6.92}$$

$$f = \frac{(Y_{SZ} - Y_{GZ})}{(Y_{WZ} - Y_{GZ})} = \frac{(Y_{SZ} - Y_{GZ})}{(Y_{SO} - Y_{GZ})} \tag{6.87}$$

the latter equation being the same. From these foregoing equations one finds

$$(1-\varepsilon) f \exp[P] = f \tag{6.93}$$

which simplifies to the sought expression for ε

$$\varepsilon = 1 - \exp[-P] \tag{6.94}$$

where [P] is given by equation 6.84 above. When the solids remain entirely in the first drying period, then equation 6.94 reduces to

$$\varepsilon = 1 - \exp[- N_1] \tag{6.95}$$

which is identical to equation 6.90 for co-current working. This result is expected, since the extent of drying ought to be independent of the solids-flow direction when only unbound moisture is being driven off.

In general, the loading ratio will differ according to the mode of working, for a given number of transfer units and outlet-solids moisture content, and thus f. From equations 6.89 and 6.94, one finds

$$\frac{\varepsilon_{ctr}}{\varepsilon_{co}} = \{1 - \exp[-P]\} / \{\frac{1 - \exp[-P]}{1 - (1-f)\exp[-P]}\}$$

$$= 1 - (1-f)\exp[-P] \qquad (6.96)$$

Thus the loading ratio for the countercurrent case (ε_{ctr}) is always smaller once hindered drying begins, reflecting the more effective use of equipment. When drying down to low moisture levels, this difference in performance can be very marked. Plots of the moisture-loading capacity of dryers in Fig. 6.29 illustrates this effect for continuously worked plant.

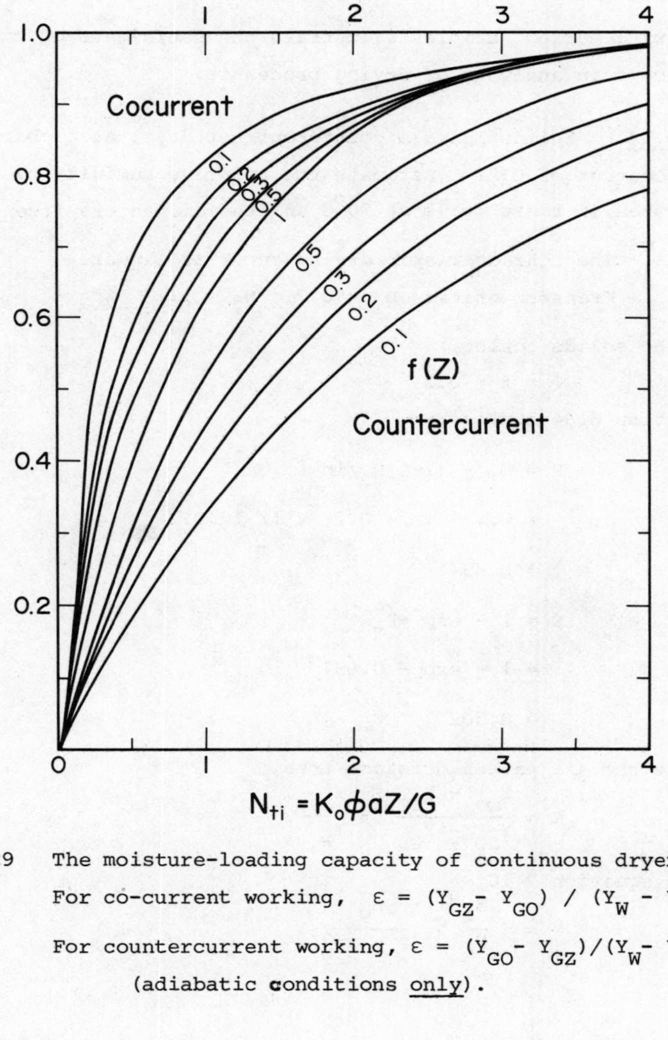

Fig. 6.29 The moisture-loading capacity of continuous dryers.
For có-current working, $\varepsilon = (Y_{GZ} - Y_{GO}) / (Y_W - Y_{GO})$
For countercurrent working, $\varepsilon = (Y_{GO} - Y_{GZ})/(Y_W - Y_{GZ})$.
(adiabatic conditions only).

In Fig. 6.29, the curve for f = 1 gives the loading ratio, ε_1 say, for that portion of the dryer where unhindered drying prevails. The other curves yield the loading ratio, ε_2 say, for hindered-drying conditions. For dryers in which both regimes are found, the overall loading ratio ε is given by

$$\varepsilon = 1 - (1-\varepsilon_1)(1-\varepsilon_2) \qquad (6.97)$$

for countercurrent drying on considering the form of equation 6.95. There is no simple way of compounding loading ratios for co-current conditions, but equation 6.97 may be used provided the hindered-drying zone is not too extensive.

The following worked examples illustrate the usefulness of the loading-ratio concept in analysis of drying processes.

Example 6.14. Solids leave a countercurrent dryer at a characteristic moisture content of 0.2. Estimate the relative humidity of the exhaust gases if these leave at 70°C and the air enters free of moisture.

Data: The characteristic drying curve is linear.

Transfer units: $N_1 = 0.2$; $N_2 = 1$.

At the solids outlet,

$$\Phi = f = 0.2$$

Equation 6.94 applies:-

$$P = N_1 - (1-f)N_2/\ln f$$

$$= 0.2 - (1 - 0.2) \times 1/ \ln 0.2$$

$$= 0.697$$

$$\varepsilon = 1 - \exp[-P]$$

$$= 1 - \exp[- 0.697]$$

$$= 0.502$$

Since the air enters moisture-free,

$$\varepsilon = \frac{Y_{GO} - Y_{GZ}}{Y_{SO} - Y_{GZ}} = \frac{Y_{GO}}{Y_W}$$

From equation 2.10,

$$Y = \left(\frac{M_W}{M_G}\right) \frac{\psi p_W^o}{(P - \psi p_W^o)}$$

it follows that

$$\varepsilon = \frac{Y_{GO}}{Y_W} = \psi \cdot \frac{(1 - p^o/P)}{(1 - \psi p^o/P)} \qquad \text{I}$$

Let the pressure ratio $p^o/P = \beta$, then by re-arrangement of equation I, we get

$$\psi = \frac{\varepsilon}{1 - \beta(1-\varepsilon)} \qquad \text{II}$$

At 70°C, $\beta = p_W^o/P = 31.16 / 101.3 = 0.308$.

Thus from equation II

$$\psi = \frac{0.502}{1 - 0.308(1 - 0.502)} = 0.593$$

The relative humidity of the outlet gases is thus 59 per cent.

Suppose this drying operation is to be accomplished by the co-current movement of air and the material. For the same performance, that is the same outlet-solids moisture content and transfer units in the dryer, equation 6.96 predicts that the loading ratio would have to be 66 per cent greater. Or, if the air conditions were not to be changed, then the co-current dryer would have to be more extensive. Methods of calculating the change in length needed will be given in Chapter 7. Co-current drying uses plant ineffectively.

Despite this penalty, there are times when co-current drying is preferred, especially when handling awkward materials. For instance, refined lactose is processed in rotary dryers with parallel movement of the solids and the hot air.[27] As the material becomes less tolerant of being heated as it dries out, so the air temperature falls to safer levels. Moreover, should the feed through a process upset enter wetter than normal, there is some degree of insurance with co-current working. The overwet crystals stick together and the lumps of material are shifted more slowly than individual particles along the dryer so that the extended dwell-time enables more moisture to be driven off. In counter-current operation, a sudden process upset could cause enough initial evaporation to humidify the outlet air beyond the dewpoint, to induce local condensation which would make the feed wetter still. The operation becomes unstable rather than self-regulating.

The minimum air-rate in countercurrent operation, when the outlet air is just saturated, corresponds to a loading ratio of 1. Thus the loading ratio in any instance gives a measure of the amount of air above this minimum that is being used. The following worked example demonstrates such a calculation and shows how one can consider the case when the solids enter the dryer below their critical moisture content.

Example 6.15. Material enters a progressive, countercurrently worked textile dryer at a moisture content of 0.45 kg kg^{-1} and the stuff leaves at a moisture content of 0.15 kg kg^{-1}. The critical moisture content is 0.50 kg kg^{-1} and the equilibrium-moisture content is 0.12 kg kg^{-1}. If there are 2.5 internal transfer units in the dryer, estimate by how much the air-rate exceeds the minimum value and thus determine what change in moisture content of the material leaving the dryer accompanies a ± 10 per cent fluctuation in feed rate.

Data: The characteristic drying curve is given by

$$f = \Phi^{1.2}$$

Fig. 6.30 Progressive countercurrent dryer.

(a) Since $X_o < X_{cr}$ the stuff is wholly within the falling-rate period of drying. We consider three drying operations:

1. A fictitious dryer accepting material at the critical point and drying it to the given outlet moisture content.
2. Another fictitious dryer accepting material at the critical point and drying it to the given inlet moisture content.
3. The drying operation which represents the difference between the foregoing cases, being the actual one.

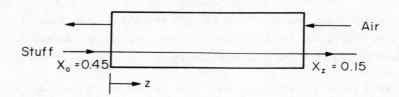

At the solids outlet:

$$\Phi = \frac{(0.15 - 0.12)}{(0.50 - 0.12)} = 0.0789$$

$$f = \Phi^{1.2} = (0.0789)^{1.2} = 0.0475$$

At the solids inlet:

$$\Phi = \frac{(0.45 - 0.12)}{(0.50 - 0.12)} = 0.868$$

$$f = \Phi^{1.2} = (0.868)^{1.2} = 0.844$$

For the hypothetical extension of the dryer to an inlet moisture content of $\Phi = 1$, it is assumed that

$$N_{ti} \;\alpha\; \ln(\Phi_o/\Phi)$$

- on the basis of the exponential decay of drying rates with distance. Thus for this extension

$$N_{ti} = 2.5 \times \frac{\ln(1/0.868)}{\ln(0.868/0.0789)}$$

$$= 0.148$$

Therefore, for each of the drying operations we have:

case	Φ_{in}	Φ_{out}	N_{ti}
1	1	0.0789	2.5 + 0.148 = 2.648
2	1	0.868	0.148
3	0.868	0.0789	2.5

Let the loading ratio for each of these cases be ε_1 , ε_2 and ε_3 respectively. Then, by analogy with the development of equation 6.97,

$$(1-\varepsilon_1) = (1-\varepsilon_2)(1-\varepsilon_3)$$

$$\therefore \; \varepsilon_3 = 1 - (1-\varepsilon_1) \;/\; (1-\varepsilon_2)$$

Now from equation 6.94

$$\varepsilon_1 = 1 - \exp + [(1-f)N_2/\ln f]$$

$$= 1 - \exp + [(1 - 0.0475) \times 2.648/\ln 0.0475]$$

$$= 0.563$$

$$\varepsilon_2 = 1 - \exp + [(1 - 0.844) \times 0.148/\ln 0.844)]$$
$$= 0.127$$

Thus $\varepsilon_3 = 1 - (1 - 0.563) / (1 - 0.127)$

$$= 0.499 \text{ say } 0.5$$

The minimum air-rate occurs at a loading ratio of 1, so the actual air-rate is 1/0.5 = 2 times the minimum.

(b) For a 10 per cent fluctuation in feed-rate, all loading ratios will vary by ± 10%. We may thus calculate the outlet moisture content for the values of ε_1 at the extrema of this range.

For $\varepsilon_1 = 1.1 \times 0.563 = 0.619$,
from equation 6.94,

$$\frac{(1-f)}{\ln f} = \frac{\ln(1-\varepsilon_1)}{N_2} = \frac{\ln(1 - 0.619)}{2.648} = -0.3647$$

The equation $g(f) = (1-f)/\ln f = -0.3647$ can be solved interatively.

f	g
0.08	- 0.3642
0.081	- 0.3657
0.0803	- 0.3647 close

$$\Phi = f^{1/1.2} = (0.0803)^{1/1.2} = 0.1223$$

$$X = 0.1223 \times (0.50 - 0.12) + 0.12 = 0.166 \text{ kg kg}^{-1}$$

In a similar way for $\varepsilon_1 = (0.9 \times 0.563) = 0.5067$ and
$g(f) = \ln(1 - 0.5067) / 2.648 = -0.2669$

f	g
0.03	- 0.2766
0.025	- 0.2643
0.0260	- 0.2669 close

$$\Phi = f^{1/1.2} = (0.0260)^{1/1.2} = 0.0478$$

$$X = 0.0478 \times (0.50 - 0.12) + 0.12 = 0.138 \text{ kg kg}^{-1}$$

Thus the moisture content of the dried material varies between 0.138 and 0.166 kg kg^{-1}.

Batch drying. In batch drying, the loading ratio falls from a maximum value when the material is uniformly wet above the critical point to a progressively smaller value as the stuff dries out. This maximum value of ε is given by equation 6.90 (or 6.95), namely

$$\varepsilon_{max} = 1 - \exp(- N_1) \quad (6.98)$$

where N_1 is the number of internal transfer units. The number of transfer units for batch dryers such as tray dryers and timber kilns is about 1, but is much larger for deep, slowly through-circulated beds such as grain bins. Under these circumstances, ε_{max} approaches 1 and a distinct drying zone is witnessed within the bed, as described in Chapter 8.

REFERENCES

1. Becker, F.C., C. Janholm and R.E. Raaschou, "Om Tørringprocessen" Danmarks Naturvidensk. Samfund No.9, København (1925).
2. Beek, W.J. and K.M.K. Muttzall, "Transport Phenomena", p.125-140, Wiley Interscience, London (1975).
3a. Cammerer, J.S., "Die Wärme und Kälteschutz in der Industrie", 3/e, Springer, Berlin/Göttingen/Heidelberg (1952).
4. Catherall, N.F., The Process Design of Continuous Dryers, B.E. Report, Univ. Canterbury, N.Z. (1970).
5. Chilton, J.H. and A.P. Colburn, Mass transfer (absorption) coefficients. Prediction from data on heat transfer and fluid friction, Ind.Eng.Chem., 26, 183-7 (1934).
6. Chilton, J.H. and A.P. Colburn, Distillation and absorption in packed columns, Ind.Eng.Chem., 27, 255 (1935).
7. Dankwearts, P.V. and C. Anolick, Mass Tranfer from a Grid Packing in an Airstream, Trans.Instn.Chem.Engrs., 40, 203-213 (1962).
8. Eckert, E., Messung der Gesamtstrahlung von Wasserdampf, VDI Forschungsheft B 8,387 (1937).
9. Gardon, R. and R.C. Akfirat, Heat-transfer characteristics of impinging two-dimensional airjets, Trans. ASME, J.Heat Transfer, 88, 101-8 (1966).
10. Glover, H.C. and A.A.H. Moss, A method of design of continuous through-circulation dryers, Trans.Instn.Chem.Engrs., 35, 208-218 (1957).
11. Herminge, L., Heat Transfer in Porous Bodies at Various Temperatures and Moisture Contents, TAPPI, 44, 570-5 (1961).
12a. Hsu, S.T., "Engineering Heat Transfer", Von Nostrand, Princeton N.J. (1963).
13. Jakob, M., "Heat Transfer", 1, p.94, Wiley, New York (1949).

14. Keey, R.B., "Drying Principles and Practice", p.99 f, Pergamon, Oxford (1972).
15. Keey, R.B., ibid. p.111 f.
16. Keey, R.B., ibid. p.249 f.
17. Keey, R.B., ibid. p.344.
18. Kerkhof, P.J.A.M. and H.A.C. Thijssen, A simple model describing the effect of axial mixing on countercurrent mass exchange, Chem.Eng. Sci., 29, 1427-1434 (1974).
19. Kerscher, E., G. Böhner and A. Schneider, Beitrag zur Wärmeübertragung bei der Furniertrocknung mit Düsenbelüftung, Holz als Roh- und Werkstoff, 26, 19-28 (1968).
20. Kmiec, A. Simultaneous Heat and Mass Transfer in Spouted Beds, Can.J.Chem.Eng., 53, 18-24 (1975).
21. Krischer, O., "Die wissenschaftlichen Grundlagen der Trocknungstechnik", 2/e, p.131 f, Springer, Berlin/Göttingen/Heidelberg (1963).
22. Krischer, O. and G.Loos, Wärme und Stoffaustausch bei erzwungener Strömung an Körpern verschiedener Form, Chem.Ing.Techn., 30, 31-9 (195
23. Kröll, K., "Trockner und Trocknungsverfahren", p.411-5, Springer, Berlin/Göttingen/Heidelberg (1959).
24. Kröll, K., ibid. p.540.
25. Luikov, A.V., "Heat and Mass Transfer in Capillary-Porous Bodies", Chapter 4, Pergamon, Oxford (1966).
26. Mayhew, Y.R. and G.F.C. Rogers, "Thermodynamic and Transport Properties of Fluids", 2/e, Blackwell, Oxford (1968).
27. McGimpsey, J.R. in "Drying Principles and Practice in the Process Industries", p.109, Caxton, Christchurch, N.Z. (1966).
28. Miller, W.R., Mass Transfer within Arrays, B.E. Report, Univ.Canterbury, N.Z. (1973).
29. Mosberger, E., Über den Wärme- und Stoffaustausch zwischen Partikel und Luft in Wirbelschichten, sowie über deren Ausdehnungsverhalten, Dissertation J.H. Darmstadt (1964).
30. Nonhebel, G. and A.A.H. Moss, "Drying of Solids in the Chemical Industry", p.23, Butterworths, London (1971).
31. Nonhebel, G. and A.A.H. Moss, ibid. p.168.
32. Nonhebel, G. and A.A.H. Moss, ibid. p.289.
33. Shepherd, C.B., C. Haddock and R.C. Brewer, Drying Materials in Trays. Evaporation of Surface Moisture, Ind.Eng.Chem., 30, 388-397 (1938).
34. Sherwood, T.K., The Drying of Solids II, Ind.Eng.Chem., 21, 976-984 (1929).
35. Sherwood, T.K., R.L. Pigford and C.R. Wilke, "Mass Transfer, p.242, McGraw Hill, New York (1975).
36. Sinning, B., "Trocknung durch Infrarotstrahlung von Strahlwänden", Verfahrenstechn., 7, 372-5 (1973).
37. Slessor, C.G.M. and D. Cleland, Surface Evaporation by Forced Convection, I. Simultaneous Heat and Mass Transfer, Int.J.Heat Mass Transfer, 5, 735-749 (1962).
38. Spiers, H.H. ed. "Technical Data on Fuel", 6/e, p.59, Brit.National Committee World Power Conf., London (1961).
39. Sørensen, A., Mass transfer coefficients on truncated slabs, Chem.Eng. Sci., 24, 1445-1460 (1969).
40. Thijssen, H.A.C., "Stofoverdrachtsprocessen", Diktaat No.6.605, p.8.1, T.H. Eindhoven (1972).
41. Toei, R., M. Okazaki, M. Kimura and H. Ueda, Heat Transfer Coefficients on the Superheated Steam Drying of Porous Solids, Kagaku Kogaku, 30, 947-9 (1966).

42. Treybal, R. E., "Mass Transfer Operations", 2/e p. 388, McGraw Hill, New York (1968).
43. Treybal, R. E., ibid p. 250-3.
44. Weast, R. C. (ed.) "Handbook of Physics and Chemistry," 48/e, E 63, Chemical Rubber Co., Cleveland, Ohio 1967/8.
45. Wilke, C. R. and O. A. Hougen, Mass Transfer in Flow of Gases Through Granular Solids Extended to Low Modified Reynolds numbers, Trans. AIChE, 41, 445-51 (1945).

PLATE 6. Tobacco slurry is fed onto a stainless-steel belt to be dried and peeled off as a 0.08 mm thin sheet, 1 m wide. View from discharge end. [Sandvik (UK) Ltd., with permission.]

Chapter 7

CONTINUOUS DRYING

7.1 State Paths in Progressive Dryers

Any drying process will be normally run as a continuous operation, unless the stuff to be dried is small in quantity or bulky in size. A continuously worked dryer will need less labour, fuel and floor space, and will discharge a more uniformly dried product than a batch dryer of similar capacity. As noted in Section 6.5, the solids may move progressively through the dryer, or they may be mixed up in the drying process. In the former case, the driving force for drying is held everywhere at its greatest possible value for the given set of process conditions. The variation in magnitude of the driving force or drying potential is given by the separation of the state paths for the air and the solids in their progress through the dryer. These state paths can be conveniently displayed on an enthalpy-humidity chart.[9]

We would expect the drying medium to cool itself as it picks up moisture, while the temperature of the solids to rise on drying. Nevertheless, in some cases, it may be useful to work the dryer with a different air-temperature profile, particularly if the thermal sensitivity of the solids varies with moisture content. It is thus worthwhile analysing the more general, non-adiabatic case.

Enthalpy gradient. Let us consider the moisture and enthalpy change over an infinitesimally short zone of length dz in a progressive dryer. The change in bulk-air humidity dY_G may be related to the evaporation from an incremental surface of extent $d(a_oSz)$, where a_o is the exposed surface per unit volume of air-space which has a cross-sectional area S (as shown in Fig. 7.1).

A mass balance thus yields

$$\pm\ (G_oS)dY_G - N_V d(a_oSz) = 0 \qquad (7.1)$$

where the positive sign is taken in co-current flow of solids and air,

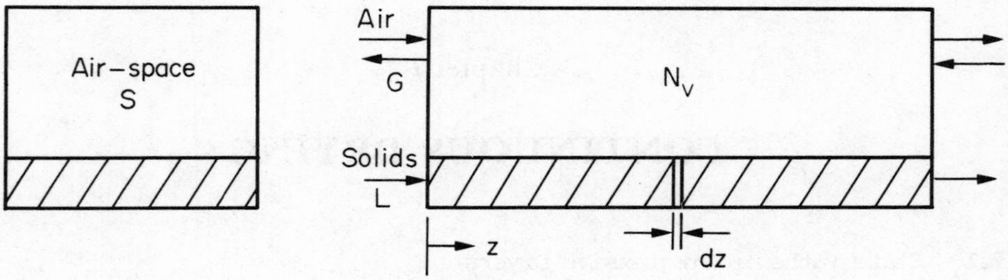

Fig. 7.1 Schematic view of progressive dryer

and the negative sign in countercurrent flow. (The negative sign arises in the counterflow of solids and air, since the bulk-air humidity decreases with distance from the solids-inlet position). On removing the common parameter S from equation 7.1, one gets

$$\pm G_o dY_G - N_V d(a_o z) = 0 \tag{7.2}$$

In equation 7.2, the drying rate N_V over a unit surface exposed to the airstream is given by equation 5.5, that is

$$N_V = f\ K_o \phi (Y_W - Y_G) \tag{5.5}$$

Equation 5.5 can also be written in terms of the surface-air humidity Y_S , rather than the saturation humidity at the wet-bulb humidity Y_W :

$$N_V = K_o \phi (Y_S - Y_G) \tag{7.3}$$

Over the considered zone, there is a total heat flux to the surface of q_T , say, the majority of which is dissipated through the evaporation of moisture. Suppose the dryer is fitted internally with heaters, which have a uniform output over the whole length of the dryer. Let this heat flux be q_H, measured in terms of unit area of the shell of the air-space in the dryer. There will be also a heat loss q_{LG} to the surroundings from this area. Let s be the ratio of the area of air-space shell to that of the solids exposed to the airstream. These various heat fluxes are sketched in Fig. 7.2.

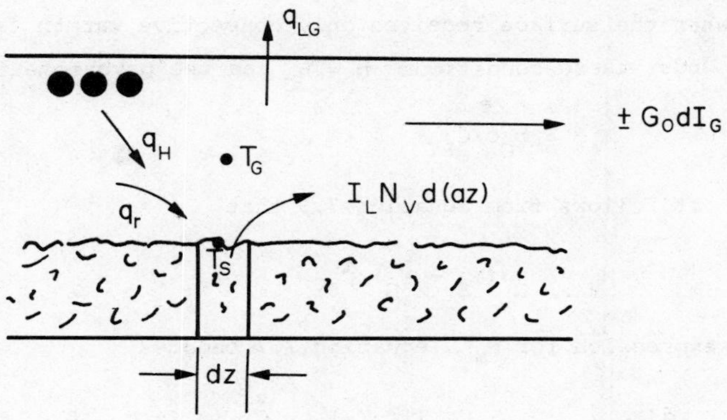

Fig. 7.2 Heat flows over an infinitesimally short zone in a progressive dryer.

By reference to Fig.7.2, a heat balance yields

$$\pm G_o dI_G = \{I_L N_V - q_T + (q_H - q_{LG})s\} \; d(a_o z) \qquad (7.4)$$

where I_L is the added enthalpy of the vaporized moisture. This enthalpy is equal to H_{LW}, the enthalpy of saturated liquid moisture at the surface temperature T_S. Again, the positive sign is taken in co-current flow, and the negative sign in the countercurrent movement of solids and air, for the reasons adduced before. If the very small Ackermann correction is ignored, then the total heat flux is the vector sum of the heat transferred to the surface $h(T_G - T_S)$ and the enthalpy convected away in the evaporation $\Delta H_{VS} N_V$. These considerations lead to the expression

$$\pm G_o dI_G = \{H_{LW} N_V - h(T_G - T_S) + \Delta H_{VS} N_V + (q_H - q_{LG})s\} d(a_o z) \qquad (7.5)$$

Since the sum of H_{LW} and ΔH_{VS} is the enthalpy of saturated moisture vapour at temperature T_S, equation 7.5 simplifies to

$$\pm G_o dI_G = \{H_{GW} N_V - h(T_G - T_S) + (q_H - q_{LG})s\} \; d(a_o z) \qquad (7.6)$$

Division of equation 7.6 by the mass-balance expression (eq. 7.2) yields the enthalpy gradient with humidity:

$$\frac{dI_G}{dY_G} = H_{GW} - \frac{h(T_G - T_S)}{N_V} + \frac{(q_H - q_{LG})s}{N_V} \qquad (7.7)$$

This equation may be used to plot the state path of the air on the enthalpy-humidity diagram to illustrate the way it is humidified in the dryer.

Equation 7.7 can, however, be put into a more convenient form for plotting when the surface receives only convective warmth from the heater. Under these conditions, $h = h_C$ and the psychrometric identity

$$K_o \phi \equiv h_C \sigma / \bar{C}_{PY} \tag{6.10}$$

applies. It follows from equation 7.3 that

$$N_V = h_C \sigma (Y_S - Y_G) / \bar{C}_{PY} \tag{7.8}$$

With this expression for N_V, equation 7.7 becomes

$$\frac{dI_G}{dY_G} = H_{GW} - \frac{\bar{C}_{PY}(T_G - T_S)}{\sigma(Y_S - Y_G)} + \frac{\bar{C}_{PY}(q_H - q_{LG})s}{h_C \sigma (Y_S - Y_G)} \tag{7.9}$$

It is also convenient to introduce the ratios:

1. heat input to convective heat

$$\eta_H = q_H s / h_C (T_G - T_S) \tag{7.10}$$

2. heat loss to convective heat

$$\eta_{LG} = q_{LG} s / h_C (T_G - T_S) \tag{7.11}$$

With these ratios, equation 7.9 transforms into the working relationship:

$$\frac{dI_G}{dY_G} = H_{GW} - \frac{\bar{C}_{PY}(T_G - T_S)}{\sigma(Y_S - Y_G)} [1 - \eta_H + \eta_{LG}] \tag{7.12}$$

Two simple forms of equation 7.12 arise for the particular cases of adiabatic and isothermal operation respectively.

In adiabatic working, $q_T = 0$ and $(q_H - q_{LG}) = 0$; thus equation 7.4 reduces to

$$\pm G_o dI_G = I_L N_V d(a_o z) \tag{7.13}$$

Division of equation 7.13 by equation 7.2, as before, yields the enthalpy gradient

$$\frac{dI_G}{dY_G} = I_L = H_{LW} \tag{7.14}$$

Equation 7.14 gives the gradient of the adiabatic-saturation path, derived earlier as equation 4.5, which intersects the moisture-saturation curve at a temperature T_S.

In <u>isothermal working</u>, $h(T_G - T_S) = (q_H - q_{LG})s$, and equation 7.4 now becomes

$$\pm G_o dI_G = H_{GW} N_V d(a_o z) \tag{7.15}$$

and the enthalpy gradient

$$\frac{dI_G}{dY_G} = H_{GW} \tag{7.16}$$

If account is taken of the small amount of sensible heat needed to warm up the moisture vapour from T_S to T_G , then H_{GW} is read as the moisture-vapour enthalpy at the temperature T_G. Equation 7.16 then represents the gradient of the isotherm for the bulk-air temperature T_G which is held constant in this case.

Fig. 7.3 shows the orientation of the various humidification paths on an enthalpy-humidity chart. Pathways of greater slope than the adiabatic line represent heat added to the system, whereas pathways of lesser slope reflect a nett heat loss.

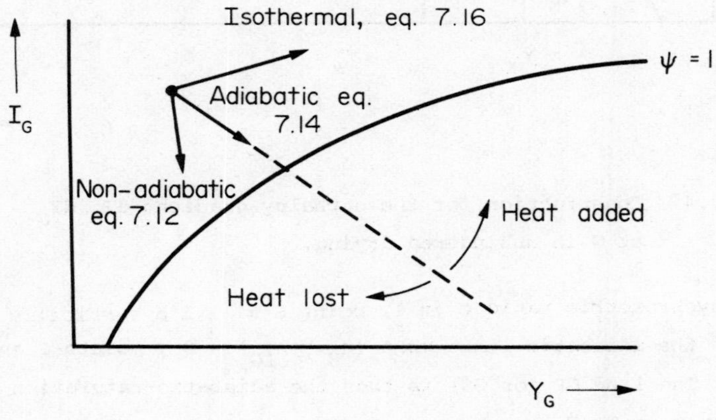

Fig. 7.3 Air-humidification pathways

The graphical construction for the enthalpy gradient dI_G/dY_G is shown in Fig. 7.4 for the case of unhindered drying when f = 1 and the surface is at the wet-bulb temperature. Point G represents the dry-bulb conditions and P the wet-bulb conditions on the enthalpy-humidity chart, while point S' lies on the wet-bulb isotherm such that the horizontal separation between R and S' is $\sigma(Y_S - Y_G)$. A vertical line segment from G through R is cut off at a point Q at a length $\bar{C}_{PY}(T_G-T_S)(1-\eta_H+\eta_{LG})$. From the point Q a line parallel to the wet-bulb isotherm is drawn to a point S which also is a horizontal distance $\sigma(Y_S - Y_G)$ away from the vertical line GRQ. Since QS has a slope of $+ H_{GW}$, it follows from equation 7.12 that the slope of the line GS gives the required enthalpy gradient at point G.

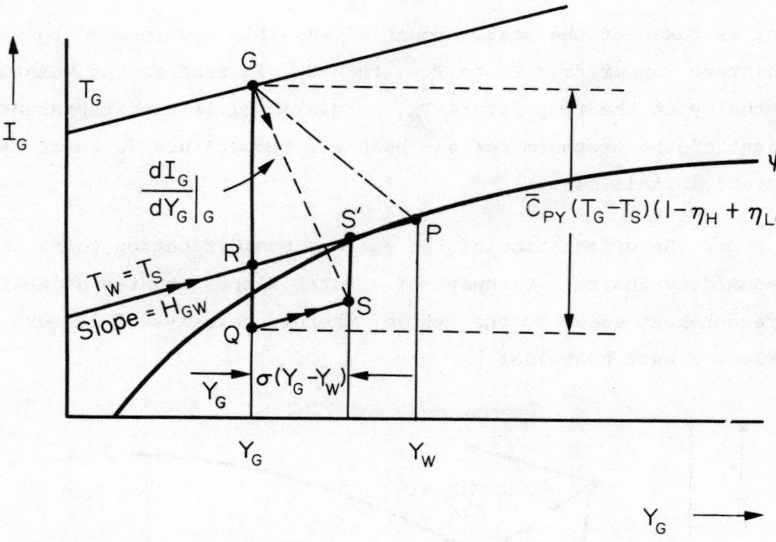

Fig. 7.4 Construction for the enthalpy gradient dI_G/dY_G at G in unhindered drying.

When the psychrometric ratio σ is 1, point S will lie vertically under P. Further, in the adiabatic limit when $(\eta_H - \eta_{LG}) = 0$, points P and S coincide. The line GP (or GS) is then the adiabatic-saturation line for a temperature $T_S(T_W)$.

In hindered drying, we find the surface humidity from the identity

$$(Y_S - Y_G) \equiv f(Y_W - Y_G) \tag{7.17}$$

The corresponding construction is shown in Fig. 7.5. Point S' is situated a horizontal distance $f(Y_W - Y_G)$ or $(Y_S - Y_G)$ away from the line PRQ, while S is a distance $\sigma(Y_S - Y_G)$ away. Also, the line GRQ has a length $\bar{C}_{PY}(T_G - T_S)(1 - \eta_H + \eta_{LG})$. From this diagram, we see that, as drying proceeds into the falling-rate region and the relative drying rate f correspondingly dwindles, point S' shifts both higher and leftwards on the chart. This curling back of the surface-state point has important consequences for the ease of drying which will be discussed later.

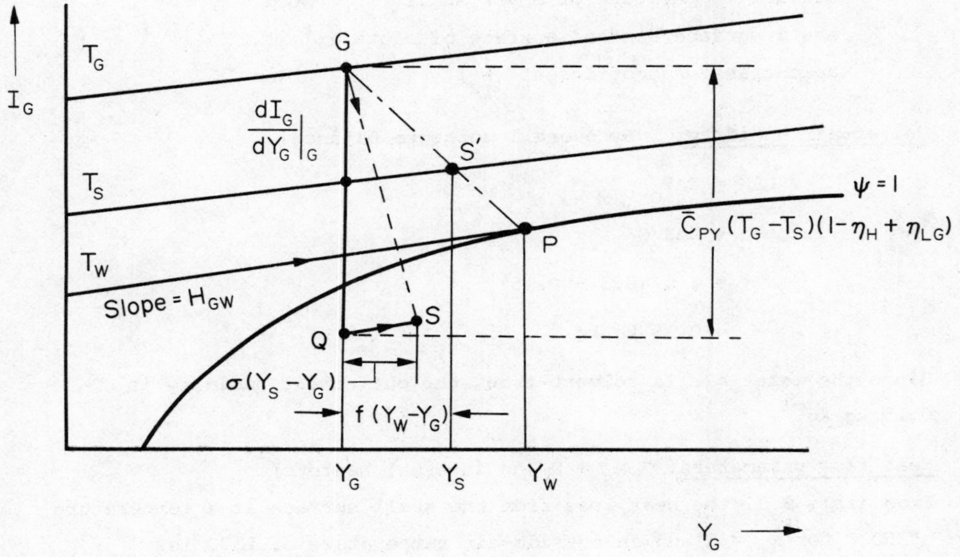

Fig. 7.5 Construction for the enthalpy gradient dI_G/dY_G at G in hindered drying.

Although equation 7.12 has been derived for evenly spread heat flows, the expression may be used when the heat fluxes vary with distance. Under these circumstances, the parameters η_H and η_{LG} become limiting "point" values and equation 7.12 is evaluated numerically by a marching solution from the inlet-air conditions. The following worked example illustrates the numerical solution for unhindered drying in the absence of internal heaters.

Example 7.1. A cloth soaked in carbon tetrachloride is being dried in a co-current dryer. Fresh, solvent-free air enters at 90°C with cloth of moisture content 0.3 kg kg^{-1}. Estimate the enthalpy and humidity of the air when the moisture content of the cloth has been reduced by one half.

Data:

air/cloth ratio (dry-substance basis)	1
critical moisture content	0.11 kg kg^{-1}
psychrometric coefficient (σ)	0.57
wet-bulb temperature	20°C
saturation humidity at 20°C	0.7 kg kg^{-1}
surface temperature of dryer shell	50°C
shell surface/exposed surface of cloth (s)	1.2
heat-transfer coefficient (h_c)	20 W m^{-2}K^{-1}

Outlet-air humidity. By overall moisture balance:

$$L\Delta X = G\Delta Y_G$$

$$\therefore \Delta Y_G = L\Delta X/G$$

$$= 1 \times (0.3 - 0.15)$$

$$= 0.15 \text{ kg kg}^{-1}$$

Since the inlet air is solvent-free, the outlet air humidity is 0.15 kg kg^{-1}.

Heat flow parameters. $\eta_H = 0$ (no internal heaters)

From Table 3.1, the heat loss from the shell surface at a temperature of 50°C for $\varepsilon_S \simeq 1$ and an outside-air temperature of 15°C is 392 W m^{2}K^{-1}. Now, from eq. 7.11

$$\eta_{LG} = q_{LG}\, s/h_C(T_G - T_S)$$

$$= 392 \times 1.2 / 20 \times (90 - 20)$$

$$= 0.336$$

This parameter changes as the air temperature within the dryer falls, but the effect is small and can be ignored in this case.

Outlet-air enthalpy. H_{GW} is the enthalpy of moisture vapour at 90°C. See Table 2.5 for data. i.e. $H_{GW} = \bar{C}_{PV}\theta + \Delta H_{VS} = (0.552 \times 90) + 217.8$

$$= 267.5 \text{ kJ kg}^{-1}$$

At the inlet, $\bar{C}_{PY}$ is the specific heat of dry air, as the humidity of the fresh air is zero, i.e. $\bar{C}_{PY} = \bar{C}_{PG} = 1.011$ kJ kg^{-1}K^{-1}.

From equation 7.12

$$\frac{dI_G}{dY_G} = 267.5 - \frac{1.011(90 - 20)}{1 \times (0.7 - 0)}[1 + 0.336]$$

$$= 132.4 \text{ kJ kg}^{-1}$$

$$\therefore \Delta I_G = 132.4 \times 0.15 = 19.9 \text{ kJ kg}^{-1}$$

The enthalpy of the dry inlet air H_{GO} is $C_{PG}\theta$

$$\therefore H_{GO} = 1.011 \times 90 = 91.0 \text{ kJ kg}^{-1}$$

Thus the outlet-air enthalpy is $91.0 + 19.9 = 110.9 \text{ kJ kg}^{-1}$

Surface-temperature gradient. Let us now consider the enthalpy changes over an infinitesimally short zone in a progressive dryer, as sketched in Fig. 7.6.

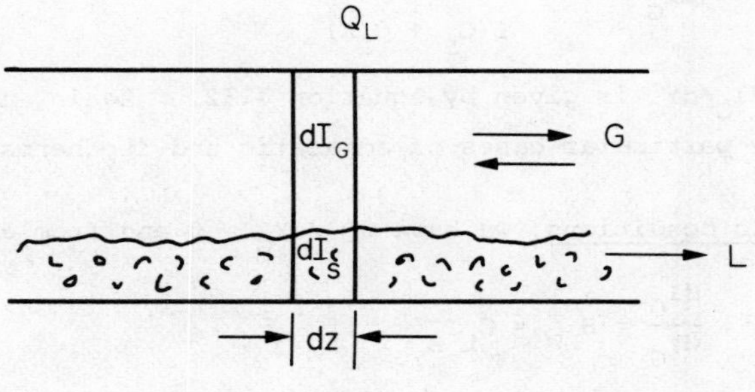

Fig. 7.6 Enthalpy changes over an infinitesimally short zone in a progressive dryer.

An enthalpy balance over this zone yields

$$\pm\, GdI_G - LdI_S - dQ_L = 0 \qquad (7.18)$$

where Q_L is the total heat loss from the dryer. The positive sign in the first term is taken when the solids and air move in the same direction, and the negative sign when countercurrent movement prevails. The negative sign appears in the second term, since the enthalpy of the solids diminish as moisture evaporates. The humid enthalpy of the solids is given by equation 2.36, which becomes

$$I_S = [C_S + C_L X\,]T_S \qquad (7.19)$$

when the very small enthalpy of wetting is ignored. On substituting equation 7.19 into 7.18, one finds that

$$\pm\, GdI_G - LC_S dT_S - LC_L d(XT_S) - dQ_L = 0 \qquad (7.20)$$

which, on re-arrangement, leads to the expression

$$\pm\, G dI_G - L(C_S + C_L X) dT_S - LC_L T_S dX - dQ_L = 0 \tag{7.21}$$

On noting that $LdX = \pm\, GdY_G$ (from a moisture balance over the considered zone), equation 7.21 may be rewritten as

$$\pm\, G dI_G - L(C_S + C_L X) dT_S \mp GC_L T_S dY_G - dQ_L = 0 \tag{7.22}$$

which itself can be re-cast to yield an expression for the surface-temperature gradient of humidity dT_S/dY_G :

$$\frac{dT_S}{dY_G} = \frac{\pm\, G\left[\dfrac{dI_G}{dY_G} - C_L T_S - \dfrac{1}{G}\dfrac{dQ_L}{dY_G}\right]}{L(C_S + C_L X)} \tag{7.23}$$

The gradient dI_G/dY_G is given by equation 7.12. Again, it is illustrative to look at the particular cases of adiabatic and isothermal operation.

Under adiabatic conditions, we have $dQ_L/dY_G = 0$ and from equation 7.14

$$\frac{dI_G}{dY_G} = H_{LW} = C_L T_S \tag{7.24}$$

so that equation 7.23 reduces to

$$\frac{dT_S}{dY_G} = 0 \tag{7.25}$$

The surface temperature remains constant throughout the drying process. It also follows that, if the drying in the "constant-rate" period is maintained at a constant temperature under adiabatic conditions, the temperature cannot remain so any longer once the drying enters the falling-rate period as the heat demand becomes less.

When the drying is conducted under isothermal conditions in a well-insulated chamber, we have $dQ_L/dY_G \simeq 0$ and from equation 7.16

$$\frac{dI_G}{dY_G} = H_{GW} = C_L T_S + \Delta H_{VS} + C_{PW}(T_G - T_S) \tag{7.26}$$

On substituting equation 7.26 into the expression for the temperature gradient (eq. 7.23), we obtain

$$\frac{dT_S}{dY_G} = \frac{\pm\, G[\Delta H_{VS} + C_{PW}(T_G - T_S)]}{L(C_S + C_L X)} \tag{7.27}$$

when the small term in dQ_L/dY_G is ignored. It thus follows from equation 7.27 that the temperature of the surface rises in co-current flow as the air above becomes damper, and falls with humidification when the solids and air move in opposite directions.

If the heat loss is high in an unheated dryer, and the enthalpy gradient thereby negative, as sometimes occurs, then the surface-temperature gradient will take a slope of opposite sign from the foregoing isothermal case. There is a negative gradient from the air intake onwards.

State paths. Equation 7.12 for the humidity gradient and equation 7.23 for the surface-temperature gradient may be invoked to follow the progress of both the air and solids through the dryer. The humidification path for the air may be traced from known air-inlet conditions with the humidity gradient so determined, while the surface-temperature profile, which reflects how moist the surface is, follows directly from the corresponding temperature gradient. As long as unhindered drying takes place, this profile will lie on the moisture-saturation curve. However, if the surface is hygroscopic and the relative humidity of the adjacent air-skin less than 1, then the surface state will lie in the unsaturated region of the humidity chart. The equilibrium relative humidity can be found from the moisture isotherm for a given moisture content, as described in Section 2.8, and thus the state point for the surface can be located at a point (ψ_S , T_S).

The state paths for the co-current operation of a dryer are sketched on an enthalpy-humidity chart in Fig. 7.7. The humidification path is shown for the general, non-adiabatic case, for which the gradient dI_G/dY_G falls off somewhat with humidity as the parameter η_{LG} becomes less owing to the self-cooling of the air being humidified. Initially, the surface conditions shift from point A to point B on the moisture-saturation curve as the wet-bulb temperature rises due to the heating of the surface. Beyond point B, the state path for the surface diverges from the moisture-saturation curve, as the surface temperature climbs above the wet-bulb value. The surface-state path begins to curl back on itself since the drying potential $\Pi = (Y_S - Y_G)$ weakens and the two state paths converge towards a common point, which represents equilibrium of the bulk of the air with the surface material. This condition defines the end of drying, and clearly can only be reached with an infinitely long dryer. In practice, however, even with a dryer of modest dimensions, the final driving force (Π_Z) will be quite small for co-current operation.

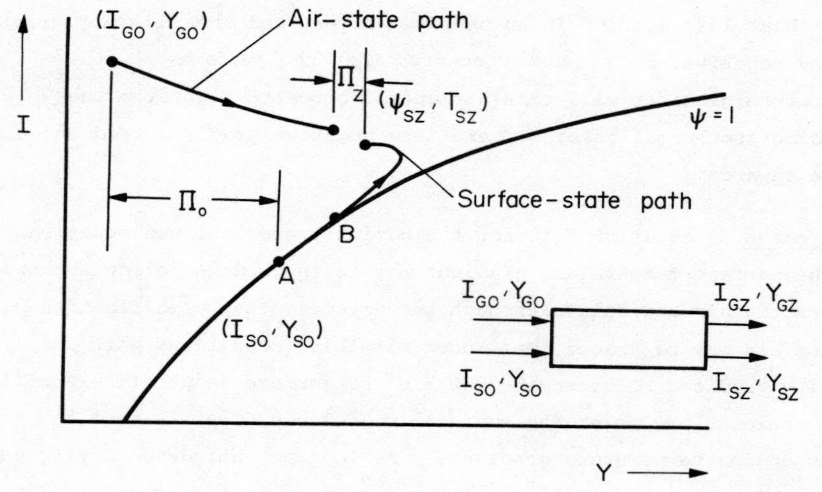

Fig. 7.7 State paths for the co-current operation of a dryer.

$$\Pi_o = (Y_{SO} - Y_{GO})$$

$$\Pi_Z = (Y_{SZ} - Y_{GZ})$$

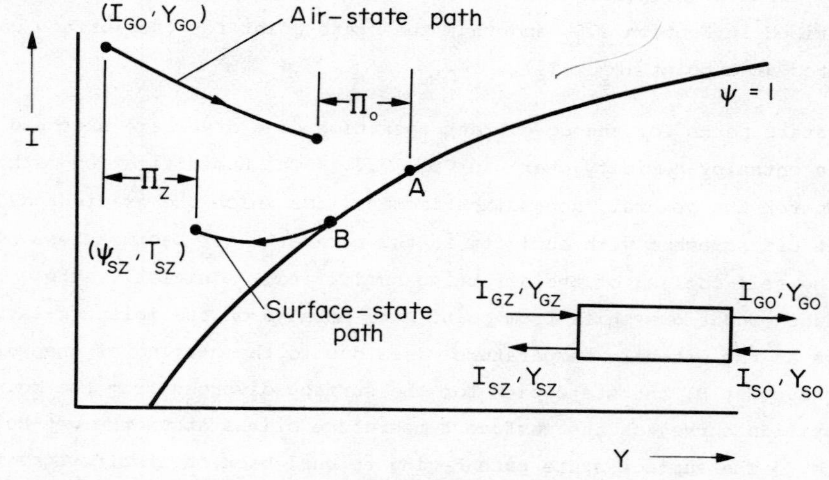

Fig. 7.8 State paths for the countercurrent operation of a dryer.

$$\Pi_o = (Y_{SO} - Y_{GO})$$

$$\Pi_Z = (Y_{SZ} - Y_{GZ})$$

The corresponding state paths for countercurrent operation are shown in Fig. 7.8. The humidity of the air now rises from the solids inlet, still with a declining gradient associated with the cooling of the air. The temperature of the solids falls as the solids move through the dryer. As before, the surface-state path initially lies on the moisture-saturation curve before it bends away into the unsaturated region when hindered drying sets in. The diagram clearly shows the advantages of countercurrent operation in maintaining a more uniform drying potential throughout the course of drying.

Minimum air-rate. By inspecting Fig. 7.8, we see that the air-state path may intersect the surface-state path should the air-rate be restricted. Under these conditions, the outlet air will be totally saturated, and the desired degree of drying cannot be undertaken. There is thus a minimum air-rate in countercurrent drying, which corresponds to the situation where the outlet-air is just saturated, as depicted in Fig. 7.9.

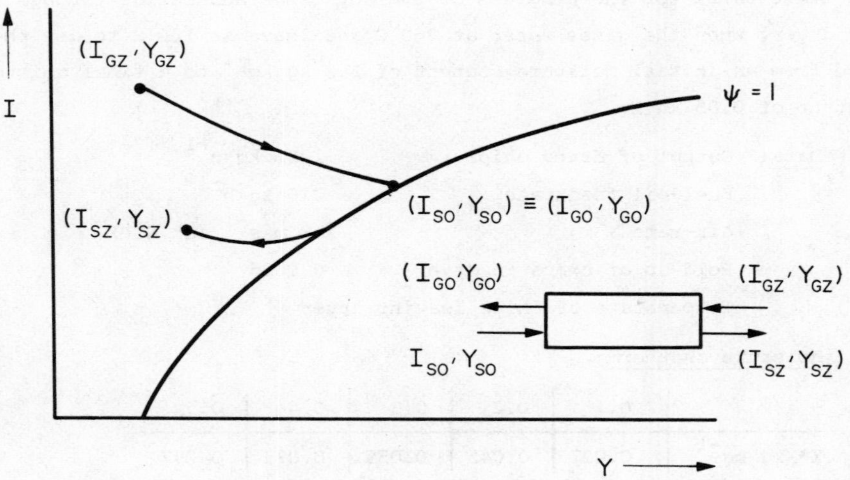

Fig. 7.9 State paths for the minimum air-rate in countercurrent operation.

This minimum air-rate can be easily found by writing a mass balance across the dryer:

$$L(X_O - X_Z) = -G(Y_{GZ} - Y_{GO}) \tag{7.28}$$

When the air leaves at the dewpoint, Y_{GO} is equal to Y_{SO}. Often this humidity will be equal to the saturation humidity at the wet-bulb temperature, Y_{WO}, at the solids-inlet position, since in most cases the solids enter with some unbound moisture to be driven off. Re-arrangement of equation 7.28 yields the minimum air/solids ratio:

$$\left(\frac{G}{L}\right)_{min} = \frac{(X_O - X_Z)}{(Y_{SO} - Y_{GZ})} \qquad (7.29)$$

For a change in moisture content of 1 kg kg^{-1}, moisture-free fresh air, and a wet-bulb temperature of 50°C at the solids inlet, the minimum air/solids ratio is 1/0.0875 = 11.4. Countercurrent operation may thus need considerable quantities of air to sustain it.

Example 7.2. Wood chips, 20 mm x 10 mm x 0.38 mm, for the manufacture of particleboard are to be dewatered in a progressive dryer 2.5 m in diameter and 8 m long. The dryer is oil-fired, the combustion gases and the wooden chips passing co-currently through the plant. Draw up the state paths for the progress of the hot gases and solids through the dryer, when the gases enter at 260°C and leave at 160°C to dry the wood from an initial moisture content of 1.3 kg kg^{-1} to a final moisture content of 0.05 kg kg^{-1}.

Data:

Output of dried chips	1820 kg h^{-1}
Fuel-oil feed rate	210 kg h^{-1}
Air-rate	18 m^{3}s^{-1} (at NTP)
Hold-up of chips in dryer	0.0025
Temperature of chips leaving dryer	72°C

Moisture isotherm:

ψ	0.1	0.2	0.3	0.4	0.5
X*/kg kg^{-1}	0.027	0.045	0.059	0.072	0.087

ψ	0.6	0.7	0.8	0.9	1.0
X*/kg kg^{-1}	0.103	0.124	0.150	0.200	0.300

Inlet-gas humidity, Y_{GO}. From equation 3.31, the water vapour produced on combustion is given by

$$g_V = 8.937\ h + w$$

where h = hydrogen content and w = water content. For dry medium fuel-oil, h = 0.112 and w = 0, thus

$$g_V = 8.937 \times 0.112 = 1.00\ \text{kg kg}^{-1}$$

If fresh air is drawn at 15°C and 60 per cent relative humidity, its humidity is 0.00643 kg kg^{-1} (see example 3.7).
The air/fuel ratio is equal to

$$18 \times 3600 \times 1.284/210 = 396$$

$$\therefore\ \text{inlet-gas humidity} = 0.00643 + 1.00/396$$

$$= 0.00896\ \text{kg kg}^{-1}$$

At a dry-bulb temperature of 260°C, air of this humidity will have a wet-bulb temperature of 52°C, which is the solids-inlet temperature also.

Inlet-gas enthalpy, I_{GO}. For $\bar{C}_{PG} = 1.014$ kJ kg^{-1}K^{-1} and $\bar{C}_{PW} = 1.902$ kJ kg^{-1}K^{-1} and $\Delta H_{VO} = 2501$ kJ kg^{-1},

$$I_{GO} = (1.014 \times 260) + [(1.902 \times 260) + 2501] \times 0.00896$$

$$= 290.5\ \text{kJ kg}^{-1}.$$

Shell/particle surface ratio, s. For a cuboid chip of sides of length a, b and c, the surface/volume ratio is

$$\frac{2ab + 2bc + 2ca}{abc} = 2\left[\frac{1}{c} + \frac{1}{a} + \frac{1}{b}\right]$$

Thus, in this example, the surface area of chips in unit volume of dryer is

$$2\left[\frac{1}{0.02} + \frac{1}{0.01} + \frac{1}{0.00038}\right] \times 0.0025 = 14\ \text{m}^{-1}$$

In a cylindrical dryer of diameter D and length L, the ratio of the shell area to the volume contained is $\pi DL/\frac{\pi D^2}{4} L = 4/D$.
In this case, therefore, the ratio is 4/2.5 = 1.6 m^{-1}

$$s = \frac{\text{shell surface}}{\text{particle surface}} = \frac{1.6}{14} = 0.114$$

Heat-loss parameter, η_{LG}. There is a substantial wall-temperature profile in the dryer, and consequently the state paths will have large changes in gradient when the particles move into the falling-rate region of drying. The dryer is thus broken up into two zones for the unhindered-drying section, and five zones for the hindered-drying part, as shown in Fig. 7.10. It is assumed, roughly, that the air-temperature changes are proportional to the moisture-content and thus humidity changes. (This is Grosvenor's old observation[3]). Since the dryer is uninsulated, the shell temperature is equated to the corresponding bulk-air temperature. The heat-transfer coefficient for the convective heating of the particles is taken to be 3.125 W $m^{-2}K^{-1}$ (see example 6.9). The solids temperatures are linearly interpolated in the falling-rate region in the absence of other information. The humidity Y_S of the air adjacent to the particles follows from the known relative humidity ψ_S, a function of the moistness of the chips, and the local wet-bulb temperature, assumed constant. The temperature T_S then follows, being uniquely determined by ψ_S and Y_S. While the foregoing procedure is very rough in a number of places, the consequential estimate of η_{LG} is still probably good enough.

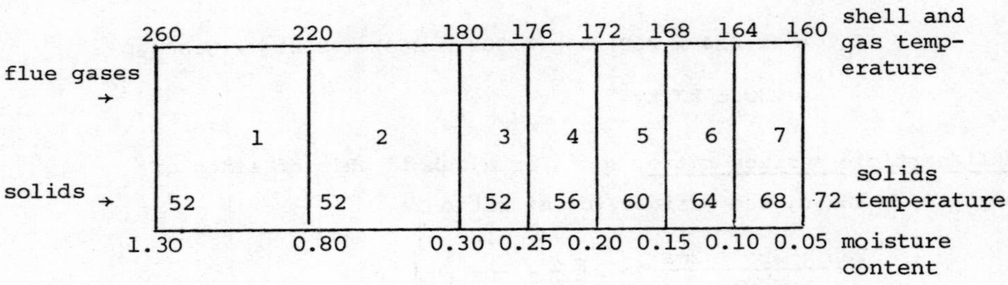

Fig. 7.10. Zones in progressive dryer for example 7.2

zone	$\bar{T}_G$/ °C	heat loss/ W m^{-2}	$\bar{X}$/ kg kg^{-1}	$\bar{T}_S$/ °C	$\eta_{LG} = q_L s/h(\bar{T}_G - \bar{T}_S)$
1	240	6400	1.05	52	1.241
2	200	4070	0.55	52	1.003
3	178	3100	0.275	54	0.912
4	174	3020	0.225	58	0.950
5	170	2920	0.175	62	0.986
6	166	2806	0.125	66	1.024
7	162	2670	0.075	70	1.059

To calculate the humidity potential, one obtains:

(i) <u>the bulk-air humidity</u> Y_G by moisture balance,

$$\Delta Y_G = \frac{L}{G}\,\Delta X$$

$$= \frac{1820}{1.05 \times 18 \times 3600 \times 1.284} = 0.0208\ \Delta X$$

(ii) the <u>surface humidity</u> Y_S from the known values of $\psi_S(T_S)$ and T_S through equation 2.26.

i.e. $$Y_S = \frac{0.622\ p_W^o}{1/\psi - p_W^o}$$

where p_W^o is the saturation vapour pressure at T_S in <u>bar</u>.

The enthalpy-gradient profile now follows through successive evaluation of equation 7.12 for each of the zones:-

zone	H_{GW}	$\bar{C}_{PY}$	$(\bar{T}_G-\bar{T}_S)$	$\bar{X}$	ψ_S	$\bar{Y}_S$	$\bar{Y}_G$	$(\bar{Y}_S-\bar{Y}_G)$	$(1+\eta_{LG})$	$\Delta I_G/\Delta Y_G$
1	2995	1.048	188	1.05	1	0.0980	0.0142	0.0838	2.061	- 2274
2	2817	1.059	148	0.55	1	0.0980	0.0246	0.0734	2.043	- 1575
3	2840	1.075	124	0.275	0.985	0.1078	0.0303	0.0775	1.912	- 449
4	2832	1.076	116	0.225	0.94	0.1279	0.0313	0.0966	1.950	+ 312
5	2825	1.077	108	0.175	0.86	0.1438	0.0324	0.1114	1.986	+ 751
6	2817	1.079	100	0.125	0.71	0.1418	0.0334	0.1084	2.021	+ 802
7	2810	1.080	92	0.075	0.42	0.0936	0.0345	0.0591	2.040	- 652

The enthalpy profile is then calculated from the values of $\Delta I_G/\Delta Y$ with the known humidity changes over each zone:-

zone	$Y_G\|_{in}$	ΔY_G	$I_G\|_{in}$	ΔI_G
1	0.00896	0.0104	290.5	- 23.6
2	0.01936	0.0104	266.9	- 16.1
3	0.02976	0.001	250.8	- 0.4
4	0.03076	0.001	250.4	+ 0.3
5	0.03176	0.001	250.7	+ 0.8
6	0.03276	0.001	251.5	+ 0.8
7	0.03376	0.001	252.3	- 0.7
out	0.03476		251.6	

The state paths may now be plotted (Fig.7.11). Since the gradient dQ_L/dY_G is unknown precisely, there is no value in "checking" the surface temperature by means of equation 7.23. The solids-state curve takes the characteristic form of curling back towards the air-state path. The effect of the heat loss is to reduce the wet-bulb temperature as drying proceeds, there being a drop of 3°C

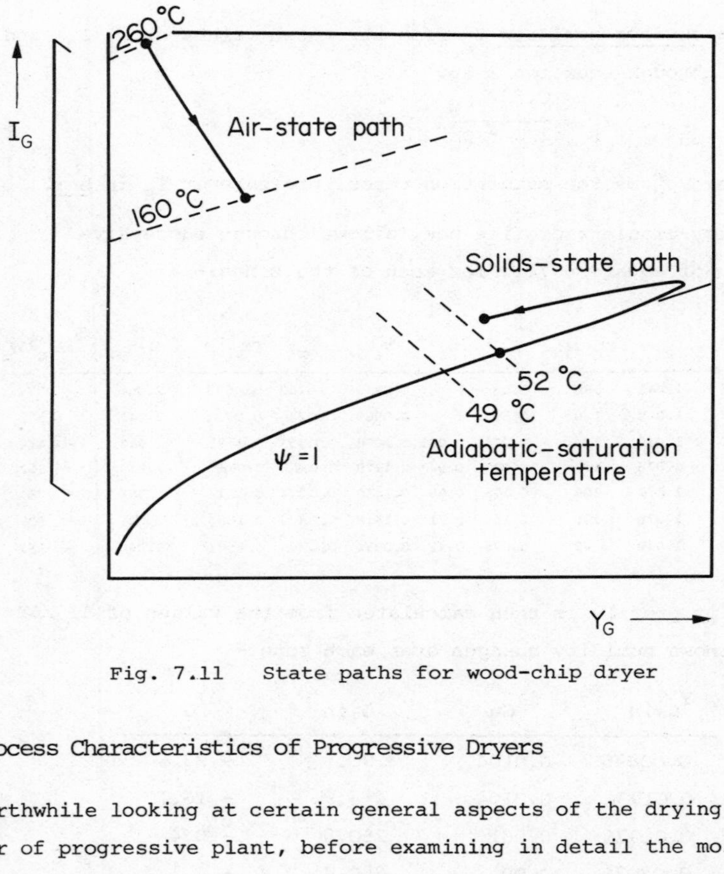

Fig. 7.11 State paths for wood-chip dryer

7.2 Process Characteristics of Progressive Dryers

It is worthwhile looking at certain general aspects of the drying behaviour of progressive plant, before examining in detail the moisture-content profiles found therein as a basis for estimating the duration of drying consequent on adopting a particular process strategy. It is assumed that cross-circulation of the solids occurs, so that the air flows parallel to the movement of solids, either in the same or opposite direction.

Firstly, let us consider how drying rates vary with distance for counter-current operation. Now, the ratio of the local drying rate N_V at some point within the dryer compared with that N_{VZ} at the air inlet is given by

$$\frac{N_V}{N_{VZ}} = \frac{f\,K_o\phi(Y_W - Y_G)}{f_Z K_o \phi_Z (Y_{WZ} - Y_{GZ})} \tag{7.30}$$

where the subscript Z stands for the solids-outlet position, and thus the air inlet, as shown in Fig. 7.12

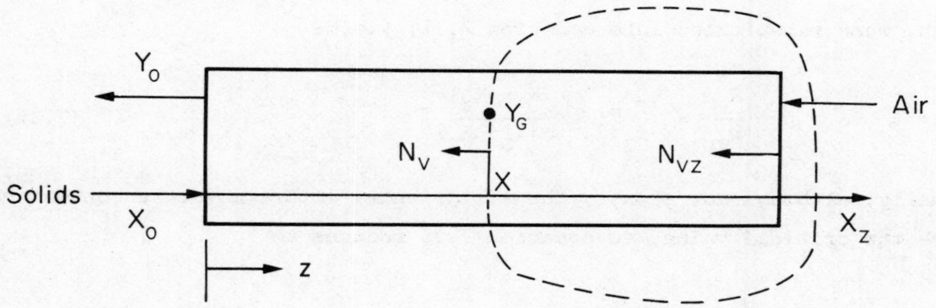

Fig. 7.12 Countercurrently worked progressive dryer

Further, let us suppose that the drying is conducted adiabatically, then the wet-bulb humidity Y_W and the humidity-potential coefficient $[\phi \simeq D/(D + Y_W)]$ remain constant throughout the dryer. Under these conditions, equation 7.30 reduces to

$$\frac{N_V}{N_{VZ}} = \frac{f(Y_W - Y_G)}{f_Z(Y_W - Y_{GZ})} = \frac{f}{f_Z}\left[1 - \frac{(Y_G - Y_{GZ})}{(Y_W - Y_{GZ})}\right] \tag{7.31}$$

The humidity change in the bulk of the air depends upon the relative quantities of air and solids passing through the dryer. The minimum air/solids ratio for countercurrent working is given by equation 7.29, so that if α times this minimum is used in practice, then

$$\frac{G}{L} = \frac{\alpha(X_O - X_Z)}{(Y_W - Y_{GZ})} \tag{7.32}$$

on noting that $Y_W = Y_{SO}$ in this case. From a moisture balance over the dotted envelope in Fig. 7.12, another equation for this flow ratio may be found, namely:

$$\frac{G}{L} = \frac{(X - X_Z)}{(Y_G - Y_{GZ})} \tag{7.33}$$

On eliminating the flow ratio G/L from the preceding two expressions (eq. 7.32 and 7.33), we can compare the humidity drop $(Y_G - Y_{GZ})$ with the humidity potential at the air inlet, $(Y_W - Y_{GZ})$:

$$\frac{(Y_G - Y_{GZ})}{(Y_W - Y_{GZ})} = \frac{(X - X_Z)}{\alpha(X_O - X_Z)} \tag{7.34}$$

which, when substituted into equation 7.31, yields

$$\frac{N_V}{N_{VZ}} = \frac{f}{f_Z}\left[1 - \frac{(X - X_Z)}{\alpha(X_O - X_Z)}\right] \tag{7.35}$$

Normally, although not always, the solids enter with a moisture content above the critical value, so equation 7.35 reduces to

$$\frac{N_{VO}}{N_{VZ}} = \frac{1}{f_Z}\left[\frac{\alpha-1}{\alpha}\right] \tag{7.36}$$

for the relative rate at the solids inlet. From equation 7.35, we see that the relative rate rises from the solids-inlet position linearly with decreasing moisture content until the solids are dried below their critical moisture content. From that point onwards in the dryer, the increasing difficulty of drying counteracts the improving humidity potential, as witnessed by the rising rates in the first drying period, and the rates start to diminish. In some cases, the maximum is close to the position where the critical point lies, but in isothermal drying the effect of the humidity potential can prevail for longer, so the maximum appears well into the falling-rate period.[6]

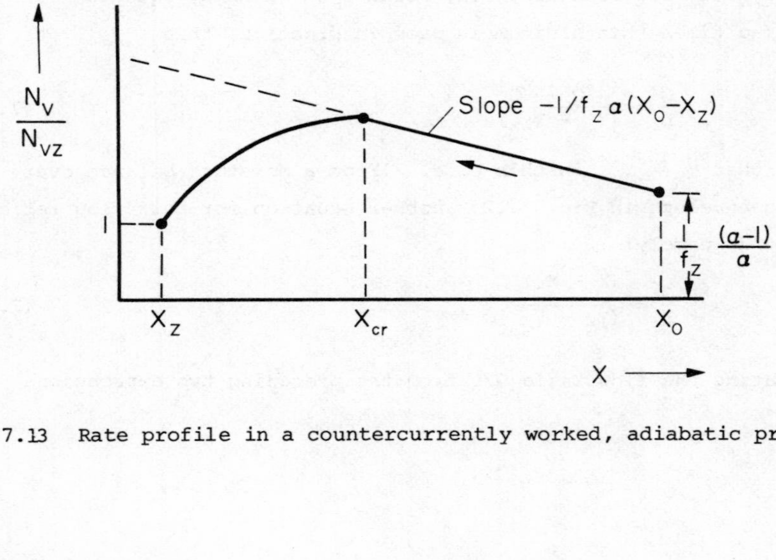

Fig.7.13 Rate profile in a countercurrently worked, adiabatic progressive dryer.

Example 7.3. Essentially non-hygroscopic solids are being dried in a countercurrent progressive chamber from an initial moisture content of 0.75 kg kg^{-1} to a final moisture content of 0.1 kg kg^{-1}. Estimate the moisture content where the drying rate is a maximum when (a) 20 per cent and (b) 50 per cent more air than the minimum is used.

Data. The falling-rate period may be assumed to be linear with a critical moisture content of 0.5 kg kg^{-1}.

From equation 7.35 for the rate profile we have

$$R = \frac{N_V}{N_{VZ}} = \frac{f}{f_Z}\left[1 - \frac{(X - X_Z)}{\alpha(X_O - X_Z)}\right]$$

For $X = \Phi X_{cr}$ and $f = \Phi$ (linear falling rate), this expression becomes

$$R = \frac{\Phi}{\Phi_Z}\left[1 - \frac{(\Phi - \Phi_Z)}{\alpha(\Phi_O - \Phi_Z)}\right] \qquad \Phi < 1$$

$$= \frac{[\alpha\Phi_O - (\alpha-1)\Phi_Z]\Phi - \Phi^2}{\Phi_Z\alpha(\Phi_O - \Phi_Z)}$$

$$\therefore \frac{dR}{d\Phi} = \frac{[\alpha\Phi_O - (\alpha-1)\Phi_Z] - 2\Phi}{\Phi_Z\alpha(\Phi_O - \Phi_Z)}$$

The rate is a maximum when $dR/d\Phi = 0$; that is, when

$$\alpha\Phi_O - (\alpha-1)\Phi_Z = 2\Phi \qquad \text{(A)}$$

In this case, $\Phi_O = 0.75/0.5 = 1.5$ and $\Phi_Z = 0.1/0.5 = 0.2$

(a) For $\alpha = 1.2$, from equation A

$$\Phi = \frac{1}{2}[(1.2 \times 1.5) - (1.2 - 1) \times 0.2] = 0.88$$

$$\text{i.e. } X = \Phi X_{cr} = 0.88 \times 0.5 = 0.44 \text{ kg kg}^{-1}$$

(b) For $\alpha = 1.5$, we have

$$\Phi = \frac{1}{2}[(1.5 \times 1.5) - (1.5 - 1) \times 0.2] = 1.075$$

This violates the condition that $\phi < 1$ and thus the maximum appears at the critical point in this case. ($X = 0.5$ kg kg^{-1}).

The equivalent expression[8] to equation 7.35 for co-current working becomes

$$\frac{N_V}{N_{VO}} = f\left[1 - \frac{(X_O - X)}{\alpha(X_O - X_Z)}\right] \qquad (7.37)$$

Equation 7.37, like equation 7.35, is normalized with respect to the rate at the air inlet, which is also the solids-inlet position in this mode of operation. Inspection of equation 7.37 shows that the rates steadily diminish as the solids go through the dryer, due to the adverse changes in humidity potential, the effect becoming more pronounced once the solids enter the falling-rate period. These changes are illustrated in Fig. 7.14. The relative rate at the solids outlet is $f(\alpha-1)/\alpha$. At the minimum flow ratio $(\alpha-1)$ the air leaves saturated with moisture and the relative rate N_V/N_{VO} is zero.

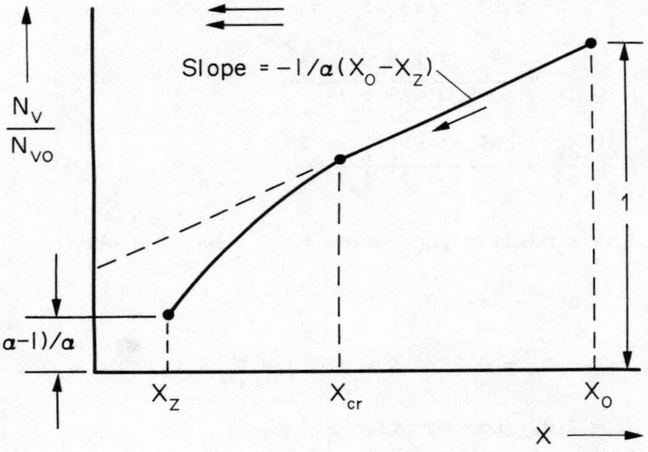

Fig. 7.14 Rate profile in a co-currently worked adiabatic progressive dryer.

The rate profiles (eq. 7.35 and 7.37) have one important property: they are independent of wet-bulb depression. Manipulation of the drying potential will influence the absolute rate of drying, but the relative magnitude of the local rates within the dryer cannot be influenced in this way. This conclusion, however, is restricted to adiabatic variations only, since adding or withdrawing further heat during the course of drying will modify both the drying potential and local rates, as noted earlier.

This conclusion also has another, and implicit, restriction. The analysis presumes that a single characteristic drying curve applies. The variation in rates can also depend upon the intensity of drying to judge from analyses[8] of the progressive drying of non-hygroscopic slabs in terms of the receding-plane theory of moisture movement. Under conditions of high-intensity drying, when the relative intensity $\mathcal{N} = N_V b/\rho_S X \mathcal{D}_a$ is greater than 2, drying takes place within a narrow zone which progressively shifts further and further away from the exposed surface with time. On the other hand, when low-intensity drying pertains, the moisture contents throughout the whole body of the material are reduced below their initial values. As long as the material ramains in the first drying period, the drying rates are independent of the relative intensity, since the solids do not hinder drying therein. In the falling-rate period, however, the drying rates at each place depend not only on the averaged degree of moistness, which is given by the characteristic moisture content, but also on the distribution of moisture for which the relative intensity is an index. From the examples studied,[8] it appears that the variation of drying rates becomes wider as the relative intensity increases above 2 and thus as the drying zone narrows. The effect is fairly small, and becomes less significant as the moisture-transfer resistance of the solids increases relative to that of the air above. Normally then, the use of a single characteristic drying curve will be good enough to fit the drying-rate profile.

Let us consider the case of co-current drying by way of illustration. From equation 7.37, the normalized drying rate at the place where the solids have reached the critical moisture content is given by the expression

$$\left.\frac{N_V}{N_{VO}}\right|_{X_{cr}} = \left[1 - \frac{(X_O - X_{cr})}{\alpha(X_O - X_Z)}\right] \tag{7.38}$$

For non-hygroscopic material, $\Phi = X/X_{cr}$ and equation 7.38 may be re-written in terms of the characteristic moisture content as

$$\left.\frac{N_V}{N_{VO}}\right|_{1} = \left[1 - \frac{(\Phi_O - 1)}{\alpha(\Phi_O - \Phi_Z)}\right] \tag{7.39}$$

Now, high-intensity drying conditions will just appear when $\mathcal{N} = 2$, so low-intensity drying conditions will prevail as long as the initial intensity $\mathcal{N}_o$ at the inlet does not exceed a value given by

$$\frac{2}{\mathcal{N}_o} = \left[1 - \frac{(\Phi_o - 1)}{\alpha(\Phi_o - \Phi_Z)}\right] \tag{7.40}$$

that is

$$\mathcal{N}_o \leq \frac{2\alpha(\Phi_o - \Phi_Z)}{1 + (\alpha-1)\Phi_o - \Phi_Z} \tag{7.41}$$

Under these conditions, the drying-rate profile is independent of relative intensity $\mathcal{N}$. When drying proceeds to low moisture levels ($\Phi_Z \to 0$), this limiting value of $\mathcal{N}_o$ will range from 2, if the feed is entering at the critical moisture content, to $2\alpha/(\alpha-1)$ at high initial moisture contents. Should copious quantities of air be used, and thus α becomes very large, then it follows from equation 7.41 that high-intensity drying prevails everywhere.

7.3 Progressive Cross-circulated Dryers

Let us consider a progressive, cross-circulated dryer as shown in Fig.7.15. The dryer is assumed to have a total cross-sectional area of S_T, of which the greater fraction ε is occupied by the air and a lesser fraction $(1-\varepsilon)$ by the solids. Again, we take the specific dry gas flow G_o to be positive in co-current movement of solids and air, and negative in countercurrent movement. The specific dry-substance flow rates L_o and G_o are defined in terms of the areas available for each stream. The specific flow rates, in terms of the dryer's cross-section area, are $G_o\varepsilon$ and $L_o(1-\varepsilon)$ respectively.

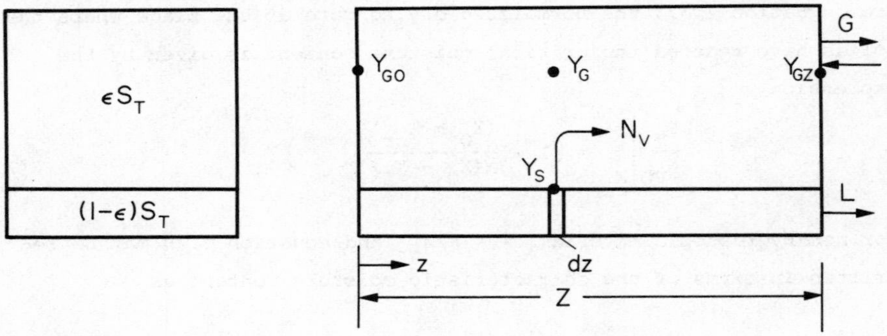

Fig. 7.15 Progressive cross-circulated dryer.

The moisture-content and drying-rate profiles along the dryer may be found by striking moisture balances from place to place in the dryer together with a suitable constitutive relationship for the local drying rate. By equating the rate of humidification to the moisture-transfer rate, we get one mass balance:

$$\pm\, G_o(\varepsilon S_T)dY_G = N_V d(aS_T z) \tag{7.42}$$

where a is the exposed area of solids per unit volume of dryer. A second mass balance is obtained by equating the rate of moisture loss to the moisture-transfer rate:

$$-\, L_o(1-\varepsilon)S_T dX = N_V d(aS_T z) \tag{7.43}$$

In each case, the drying rate over unit area (N_V) is determined by the constitutive relationship

$$N_V = fK_o\phi(Y_W - Y_G) \tag{5.5}$$

If the dryer is uniformly loaded, which is the normal practice, then the exposed area a does not alter throughout the dryer. The mass-transfer coefficient K_o is usually, but not neccessarily, taken as the distance-averaged value from measurements of the overall drying rate on the plant. However, ϕ and Y_W, besides the humidity Y_G of the bulk of the air, will vary with distance. In general, then, these foregoing equations are solved in their finite-difference form by a marching solution from known conditions at one end of the dryer. The following worked example illustrates such a procedure for the case of isothermal drying.

Example 7.4. A conveyer dryer is to dry peeled timber veneers from an initial moisture content of 1.35 kg kg^{-1} to an outlet moisture content of 0.15 kg kg^{-1}. Air enters countercurrently to the conveyed veneer at a temperature of 84°C and a relative humidity of 5.2 per cent. Estimate the variation of drying rate with moisture content if the air temperature is maintained throughout the dryer at 84°C and 58 per cent more than the minimum is used.

Data. Characteristic drying curve, $f = \Phi^{3/4}$, $0 < \Phi < 1$

Critical moisture content, $X_{cr} = 1.2$

Equilibrium moisture content, $X^* = 0.16\psi$, $0 < \psi < 0.8$

Minimum air-rate. Air at 84°C and $\psi = 0.052$ has a humidity of 0.0186 kg kg^{-1} and the corresponding wet-bulb temperature is 36°C. At the minimum air-rate, however, the air will leave the dryer saturated at 84°C, not at 36°C, since the drying is isothermal. The corresponding saturation humidity Y_G^* is 0.7788 kg kg^{-1}. By overall mass balance

$$\frac{G_{min}}{L} = \frac{\Delta X}{Y_G^* - Y_G|_{in}}$$

$$= \frac{1.35 - 0.15}{0.7788 - 0.0186} = 1.579$$

$$\therefore G/L = 1.579 \times 1.58 = 2.5$$

Drying-rate profile. Decrements of 0.1 kg kg^{-1} moisture content are taken. The corresponding decrements of humidity are thus 0.1/2.5 = 0.04 kg kg^{-1}.

X	Y_G	T_W	Y_W	$\phi = D/(D+Y_W)$	$\phi\Pi = \phi(Y_W-Y_G)$
1.35	0.4986	78.7	0.5071	0.551	0.0047
1.25	0.4586	$\frac{dY_W}{dY_G} = 1.02$	0.4664	0.571	0.00445
1.15	0.4186		0.4257	0.594	0.0042
1.05	0.3786	74.8	0.3850	0.618	0.00395
0.95	0.3386	$\frac{dY_W}{dY_G} = 1.005$	0.3448	0.643	0.0040
0.85	0.2986		0.3046	0.671	0.0040
0.75	0.2586	69.0	0.2644	0.702	0.0041
0.65	0.2186	$\frac{dY_W}{dY_G} = 0.971$	0.2254	0.734	0.0050
0.55	0.1786		0.1866	0.769	0.00615
0.45	0.1386	59.2	0.1478	0.808	0.0074
0.35	0.0986	$\frac{dY_W}{dY_G} = 0.904$	0.1116	0.848	0.0110
0.25	0.0586		0.0755	0.892	0.0151
0.15	0.0186	36.0	0.0393	0.941	0.0195

↑ airflow direction

The calculation is very sensitive to the estimate of wet-bulb temperature which must be calculated to within 0.1°C if the driving force is to be determined to within 2 per cent at the air outlet. The wet-bulb temperatures are found by following the isothermal state path for the air-inlet conditions, as shown in Fig. 7.16. To smooth out random errors, the gradient dY_W/dY_G, being almost constant, is found from somewhat gross intervals and used to interpolate local wet-bulb humidities, rather than by getting these from estimates of wet-bulb temperatures themselves.

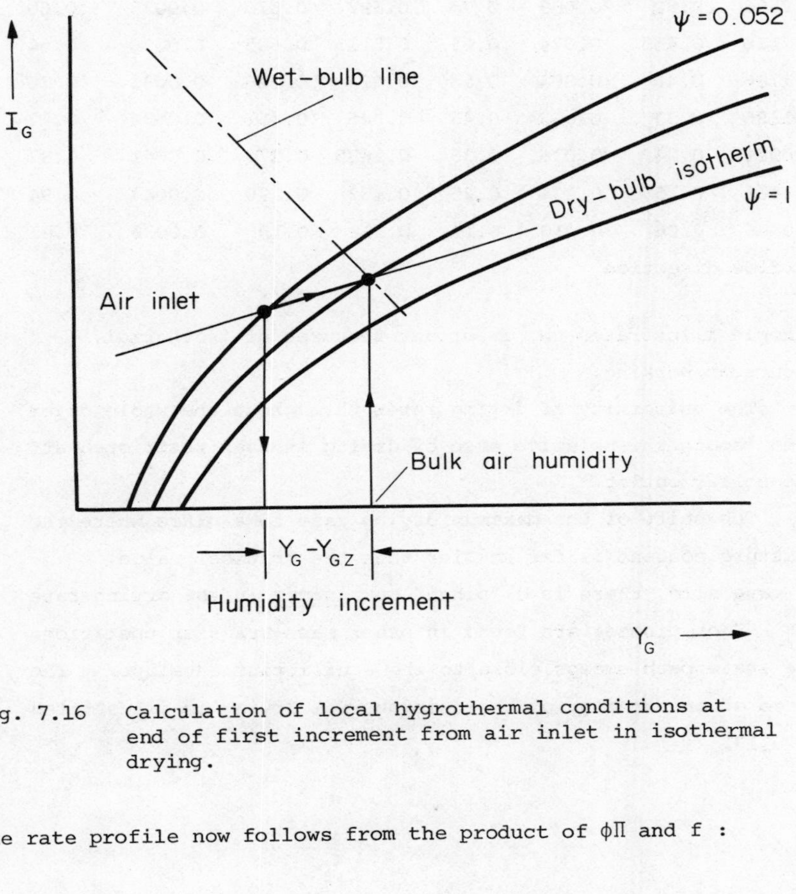

Fig. 7.16 Calculation of local hygrothermal conditions at end of first increment from air inlet in isothermal drying.

The rate profile now follows from the product of $\phi\Pi$ and f :

Y_G	ψ	X*	X	Φ	f	$f\phi\Pi$	N_V/N_{VO}
0.4986	0.80	0.128	1.35	1.140	1	0.0047	1.00
0.4586	0.76	0.122	1.25	1.046	1	0.0045	0.96
0.4186	0.715	0.114	1.15	0.954	0.965	0.00405	0.86
0.3786	0.675	0.108	1.05	0.863	0.895	0.0036	0.77
0.3386	0.625	0.100	0.95	0.773	0.824	0.0033	0.70
0.2986	0.58	0.093	0.85	0.683	0.751	0.0030	0.64
0.2586	0.52	0.083	0.75	0.597	0.679	0.0028	0.60
0.2186	0.465	0.074	0.65	0.5115	0.605	0.0030	0.64
0.1786	0.40	0.064	0.55	0.424	0.526	0.0033	0.70
0.1386	0.33	0.053	0.45	0.346	0.451	0.0034	0.72
0.0986	0.24	0.038	0.35	0.2685	0.373	0.0041	0.87
0.0586	0.15	0.024	0.25	0.192	0.290	0.0044	0.94
0.0186	0.06	0.010	0.15	0.113	0.195	0.0038	0.81

↑ airflow direction

This example illustrates two important features of isothermal, countercurrent working:

1. The uniformity of drying rates throughout the whole dryer, even though the relative ease of drying is much restricted at the solids outlet.
2. The shift of the maximum drying rate to a place where the moisture content is far smaller than the critical value.

In this case also, there is a "pinch", a minimum in the drying-rate profile. Such pinches are found in other mass-transfer operations when the state path sweeps close to the equilibrium envelope. The appearance of an enthalpy pinch in isothermal drying is illustrated in Fig. 7.17.

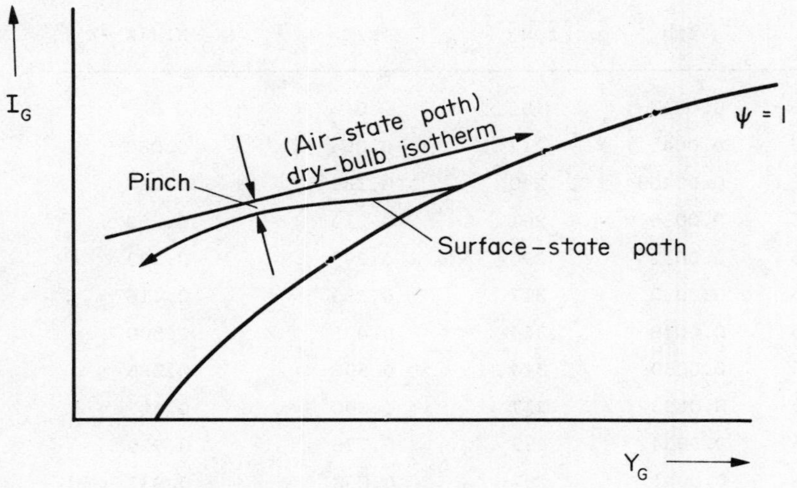

Fig. 7.17 A pinch zone in isothermal drying

Moisture-content profile. From equation 7.43

$$- L(1-\varepsilon)\delta X = \bar{N}_V \, a\delta z$$

where $\bar{N}_V = \frac{1}{2}[N_V|_z + N_V|_{z+\delta z}]$

Further, $$\frac{\delta z}{Z} = \frac{[\bar{N}_V]_i^{-1}}{\sum_1^n [\bar{N}_V]_i^{-1}}$$

Since N_V, or $f\phi\Pi$ which is proportional to it, is known as a function of X, the moisture-content profile (X vs z) follows.

Example 7.5. Evaluate the moisture-content profile along the band dryer for the conditions given in example 7.4.

The profile is evaluated, as outlined above, from known values of X and $f\phi\Pi$ at discrete points in the dryer.

X	$f\phi\Pi$	$[\overline{f\phi\Pi}]^{-1}$	z/Z	$(X_0-X)/(X_0-X_Z)$
1.35	0.0047)	0	0	0
1.25	0.0045 } →	217	0.064	0.083
1.15	0.00405	240	0.135	0.167
1.05	0.0036	261	0.213	0.250
0.95	0.0033	290	0.299	0.333
0.85	0.0030	317	0.393	0.416
0.75	0.0028	344	0.495	0.500
0.65	0.0030	344	0.596	0.583
0.55	0.0033	317	0.690	0.667
0.45	0.0034	299	0.778	0.750
0.35	0.0041	267	0.858	0.833
0.25	0.0044	235	0.928	0.917
0.15	0.0038	244	1.000	1.000
		Σ $\overline{3375}$		

The moisture content thus falls almost uniformly as the material
goes through the dryer, which is expected as the drying rates
themselves are maintained with little variation throughout.

The drying-rate profiles also lead directly to the time of drying.
Since

$$N_V = -\frac{d}{d\tau}\left(\frac{\rho_s X}{a}\right) \tag{7.44}$$

it follows that

$$\tau = -\frac{\rho_s}{a}\int_{X_o}^{X_Z}\frac{dX}{N_V} \tag{7.45}$$

Thus

$$\tau = \frac{-\rho_s}{K_o a}\int_{X_o}^{X_Z}\frac{dX}{f\phi\Pi} \tag{7.46}$$

on introducing the mass-transfer coefficient K_o.   The following
worked example illustrates the procedure.

Example 7.6.   Estimate the time to dry 2.5 mm thick veneer on a
conveying band under the isothermal conditions outlined in Example 7.4.
What band speed is needed ?

Data bone-dry density 500 kg m^{-3}

mass-transfer coefficient 0.15 kg $m^{-2}s^{-1}$

From the moisture-content profile, we find

$$\int = -\Sigma\, [\,\overline{f\phi\Pi}\,]^{-1}\, \delta X = -\delta X \Sigma\, [\,\overline{f\phi\Pi}\,]^{-1} = 0.1 \times 3375 = -337.5.$$

The exposed area per unit volume of dryer (a) is 1/0.0025 = 400 m^{-1}.

$$\therefore\ \tau = -\frac{500}{0.15 \times 400}\,[-337.5] = 2812\text{s or } 46.8 \text{ min.}$$

The band speed needed is thus 20/46.8 = 0.427 m min^{-1}.

The drying in this example is fairly gentle. Use of a higher air temperature, say up 150°C, would reduce the drying time threefold to about 15 min.[10] If the band speed were maintained at 0.4 m min^{-1}, a dryer only 6m long would be needed.

Keey[7] shows that, by repeating calculations of the sort shown in the two foregoing examples, one can explore various process options to find the preferred one. In the case studied, isothermal drying provides a quicker process compared with the adiabatic case for the same inlet-air conditions, but the cost of the extra heating can outweigh the saving in the smaller capital cost of the plant.

Under certain circumstances, it is possible to derive analytical expressions for adiabatic drying. Such expressions will now be derived.

Adiabatic drying, first period. Under these conditions, equation 7.42 with equation 5.5 becomes

$$\pm\, G\, \varepsilon\, dY_G = [K_o \phi (Y_W - Y_G)]\, d(az) \qquad (7.47)$$

The gas-rate G is defined in terms of unit area of air-space in the dryer. The corresponding rate for unit cross-sectional area is Gε. When the drying is adiabatic, both ϕ and Y_W remain constant so that equation 7.47 can be re-arranged to yield

$$\int_{Y_{GO}}^{Y_{GZ}} \frac{dY_G}{(Y_W - Y_G)} = \frac{K_o \phi a}{\pm\, G\varepsilon} \int_o^Z dz \qquad (7.48)$$

which has the solution

$$- \ln\left[\frac{Y_W - Y_{GZ}}{Y_W - Y_{GO}}\right] = \frac{K_o \phi a Z}{\pm G_o \varepsilon} \tag{7.49}$$

Since $G_o \varepsilon$ is the specific gas-rate in terms of the total area across the dryer, the right-hand side is the number of internal transfer units N_{ti} and $G\varepsilon/K_o\phi a$ is the extent of a transfer unit. Because the drying rates everywhere are directly proportional to the humidity potential $(Y_W - Y_G)$ at each place, equation 7.49 can be transformed as

$$\ln\left[\frac{N_{VZ}}{N_{VO}}\right] = \frac{K_o \phi a Z}{\mp G_o \varepsilon} \tag{7.50}$$

or

$$N_{VZ} = N_{VO} \exp(\mp K_o \phi a Z/G_o \varepsilon) \tag{7.51}$$

The local drying rates thus fall exponentially with distance from the place where the solids enter in co-current movement of solids and air, but rise exponentially in countercurrent movement. For identical air-inlet conditions, the drying-rate profiles are symmetrical as shown in Fig. 7.18. Thus, it is immaterial at which end of the dryer the air is admitted, for the average drying rate over the unit is the same in both instances.

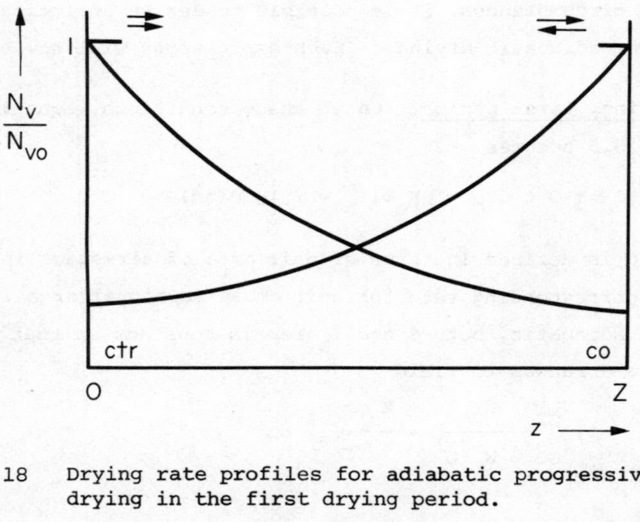

Fig. 7.18 Drying rate profiles for adiabatic progressive drying in the first drying period.
(Drying rates normalized to value 1 at air inlet)

Substitution of equation 7.51 into the equation for the moisture loss (eq. 7.43) gives the expression

$$-L(1-\varepsilon)\int_{X_O}^{X_Z} dX = \int_O^Z [N_{VO} \exp \mp(K_O\phi az/G_O\varepsilon)]\, adz \tag{7.52}$$

which, on integration, reduces to

$$-L(1-\varepsilon)(X_Z-X_O) = N_{VO}a\left[\frac{\mp G\varepsilon}{K_O\phi a}\right][\exp \mp(K_O aZ/G_O\varepsilon) - 1] \tag{7.53}$$

that is

$$(X_O - X_Z) = \mp \frac{G_O}{L_O}\cdot\frac{\varepsilon}{(1-\varepsilon)}\cdot\frac{1}{K_O\phi}[N_{VZ} - N_{VO}] \tag{7.54}$$

With parallel movement of solids and air, N_{VZ} is less than N_{VO} (see Fig. 7.18), so that the moisture contents decrease linearly with drying rate, or exponentially with distance. On the other hand, when the material moves against the direction of the airflow, N_{VZ} is now greater than N_{VO} and the drying rates rise as the moisture content is reduced. Again, equation 7.54 illustrates that it does not matter whether co- or countercurrent working is adopted when unhindered drying takes place.

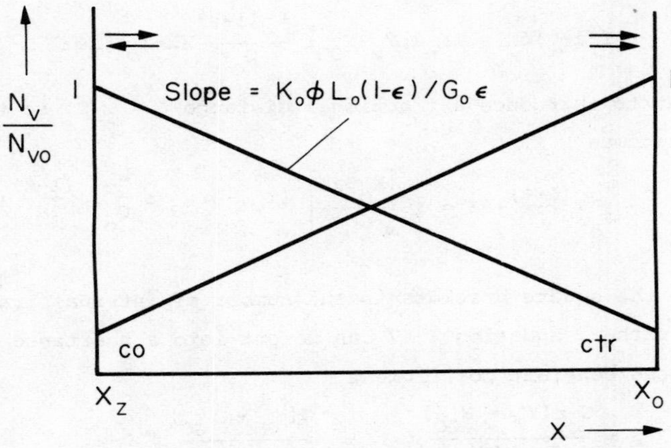

Fig. 7.19 Variation of drying rate with moisture content for adiabatic progressive drying in the first drying period.

(Drying rates normalized to value 1 at air inlet)

The slope of the profiles in Fig. 7.19 has a magnitude of $K_O\phi L_O(1-\varepsilon)/G_O\varepsilon$. Thus the determination of the drying rate at two points is sufficient to yield the mass-transfer coefficient averaged over the whole zone.

<u>Adiabatic drying, falling-rate period.</u> There is no general solution to this situation, but an analytical solution is possible if the characteristic drying curve takes a simple analytical form. However, all drying curves can be approximated by linear segments, and usually not many are needed. This technique of quasi-linearization is advocated by Schlünder[10] who notes that the method is well within the capability of modern programmable, hand-held calculators. In this book, we will consider the simpler case when the drying curve is effectively linear over the moisture-content range of interest.

The starting equations are:

(1) Moisture loss $$- L_o(1-\varepsilon)S_T dX = N_V d(aS_T z) \tag{7.43}$$

(2) Drying rate $$N_V = fK_o\phi(Y_W - Y_G) \tag{5.5}$$

(3) Local humidity $$Y_G = Y_{GO} \pm \frac{L_o(1-\varepsilon)}{G_o\varepsilon}(X-X_o) \tag{7.55}$$

(In equation 7.55, the specific flow rates L_o and G_o are those for the flow through the area available to each material. These equations can be coalesced into the single expression.

$$- L(1-\varepsilon)dX = fK_o\phi[Y_W - Y_{GO} \pm \frac{L_o(1-\varepsilon)}{G_o\varepsilon}(X-X_o)]\,adz \tag{7.56}$$

It is convenient to introduce a fractional distance $\zeta = z/Z$, so that equation 7.56 becomes

$$- \frac{L_o}{G_o}\cdot\frac{(1-\varepsilon)}{\varepsilon}\cdot dX = f\left[\frac{K_o\phi aZ}{G\varepsilon}\right]\{Y_W - Y_{GO} \pm \frac{L_o}{G_o}\cdot\frac{(1-\varepsilon)}{\varepsilon}(X-X_o)\}d\zeta \tag{7.57}$$

The term within the square brackets is the number of internal transfer units N_{ti}. Further, equation 7.57 can be put into a shortened form by introducing the constant coefficient

$$\beta = \frac{G_o\varepsilon(Y_W - Y_{GO})}{L_o(1-\varepsilon)(X_{cr}-X^*)} = \frac{G(Y_W - Y_{GO})}{L(X_{cr}X^*)} \tag{7.58}$$

and the characteristic moisture content $\Phi = (X-X^*)/(X_{cr}-X^*)$.

Equation 7.57 thus becomes

$$-d\Phi = f[N_{ti}]\{\beta \pm (\Phi-\Phi_o)\}\,d\zeta \tag{7.59}$$

whence

$$N_{ti} = N_{ti}\int_o^1 d\zeta = -\int_{\Phi_o}^{\Phi_Z}\frac{d\Phi}{f\{\beta \pm (\Phi-\Phi_o)\}} \tag{7.60}$$

Let us suppose that the falling-rate curve is linear, then equation 7.60 can be split into two parts, yielding the number of transfer units for the first and second drying periods respectively, that is

$$N_{ti} = -\int_{\Phi_o}^{1} \frac{d\Phi}{\beta \pm (\Phi - \Phi_o)} - \int_{1}^{\Phi_z} \frac{d\Phi}{\Phi(\beta \pm (\Phi - \Phi_o))} \qquad (7.61)$$

Equation 7.61 is composed of standard integrals, for which we get the final solution

$$N_{ti} = \pm \ln \left[\frac{\beta \mp (\Phi_o - 1)}{\beta}\right] - \frac{1}{\beta \mp \Phi_o} \ln \left[\frac{\Phi_z(\beta \mp (\Phi_o - 1))}{\beta \mp (\Phi_o - \Phi_z)}\right] \qquad (7.62)$$

Equation 7.62 gives an implicit expression for the moisture content Φ_z in terms of the number of transfer units, N_{ti} , and thus distance in the dryer. In this form, the expression provides a ready means of contrasting the different dryer lengths needed, according to the airflow direction chosen, whether against or with the direction of the movement of the material being dried. Such a calculation is shown in Example 7.7 below. When drying down to low moisture levels, the number of transfer units in the falling-rate zone $(\Phi_o = 1)$ is roughly $-\ln \Phi_z/(\beta \pm 1)$, which illustrates the exponential difficulty of drying.

Example 7.7. Estimate the greater length of dryer needed when co-current working is employed compared with countercurrent drying for outlet moisture contents of $\Phi_z = 0.8, 0.6, 0.4, 0.2$. The solids enter at the critical point and the coefficient β (eq. 7.58) takes a value of 2.

From equation 7.62, for transfer in the falling-rate zone alone,

$$N_{ti} = -\frac{1}{\beta \mp 1} \ln \left[\frac{\beta \Phi_z}{\beta \mp (1 - \Phi_z)}\right]$$

$$= -\frac{1}{2 \mp 1} \ln \left[\frac{2\Phi_z}{2 \mp (1 - \Phi_z)}\right]$$

The upper sign is taken in co-current, and the lower in countercurrent flow; whence we get the following values:

Φ_Z	$N_{ti}/_{co}$	$N_{ti}/_{ctr}$	Z_{co}/Z_{ctr}
0.8	0.118	0.106	1.113
0.6	0.288	0.231	1.246
0.4	0.560	0.393	1.425
0.2	1.099	0.649	1.693

The length ratio in the last column is found by dividing each figure in column 2 by the corresponding one in column 3. We see that, on drying down to $\Phi_Z = 0.2$, the co-current dryer must be 69 per cent longer.

The formal expression for the drying-rate profile is more cumbersome. However, a simpler analysis follows if we take the constitutive relationship

$$f = \exp(-\ g\xi N_2) \tag{7.63}$$

for the relative drying rate, where ξ is the fractional distance into the falling-rate zone and N_2 is the total number of transfer units therein. Originally proposed as an empirical expression, equation 7.63 has been shown to fit a linear drying curve quite well[3] and is obeyed exactly under constant external conditions. The unknown coefficient g is found from the specified conditions at the solids outlet:

$$g = -\ \ln f_Z/N_2 \tag{7.64}$$

From equation 7.42 for the rate of humidification it can be shown that the drying rate at the place where the solids emerge relative to that at the point of entry is given by

$$N_{VZ}/N_{VO} = f_Z \exp \mp [N_1 - (1-f_Z)N_2/\ln f_Z] \tag{7.65}$$

where N_1 is the number of transfer units in the unhindered-drying zone and N_2 is the number in the falling-rate zone.[6] The negative exponential form is taken in co-current movement, the positive form in countercurrent movement of air and solids. An example of the profiles predicted from equation 7.65 is worked out below.

Example 7.8. Evaluate the drying-rate profiles for co-current and countercurrent working when the outlet characteristic moisture content (Φ_z) is 0.1, and there are 0.5 transfer units in the first and 1.0 transfer units in the second (falling-rate) drying zone. The characteristic drying curve is linear.

Rate at critical point. $\Phi = 1$, $f = 1$, $N_1 = 0.5$, $N_2 = 0$
The fractional distance into the dryer is $N_1/(N_1+N_2) = 0.5/1.5 = 0.333$. From equation 7.65, the normalized rate $R = N_V/N_{VO}$ is given by

$$R_{co} = \exp - [N_1] = \exp(-0.5) = 0.607$$

$$R_{ctr} = \exp + [N_1] = \exp(+0.5) = 1.649$$

Rate at end point. $\Phi_Z = 0.1$, $f_Z = 0.1$, $N_1 = 0.5$, $N_2 = 1.0$
From equation 7.65

$$R_{co} = 0.1 \exp - [0.5 - (1-0.1)/\ln 0.1] = 0.041$$

$$R_{ctr} = 0.1 \exp + [0.5 - (1-0.1)/\ln 0.1] = 0.244$$

The intermediate drying rates may be interpolated roughly by using log.-linear paper, as shown in Fig. 7.20.

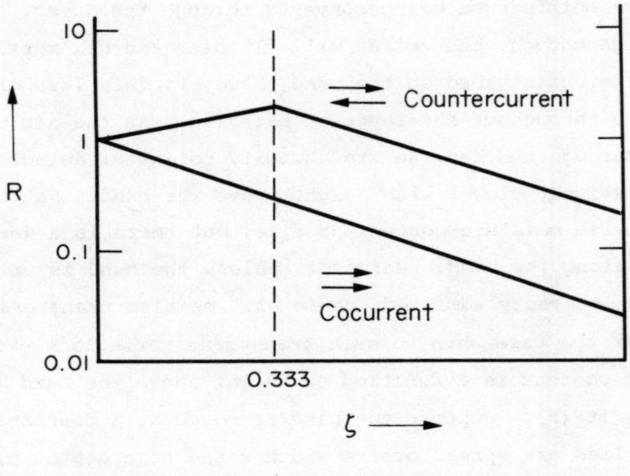

Fig. 7.20 Normalized drying-rate profiles for example 7.8

7.4 Progressive Through-Circulated Dryers

Particulate material is often through-circulated, rather than cross-circulated by the drying gas, to improve moisture-transfer rates, as noted before in considering the relative magnitude of the transfer coefficients for the two ways of drying. In through-circulation, the air passes across the direction of the solids, usually conveyed on some sort of perforate band, as sketched in Fig.7.21. For this reason, this mode of drying is sometimes called crossflow drying.

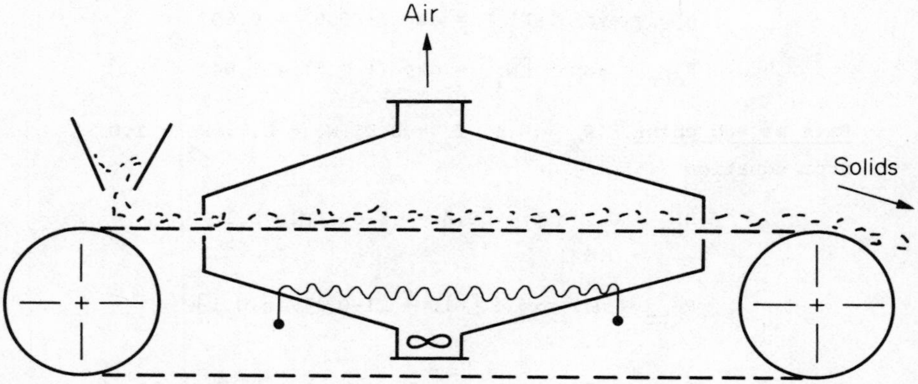

Fig. 7.21 Crossflow or through-circulation drying

The solids lose moisture on being conveyed through the dryer. If the air velocity is high enough, the solids will fluidize and mix vertically. Should the material lie undisturbed on the band, however, this loss of moisture will not be the same throughout the layer of material. As the air becomes damper on streaming through the bed, so its humidity potential dwindles and the drying becomes progressively slower with height above the band. Not only then is there a lengthwise moisture-content profile, but there is a vertical one as well at each point along the band. Moreover, unless the band is uniformly loaded and the air to it evenly admitted, there will be also transverse variations. Let us consider the case when no such transverse variations exist. The local moisture content is a function of height above the band (y) and distance along it (z). Suppose the band is moved at a constant speed of u_S, And the solids are spread over a width w and to a depth b,

as sketched in Fig. 7.22

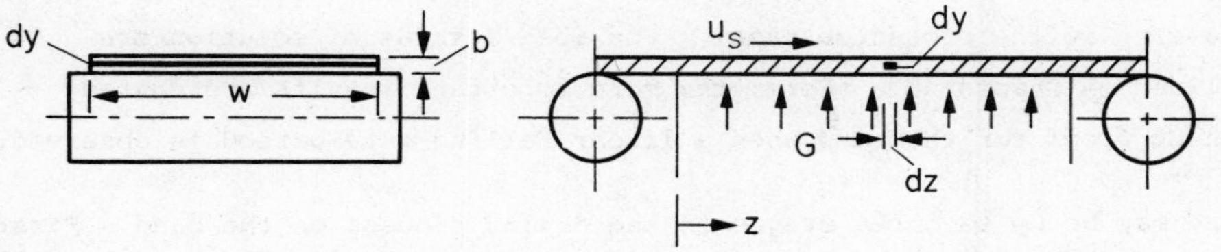

Fig. 7.22 A perforated band dryer

A moisture balance over an infinitesimally slender core of cross-section dy dz yields

$$- dX[\rho_s (1-\varepsilon)\ dy\ w\ u_s] = d(G\ Y_G)\ wdz \qquad (7.66)$$

The rate of moisture loss can also be equated to the moisture-transfer rate through equation 5.5:

$$- dX[\rho_s (1-\varepsilon) dy\ w\ u_s] = fK_o\phi(a\ w\ dy\ dz)(Y_W-Y_G) \qquad (7.67)$$

This pair of equations simplifies to:

$$- \rho_s (1-\varepsilon) u_s \frac{dX}{dz} = G \frac{dY_G}{dy} \qquad (7.68a)$$

$$- \rho_s (1-\varepsilon) u_s \frac{dX}{dz} = fK_o\phi a(Y_W-Y_G) \qquad (7.68b)$$

The pair can be further simplified by introducing dimensionless parameters which represent the extensiveness of the dryer in the vertical and horizontal directions. We have:

1. The relative distance through the bed, $\omega = K_o\phi ay/G$.
2. The relative distance along the band, $\theta = K_o\phi az/(X_{cr}-X^*)\rho_s(1-\varepsilon)u_s$

together with the humidity potential $\Pi = (Y_W-Y_G)$ and the characteristic moisture content $\Phi = (X-X^*)/(X_{cr}-X^*)$. These dimensionless parameters reduce the two equations to

$$\frac{d\Phi}{d\theta} = \frac{d\Pi}{d\omega} \qquad (7.69a)$$

$$\frac{-d\Phi}{d\theta} = \Pi f \qquad (7.69b)$$

This system of equations was first solved by Van Meel[14] for the case of batch drying, in which the time of drying is equivalent to the residence time z/u_s in the crossflow case. Possible methods of solution are outlined in Chapter 8. Here, the more important results are stated without proof for the case when a linear falling-rate period is observed.

There may be up to three stages in the drying process on the band. First, the solids are everywhere above the critical point, and the local moisture content is given by

$$\Phi = \Phi_o - \Pi_o e^{-\omega}\theta \tag{7.70}$$

where Φ_o is the initial moisture content and Π_o is the humidity potential of the air underneath the band. The end of this period is found at a distance θ_c from the inlet, where

$$\theta_c = (\Phi_o - 1)/\Pi_o \tag{7.71}$$

Afterwards, material close to the band has less moisture than the critical content, but the upper layers have more moisture. The critical moisture content thus begins to sweep upwards in this period from the base towards the top of the stuff spread out on the band. Within the upper section, at moisture contents above the critical value, the local moisture contents are given by

$$\Phi = \Phi_o - (\Phi-1)e^{-\omega}\left[\frac{\Phi_o \exp(\Pi_o\theta + 1 - \Phi_o) - 1}{\Phi_o - 1}\right]^{1/\Phi_o} \tag{7.72}$$

The distance for the critical point to sweep through the bed in this way is evaluated as

$$\theta_s = \frac{1}{\Pi_o} \ln\left[\frac{1 + (\Phi_o - 1)\exp \Phi_o N_t}{\Phi_o}\right] \tag{7.73}$$

where N_t is the total number of transfer units in the direction of the airflow ($N_t = K_o\phi ab/G$). In the final stage, the moisture contents are everywhere below the critical value. The moisture contents in this case are given by

$$\Phi = \frac{\Phi_o}{\Phi_o[\exp(\Pi_o\theta+1-\Phi_o) - 1]\exp(-\Phi\omega) + 1} \tag{7.74}$$

Equation 7.74 applies to the zone in the second stage in which moisture contents less than the critical are found.

The extent of the moisture-content variations on the band is illustrated by the following worked example which represents the case of drying a granular layer about 50 mm thick spread over a perforate band.

Example 7.9. Estimate the vertical moisture-content variations in granular material, which is being dried on a through-circulated, perforate band, in the middle of the band and at the position where the solids are discharged. The solids enter with a free moisture content $(X-X^*)$ of 0.6 kg kg^{-1}, and the air approaches the band with a humidity potential of 0.025 kg kg^{-1}.

Data

Number of transfer units (in airflow direction (N_t)	= 0.5
Band speed (u_s)	1 m min^{-1}
Band length (Z)	15 m
Mass-transfer coefficient (K_o)	120 g $m^{-2}s^{-1}$
Humidity-potential coefficient (ϕ)	0.8
Interfacial area of particles (a)	750 m^{-1}
Voidage of material on band (ε)	0.45
Bulk density of solids (ρ_s)	1200 kg m^{-3}
Free moisture content at critical point $(X-X^*)$	0.5 kg kg^{-1}

The band extends for a relative distance θ_Z, where

$$\theta_Z = \frac{K_o \phi a Z}{(X_{cr}-X^*)\rho_s(1-\varepsilon)u_s}$$

$$= \frac{0.12 \times 0.8 \times 750 \times 15}{0.5 \times 1200 \times (1-0.45) \times 1/60}$$

$$= 196, \text{ say } 200$$

The initial moisture content Φ_o is 0.6/0.5 = 1.2 .
The relative distance into the dryer at the end of the first stage is given by eq. 7.71 :

$$\theta_c = (\Phi_o - 1)/\Pi_o$$

$$= (1.2 - 1) / 0.025$$

$$= 8$$

At the end of second stage

$$\theta = \theta_c + \theta_s$$

$$= 8 + \frac{1}{\Pi_o} \ln\left[\frac{1 + (\Phi_o - 1)\exp \Phi_o N_t}{\Phi_o}\right]$$

$$= 8 + \frac{1}{0.025} \ln\left[\frac{1 + (1.2 - 1)\exp(1.2x0.5)}{1.2}\right]$$

$$= 13.1$$

At the half-way position, the stuff will be wholly in the third stage of drying ($\Phi < 1$ everywhere). Equation 7.74 applies.

Middle of dryer ($\theta = 100$).

$$\Phi = \frac{\Phi_o}{[\Phi_o \exp(\Pi_o \theta + 1 - \Phi_o) - 1]\exp - (\Phi_o \omega) + 1}$$

$$= \frac{1.2}{[1.2 \exp(0.025 \times 100 + 1 - 1.2) - 1]\exp(-1.2\omega) + 1}$$

$$= \frac{1.2}{10.97 \exp(-1.2\omega) + 1}$$

The moisture contents are evaluated at intervals of 10 mm ($\omega = 0.1$) :-

ω	Φ	
0.1	0.112	
0.2	0.125	
0.3	0.139	average = 0.140
0.4	0.154	
0.5	0.171	

Thus at the midway position, gross variations exist.

End of Dryer ($\theta = 200$)

$$\Phi = \frac{1.2}{[1.2 \exp(0.025 \times 200 + 1 - 1.2) - 1] \exp(-1.2\omega) + 1}$$

$$= \frac{1.2}{144.8 \exp(-1.2\omega) + 1}$$

whence we get the following values:-

ω	Φ	
0.1	0.009	
0.2	0.010	
0.3	0.012	average = 0.012
0.4	0.013	
0.5	0.015	

Note, that only by striving for an almost bone-dry product can the absolute differences themselves be reduced to a small level.

The drying-rate profiles can be found from the gradient $d\Phi/d\theta$, by differentiating the appropriate expressions for Φ with respect to θ Once the mean moisture contents are known, however, the relative rate may be estimated from the expression[7]

$$\frac{N_{AZ}}{N_{AO}} = \frac{1 - \exp(-N_t\bar{\Phi}_Z)}{1 - \exp(-N_t)} \qquad (7.75)$$

when the falling rate is linear ($f=\Phi$) and the solids enter above the critical point. For the worked example just given, the relative rate at the midway position, where $\bar{\Phi}_Z = 0.140$, is predicted to be 0.172 from equation 7.75.

Vertical moisture content gradients can be ignored in fluid-bed, crossflow dryers of long aspect ratio. Under these conditions, the longitudinal profile follows with ω=o. If the feedstock is very wet and sticky, it is desirable to place a well-mixed zone ahead of the plug-flow section to get smooth running. Such a dryer could be analysed by the methods outlined in the following section.

7.5 Dryers with Solids Mixing

Not always do the solids pass through a dryer without being disturbed or mixed up to some degree. Although the return of dried material towards the solids inlet results in diminished drying rates, as noted previously in Section 6.5, there is often some benefit in turning over the material at each place to bring wetter stuff to the surface, and so speed up the process.[2] This mixing may be gentle by just raking the surface, or more thorough, as in the tumbling action of rotary dryers for example. In other instances, mixing may be virtually complete and moisture-content gradients almost eliminated, as in fluid-bed drying under conditions far from incipient fluidization. The hygrothermal conditions are then essentially uniform throughout the dryer.

Let us consider the limiting behaviour of a well-mixed system first. Suppose a bed of material is fluidized so that the height of the dense phase is Z, as shown in Fig. 7.23.

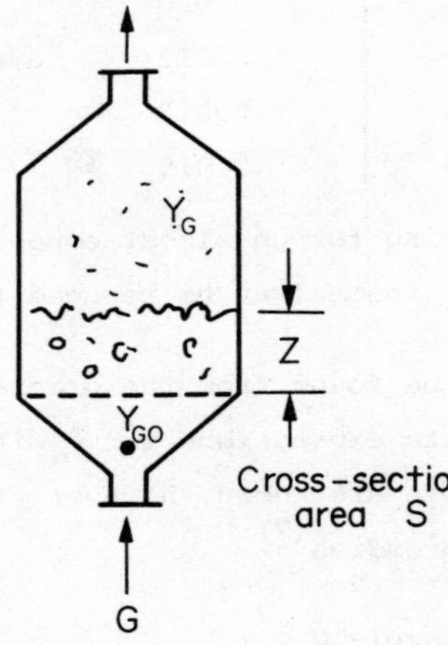

Fig. 7.23 A fluid-bed dryer

The overall drying rate for such a dryer is given by

$$N_V aSZ = GS(Y_G - Y_{GO}) \tag{7.76}$$

On noting that N_V is given by $fK_o\phi(Y_W-Y_G)$ from eq. 5.5, equation 7.76 can be re-arranged to yield the expression

$$f\left[\frac{K_o\phi aZ}{G}\right](Y_W - Y_G) = (Y_G - Y_{GO}) \tag{7.77}$$

The term in the square brackets is the number of internal transfer units N_{ti}, and equation 7.77 itself can be re-cast in terms of the humidity potentials at the inlet and outlet of the fluid bed:

$$\frac{(Y_W - Y_{GO})}{(Y_W - Y_G)} = 1 + fN_{ti} \tag{7.78}$$

Now, sometimes a small sample of material is fluidized in the laboratory to determine the drying characteristics of the stuff. The apparent relative drying rate of the material in such a test is given by

$$\bar{f} = N_V/N_V|_{f=1} \tag{7.79}$$

that is

$$\bar{f} = \frac{f\ (Y_W-Y_G)}{(Y_W-Y_G)|_{f=1}} \tag{7.80}$$

The humidity potentials in equation 7.80 can be evaluated with equation 7.78; such a calculation leads to the expression

$$\bar{f} = f\left[\frac{1 + N_{ti}}{1 + fN_{ti}}\right] \tag{7.81}$$

where f is the "true" relative drying rate for the mean moisture content of the bed. Therefore, when the bed is very extensive and N_{ti} large, $\bar{f}$ approaches 1 and the bed appears to reside in the first drying period except at the end of drying (f=0). By re-writing equation 7.81, we can get an explicit expression for the true relative rate f from the apparent value $\bar{f}$:

$$f = \bar{f}/[1 + N_{ti}(1-\bar{f})] \tag{7.82}$$

The difference between f and $\bar{f}$, as given by equation 7.82, is listed in Table 7.1 for a few values of N_{ti}.

TABLE 7.1 TRUE VALUE OF RELATIVE DRYING RATE f FROM APPARENT VALUE $\bar{f}$ IN FLUIDIZED-DRYING TESTS

$\bar{f}$	$N_{ti}=K_o\phi aZ/G$ 0.1	0.2	0.5	1.0
1	1.000	1.000	1.000	1.000
0.8	0.784	0.769	0.727	0.667
0.6	0.577	0.556	0.500	0.429
0.4	0.377	0.357	0.308	0.250
0.2	0.185	0.172	0.143	0.111
0.1	0.092	0.085	0.069	0.053

Residence-time distribution. However, even in a fluid-bed dryer, not always will complete mixing take place, particularly when coarsely particulate solids are being fluidized.[13] Also, inappropriately placed inlets and outlets can lead to some particles short-circuiting the bed, while others are held back. The moisture content of the material leaving the bed is given by

$$\Phi_Z = \int_0^\infty E(\tau)\Phi(\tau)\,d\tau \tag{7.83}$$

where $\Phi(\tau)$ is the moisture content after a given time of drying τ under the external conditions prevailing and $E(\tau)$ is the external-age distribution or "life-span" of the particles in the bed. The spread of the particles' dwell-times in the dryer is normally obtained experimentally by monitoring the progress of a tracer added to the feed material.

Should the material exhibit linear falling-rate behaviour, then

$$\frac{d\Phi}{d\tau} = - k\Phi \tag{7.84}$$

where k is an empirical constant <u>drying coefficient</u>. It thus follows that

$$[\ln \Phi]_{\Phi_o}^{\Phi} = - k\tau \tag{7.85}$$

and so

$$\Phi = \Phi_o e^{-k\tau} \tag{7.86}$$

Provided the first drying period is not too prolonged, then the value of the drying coefficient k can be adjusted to achieve a reasonable fit to the whole drying curve. When τ is very small, the higher-order terms which represent the exponential in polynomial form become negligible and thus

$$\Phi/\Phi_o = 1 - k\tau \tag{7.87}$$

The drying coefficient k can thus be determined as the limiting slope of the drying curve, as illustrated in Fig. 7.24.

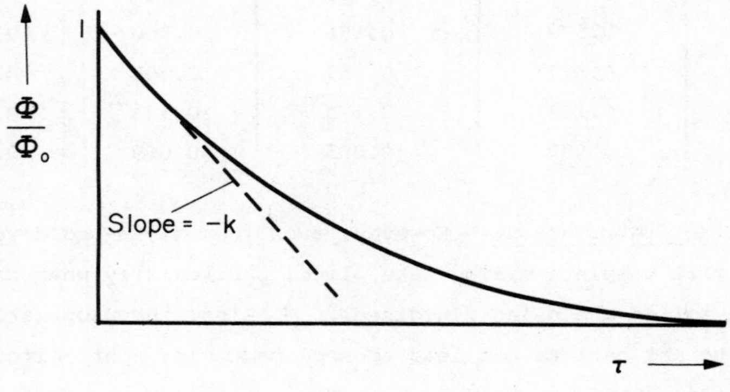

Fig. 7.24 Drying curve

The drying coefficient thus has the meaning of the ratio of the maximum rate of evaporation to the mass of drying material present.

The expression for the outlet moisture content now becomes

$$\Phi_Z = \Phi_o \int_o^\infty E(\tau) e^{-k\tau} d\tau \quad (7.88)$$

When the mixing can be described in terms of a sequence of n well-mixed zones, then

$$E(\tau/\tau_R) = \frac{n^n (\tau/\tau_R)^{n-1}}{(n-1)!} \cdot e^{-n\tau/\tau_R} \quad (7.89)$$

where τ_R is the space-time of the material in the bed, the ratio of the bed volume to the volumetric throughput. In the limit, when the whole dryer acts as a single well-mixed vessel, n=1 so

$$E(\tau) = \frac{1}{\tau_R} [e^{-\tau/\tau_R}] \quad (7.90)$$

Therefore

$$\Phi_Z = \frac{\Phi_o}{\tau_R} \int_o^\infty e^{-\tau/\tau_R} e^{-k\tau} d\tau \quad (7.91)$$

which has the solution

$$\Phi_Z = -\frac{\Phi_o}{\tau_R} . [\frac{1}{k+1/\tau_R}](-1) \quad (7.92)$$

or

$$\Phi_Z = \Phi_o/(k\tau_R + 1) \quad (7.93)$$

Thus, if complete mixing prevails, the ratio of the inlet to the outlet moisture content is linearly related to the space-time. A simple graphical test of the variation of product moistness with throughput will then show whether complete mixing can be assumed or not. This test is shown in Fig. 7.25. Data falling below the mean line may indicate short-circuiting, whereas data falling above could derive from the effects of more quiescent zones restricting the movement of material.

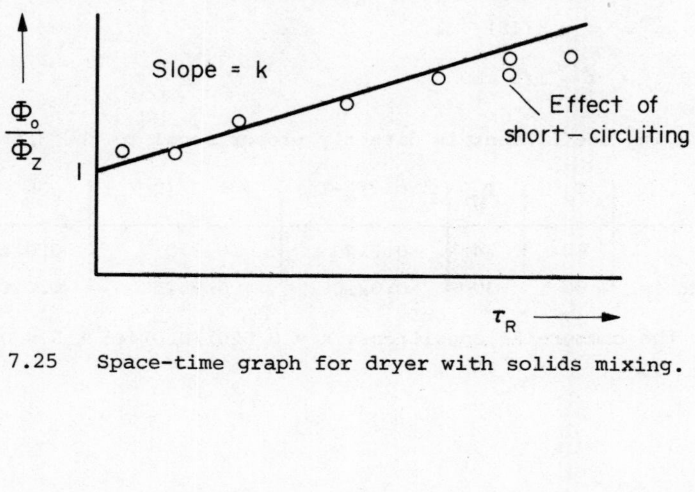

Fig. 7.25 Space-time graph for dryer with solids mixing.

The following worked examples illustrate the use of residence-time distributions in predicting the extent of drying when solids mixing occurs.

Example 7.10. The free moisture content of pharmaceutical tablets is reduced to one-quarter of its initial value after 15 min in a laboratory fluid-bed dryer in which complete mixing takes place. A tracer pulse is admitted to the inlet of a commercial dryer, and the outlet response is monitored with the following results:

time/min	tracer/g	time/min	tracer/g
0 - 2	8.5	10 - 12	0.50
2 - 4	12.2	12 - 14	0.15
4 - 6	7.5	14 - 16	0.05
6 - 8	3.2	16 - 18	0.01
8 - 10	1.13	18 - 20	0

What will be the outlet moisture content if the space-time in the commercial dryer is 20 min.? The following psychrometric readings were taken:

	dry bulb / ^oC	wet bulb / ^oC
test dryer	60	24
commercial dryer	80	30

Drying coefficient. Since complete mixing occurs in the laboratory test, from equation 7.93

$$\Phi_o/\Phi_Z = k\tau_R + 1$$

$$\therefore \quad 4 = k(15) + 1$$

$$\therefore \quad k = 1/5 \text{ min}^{-1}$$

The drying coefficient is directly proportional to the driving force $\phi\Pi$

dryer	T_G	T_W	$\Pi = (Y_W - Y_G)$	$\phi = D/(D+Y_W)$	$\phi\Pi$
test	60	24	0.0149	0.970	0.01445
commercial	80	30	0.0210	0.9575	0.0201

Under the commercial conditions, $k = 0.0201/0.01445 \times 5 = 0.278 \text{ min}^{-1}$.

Residence-time distribution. The mass of tracer appearing at the outlet can be taken directly as a measure of the life-span within the vessel, but the data must be "normalized" so that $\int_0^{\infty} E d\tau = 1$.

τ/min	$E^*(\tau)$	$E(\tau)$	$\Phi/\Phi_o = e^{-k\tau}$	$E(\tau)\Phi/\Phi_o$
0	0	0	1	0
1	8.5	0.1278	0.7573	0.0968
3	12.2	0.1835	0.4343	0.0797
5	7.5	0.1128	0.2491	0.0281
7	3.2	0.0481	0.1428	0.0069
9	1.13	0.0170	0.0819	0.0014
11	0.50	0.0075	0.0470	0.00035
13	0.15	0.0023	0.0270	0.00006
15	0.05	0.00075	0.0154	0
17	0.01	0.00015	0.0089	0
19	0	0	0.0051	0

$$\sum_0^{20} E^*(\tau)\delta\tau \simeq \int_0^{\infty} E^*(\tau)d\tau$$

$$= (8.5 \times 2) + (12.2 \times 2) + \ldots\ldots \quad = 66.5$$

To normalize the age distribution the figures in column two are divided by 66.5 to yield those in column three.

Whence

$$\Phi/\Phi_o = \sum_0^{20} E(\tau)\ (\Phi/\Phi_o)\delta\tau$$

$$= (0.0968 \times 2) + (0.0797 \times 2) + \ldots\ldots$$

$$= 0.427$$

The drying in the commercial unit is much poorer, notwithstanding the greater humidity driving force, as a considerable fraction of the material passes through very quickly and so emerges only partially dried.

Example 7.11. A powdered salt consisting of 0.5 mm granules is fluidized in a cylindrical chamber of 1.75 m diameter to dry it. Complete mixing occurs in the fluid bed which is 100 mm high. To double the throughput, a second bed of identical dimensions is set up in series with the first. What must the bed height be, if this is the same in each of the two units?

Data. Maximum specific evaporation per unit grid area (N_W) $0.15\ kg\ m^{-2} s^{-1}$

Bed porosity (ε)	0.7
Salt density (ρ_s)	$1560\ kg\ m^{-3}$
Throughput of bone-dry solids	$2000\ kg\ h^{-1}$

Drying coefficient, k. By definition of this coefficient

$$k = \frac{\text{evaporation rate}}{\text{mass of material}} = \frac{N_W S}{(1-\varepsilon) S Z \rho_s}$$

where S = cross-sectional area of the grid,

and Z = bed height

$$\text{i.e.}\quad k = N_W/(1-\varepsilon) Z\rho_s = \frac{0.15 \times 60}{(1 - 0.7) \times 0.1 \times 1560}$$

$$= 0.192\ min^{-1}$$

Space-time, τ_R. Volumetric feedrate = 2000/1560 = 1.282 $m^3 h^{-1}$

$$\text{Bed volume} = \frac{\pi}{4} \times 1.75^2 \times 0.1 = 0.2405\ m^3$$

$$\therefore\quad \tau_R = \frac{0.2405}{1.282} \times 60 = 11.3\ min.$$

Moisture-content ratio, Φ_Z/Φ_O. For complete mixing

$$\frac{\Phi_Z}{\Phi_O} = \frac{1}{k\tau_R + 1} = \frac{1}{(0.192 \times 11.3) + 1}$$

$$= 0.315$$

Two-stage dryer. From equation 7.89,

$$E(\tau/\tau_R) = \frac{2^2(\tau/\tau_R)}{1}\ e^{-2\tau/\tau_R}$$

$$\therefore\quad E(\tau) = \frac{4\tau}{\tau_R^2}\ e^{-2\tau/\tau_R}$$

and from equation 7.88,

$$\frac{\Phi_Z}{\Phi_O} = \int_0^\infty E(\tau)\Phi(\tau)d\tau$$

$$= \frac{4}{\tau_R^2} \int^{\infty} \tau e^{-2\tau/\tau_R} e^{-k\tau} d\tau$$

$$= - \frac{4}{\tau_R^2} \left[\frac{e^{-(K+2/\tau_R)\tau}}{(k + 2/\tau_R)^2} \{(k+1/\tau_R)\tau + 1\} \right]_o^{\infty}$$

$$= + \frac{4}{\tau_R^2} \left[\frac{1}{(k+2/\tau_R)^2} \right]$$

By algebraic re-arrangement

$$\tau_R = \frac{2}{k} \left[\frac{1}{\sqrt{\Phi_Z/\Phi_O}} -1 \right]$$

$$= \frac{2}{0.192} \left[\frac{1}{\sqrt{0.315}} - 1 \right]$$

$$= 8.14 \text{ min}$$

To accommodate the lesser residence time, the bed height must be lowered to a value 8.14 x 100/11.3 = 72 mm, or by 28 per cent.

In this latter example, the subdivision of the drying process into two stages improves the drying because the particles tend to bunch together to dwell in the dryer for a longer period rather than emerge at the outlet randomly, which is the case for perfect mixing. As the process is broken down into more and more stages, so the bunching becomes more pronounced until, in the limit of an infinite number of stages, the solids emerge at a single instant.

7.6 Multistaged Dryers

In the previous example, it was found that the overall time that a moist material need spend in a drying vessel could be reduced for the same product moistness by subdividing a well-mixed dryer into two stages. There are other advantages in sectioning the dryer into stages: it is possible to set the process conditions in each stage separately to take account of the changing susceptibility of the material to heat as it dries out, or to compensate for the difficulty of drying when the moisture content is low.

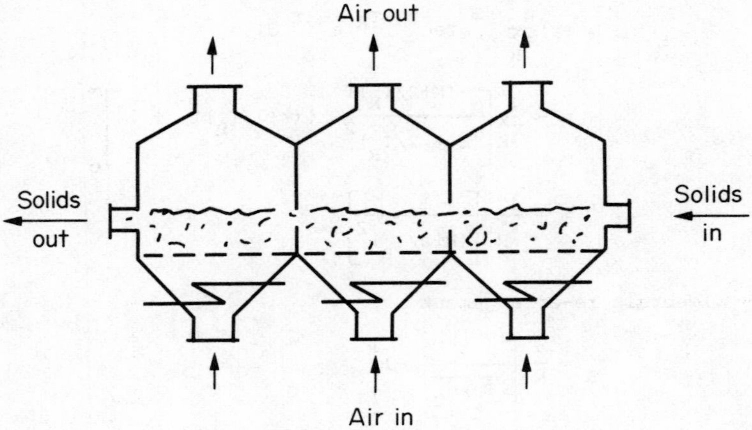

Fig. 7.26 Crossflow-type multistaged dryer.

In a multistaged fluid-bed dryer, like that shown in Fig. 7.26, the solids progress in a crossflow fashion from stage to stage, each of which having its own air inlet. Clearly, the conditions in each section can be adjusted independently of the others. Unlike the crossflow progressive dryers already considered, the moisture content of the solids changes discontinuously from stage to stage. The progressive dryer could, however, be considered as a crossflow unit having an infinite number of very small stages so that the moisture-content change is continuous.

There are other dryers in which the moisture content of the drying stuff varies continuously but the humidity of the bulk air changes in a step-wise manner. One such dryer is the multiple-drum system, used for drying scoured and greasy wool, as sketched in Fig. 7.27.

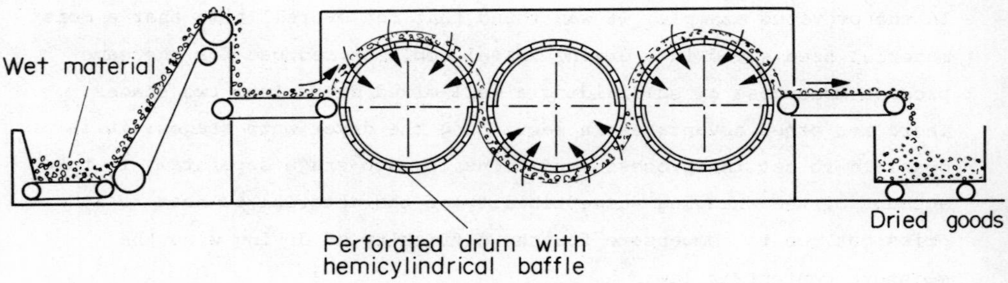

Fig. 7.27 Counterflow-type multistaged dryer.

A mat of loose fibres is drawn successively over and under a series of suction drums. Air is sucked through the wool layer by axially mounted fans, one to each rotating drum. The drums are housed in baffled chambers which permit most of the air to be recirculated, but some to counterflow against the stream of wool. The air humidity varies discontinuously from chamber to chamber, while the wool dries out progressively during its passage through the dryer.

Let X_i be the moisture content of material <u>leaving</u> the ith stage of a multistaged unit. Likewise, let Y_{Gi} be the bulk-air humidity leaving that stage, and so on. Then the air and material streams associated with multistaged dryers can be labelled as set out in Fig. 7.28. Note that the fresh air and incoming solids are labelled as if they were arriving from a fictitious stage downstream.

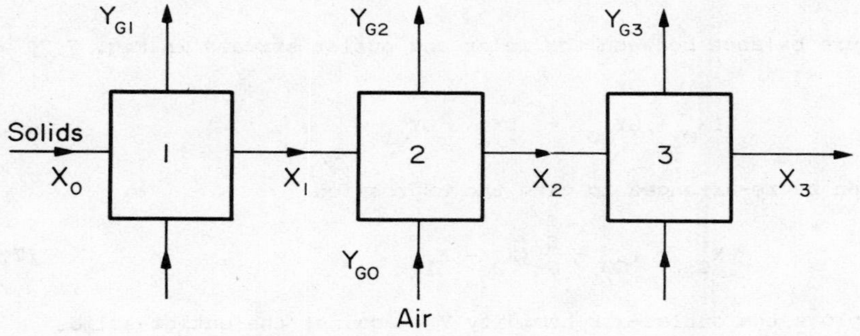

(a) Crossflow-type dryer

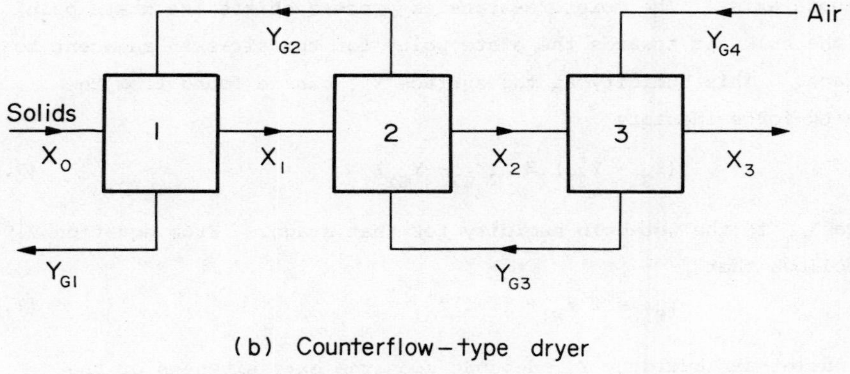

(b) Counterflow-type dryer

Fig. 7.28 Nomenclature for multistaged dryers.

Single stage. Initially, let us consider the changes in the amount of moisture in each stream over a single stage, as a basis for examining later the overall changes when a number of such stages are coupled together.

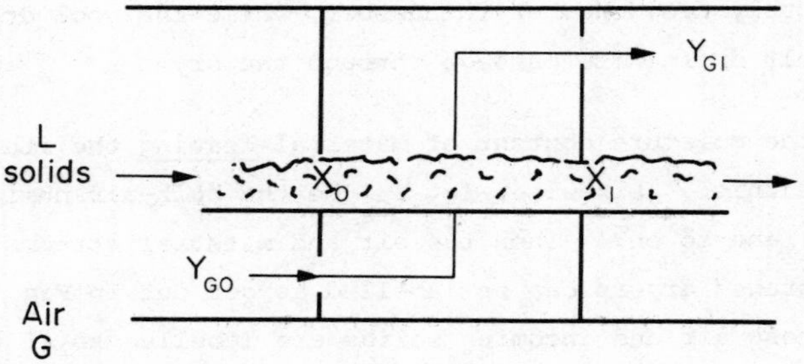

Fig. 7.29 Moisture changes over a single stage.

A moisture balance between the inlet and outlet streams in Fig. 7.29 yields

$$LX_O + GY_{GO} = LX_1 + GY_{G1} \tag{7.94}$$

which can be re-arranged to give the expression

$$Y_{G1} = Y_{GO} + \frac{L}{G}(X_O - X_1) \tag{7.95}$$

If one plots the outlet-air humidity Y_{G1} against the outlet-solids moisture content X_1 , a straight line results, the so-called operating line, which has a slope $-L/_G$ and an intercept of $Y_{GO} + LX_O/G$ on the ordinate axis. The moisture-transfer process shifts the state point for the bulk air towards the state point for the air-skin adjacent to the surface. This humidity at the surface Y_{S1} can be found from the driving-force identity

$$(Y_{S1} - Y_{G1}) \equiv f(Y_{W1} - Y_{G1}) \tag{7.96}$$

where Y_{W1} is the wet-bulb humidity for that stage. From equation 7.96, it follows that

$$Y_{S1} = f\,Y_{W1} + (1-f)Y_{G1} \tag{7.97}$$

The outlet-air humidity Y_{G1} depends upon the extensiveness of the contact of the air with the solids, for which the number of transfer units N_{te} is a measure. From equation 6.60 it follows that

$$- \ln \frac{(Y_{S1} - Y_{G1})}{(Y_{S1} - Y_{GO})} = N_{te} \quad (7.98)$$

whence

$$\frac{(Y_{G1} - Y_{GO})}{(Y_{S1} - Y_{GO})} = 1 - e^{-N_{te}} \quad (7.99)$$

which may be used to fix the state point for the outlet air on the line between the inlet and surface-air conditions

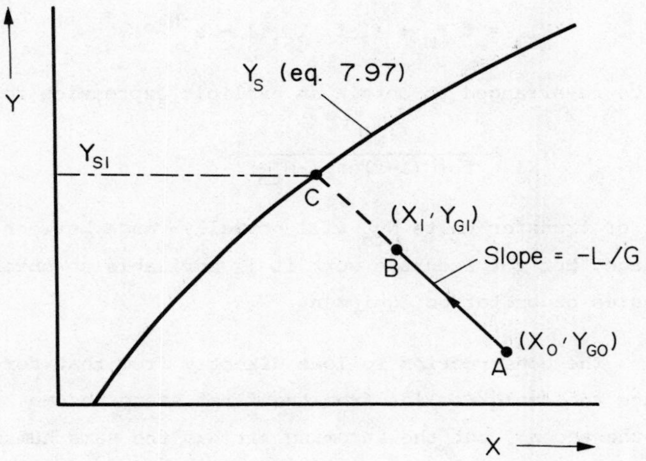

Fig. 7.30 Graphical construction for moisture-concentration changes over a single stage.

The graphical construction illustrating the moisture-concentration changes over a single stage is shown in Fig. 7.30, in which the moisture content of the solids X is plotted as abscissa and the air humidity Y as ordinate. Point A represents the moistness of the entering streams, and the line AB is the operating line, equation 7.95. The surface-air condition is located at point C on the extension of the line segment AB to the surface-air curve (eq. 7.97) such that the ratio AB/AC is given by $AB/AC = 1 - \exp(-N_{te})$, see eq. 7.99.

When the outlet-air humidity is unknown, Y_{S1} can be found from the wet-bulb humidity in the following way. Equation 7.99 can be re-arranged to give

$$Y_{G1} = Y_{GO} + (Y_{S1} - Y_{GO})\ (1 - e^{-N_{te}}) \tag{7.100}$$

or

$$Y_{G1} = Y_{S1}(1 - e^{-N_{te}}) + Y_{GO}e^{-N_{te}} \tag{7.101}$$

If one neglects the second term of equation 7.101 on the grounds of being the product of two small quantities, and substitutes the simplified expression for Y_{G1} into equation 7.97, one gets

$$Y_{S1} = f\ Y_W + (1-f)\ Y_{S1}(1 - e^{-N_{te}}) \tag{7.102}$$

which can be re-arranged to obtain an explicit expression for Y_{S1} :

$$Y_{S1} = \frac{fY_W}{f + (1-f)\exp(-N_{te})} \tag{7.103}$$

The number of transfer units N_{te} will normally range between 0.5 and 1 in each stage, but for accurate work it is advisable to obtain data from drying studies on prototype equipment.

Crossflow. The construction follows directly from that for the single stage, since the outlet-solids from the first stage become the inlet solids to the second, but the incoming air has the same humidity as that entering the first. Thus a sawtoothed construction with a common base-line, the inlet-air humidity, represents the changing moisture concentrations from stage to stage. This construction is shown in Fig. 7.31 for a two-stage unit.

If the surface-air humidity curve is almost linear, then the air should be evenly divided between stages. If the surface-air humidity is falling with decreasing moisture content of the solid, then the effect can be counteracted to some extent by increasing the amount of air (and thus decreasing the slope of the operating line). In practice, this option may not be open because of the operating characteristics of the fans used and the design of the dryer itself.

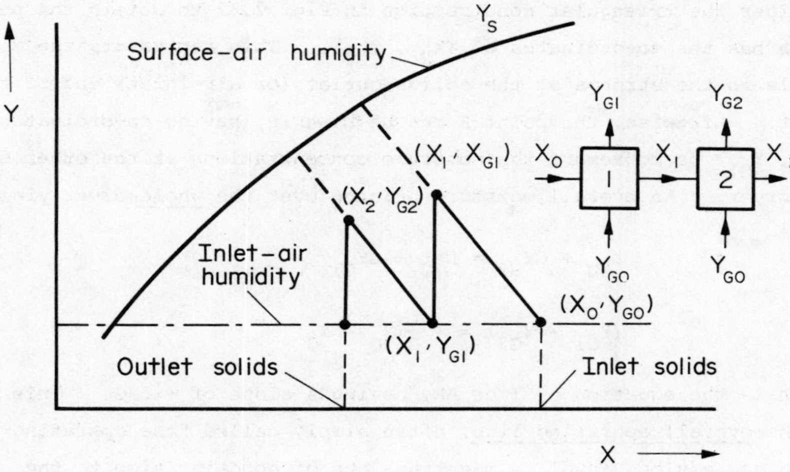

Fig. 7.31 Two-stage crossflow unit.

Counterflow. The graphical analysis for the case when air and solids move countercurrently can be the same as that for the crossflow case, however, it is not the most convenient, since the inlet streams to the first stage are no longer the terminal streams at that end. One stream enters, and a counterflowing stream emerges (see Fig. 7.32).

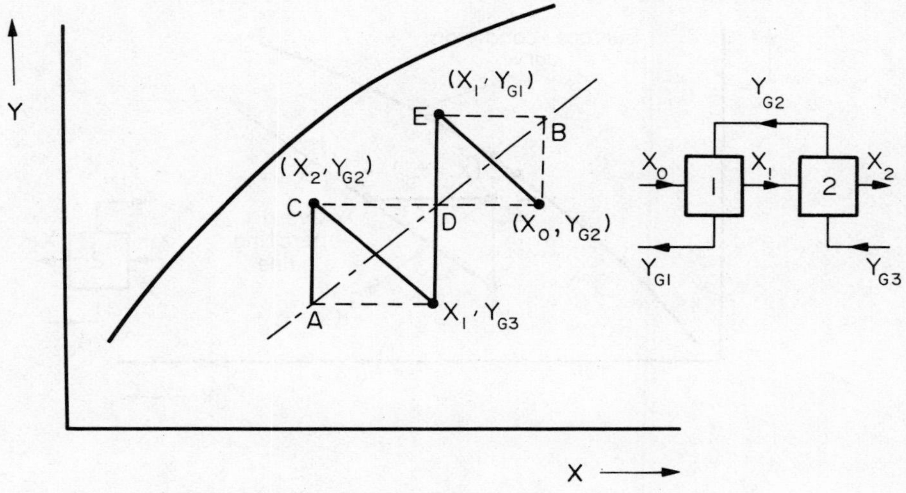

Fig. 7.32 Two-stage counterflow unit.

Consider the triangular construction in Fig. 7.32 to obtain the point A, which has the co-ordinates of (X_2 , Y_{G3}). This represents the moisture levels in the streams at the solids-outlet (or air-inlet) end of the dryer. Likewise, the point B can be drawn in, having co-ordinates of (X_O , Y_{G1}) to represent the moisture concentrations at the other end of the dryer. An overall moisture balance over the whole dryer yields

$$L_{XO} + GY_{G3} = LX_2 + GY_{G1} \tag{7.104}$$

or

$$(Y_{G1} - Y_{G3}) = + \frac{L}{G}(X_O - X_3) \tag{7.105}$$

which is the equation of line AB, having a slope of + L/G. This line is the overall operating line, often simply called "the operating line", which has a slope equal in magnitude but of opposite sign to the operating line which results from a mass balance around a single stage. The stepwise construction ACDEB between the overall operating line ADB and the points C and E, representing moisture levels in the outlet streams, gives us the number of stages from the number of steps involved.

The residual problem with this graphical analysis is the location of the outlet-stream points when only the surface-condition (Y_S vs X) and the overall operating line are known. Let us consider the concentration changes about the second stage, as shown in Fig. 7.33.

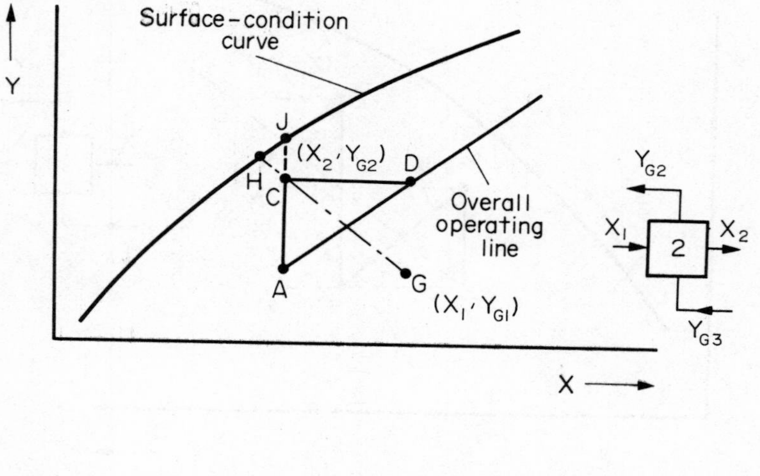

Fig. 7.33 Moisture-concentration changes about the second stage of a two-stage counterflow dryer.

Now, if the surface-condition curve and the overall operating line have already been located, the outlet concentrations can be read off once the enrichment ratio, AC/AJ is obtained. Let this ratio be e_2 and the fractional approach to saturation, GC/GH be e_1. The relationship between these ratios has been given as:[9]

$$e_1 = \frac{(1+\Lambda)e_2}{1+\Lambda e_2} \qquad (7.106)$$

where $\Lambda = \alpha G/L$, in which α is the mean slope of the surface-condition curve. Equation 7.106 can be put into another form yielding an explicit expression for the enrichment ratio e_2 , that is

$$e_2 = \frac{e_1}{1 + \Lambda\ (1-e_1)} \qquad (7.107)$$

The fractional approach to saturation e_1 can be calculated from equation 7.99 directly in terms of the number of transfer units N_{te}. With this substitution, equation 7.107 becomes

$$e_2 = \frac{1 - \exp(-N_{te})}{1 + \Lambda\exp(-N_{te})} \qquad (7.108)$$

The graphical analysis of a counterflow-type dryer is demonstrated for the case of a multiple-drum dryer in the following worked example.

Example 7.12. The following data are recorded for a five-drum dryer processing loose wool:

inlet-air humidity	0.006 kg kg^{-1}
outlet-air humidity	0.196 kg kg^{-1}
inlet moisture content	0.88 kg kg^{-1}
outlet moisture content	0.12 kg kg^{-1}
wool flow (dry-stuff basis)	10 kg min^{-1}
air flow (dry-stuff basis)	40 kg min^{-1}

The drying is adiabatic at a wet-bulb temperature of $67^{\circ}C$. Confirm that about 1 transfer unit is found in each stage.

Data Relative drying rates:

X	f	X	f
0.7370	0.823	0.3285	0.480
0.6028	0.757	0.2258	0.340
0.5391	0.727	0.1816	0.294
0.4218	0.613		

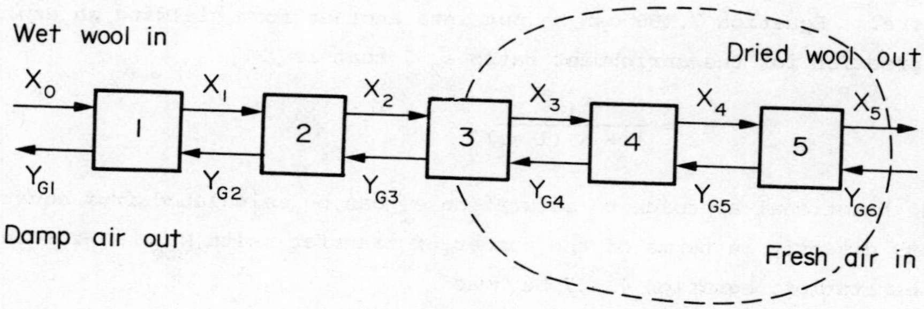

Fig. 7.34 Five-stage dryer

Overall operating line. From a mass balance over the dotted envelope in Fig. 7.34, one obtains (by analogy with the derivation of equation 7.105)

$$Y_G = Y_{G6} + \frac{L}{G}(X - X_5)$$

$$= 0.006 + \frac{10}{40}(X - 0.12)$$

$$= X/4 - 0.0024$$

This is the equation of the overall operating line.

Surface-condition curve. The wet-bulb humidity (Y_W) at 67°C is 0.234 kg kg^{-1}, whence from the data on relative drying rates one finds:

X	Y_G	f	$Y_S = fY_W + (1-f)Y_G$
0.7370	0.182	0.823	0.225
0.6028	0.148	0.757	0.213
0.5391	0.132	0.727	0.206
0.4218	0.103	0.613	0.183
0.3285	0.0797	0.480	0.154
0.2258	0.0540	0.340	0.115
0.1816	0.043	0.294	0.099

Graphical procedure. The operating line and the surface-condition curve are plotted on a common graph with rectilinear axes, see Fig. 7.35. The stepwise construction can then begin from the air-inlet end towards the wool feed.

Stage	$Y_G\|_{in}$	Y_S	Λ	$Y_G\|_{out} - Y_G\|_{in}$ (eq.7.108)
5	0.006	0.072	1.7	0.026
4	0.032	0.112	1.7	0.031
3	0.061	0.153	1.28	0.040
2	0.101	0.200	0.64	0.051
1	0.151	0.223	0.24	0.042
0	0.193 = outlet humidity			
	0.86 = inlet moisture content			

The stepwise analysis predicts that the wool enters with a moisture content of 0.86 kg kg^{-1} compared with an actual value of 0.88 kg kg^{-1}. The difference is slight, and within the error of the method and the estimation of the number of transfer units. The method confirms that the humidification of the air in each drum corresponds to 1 transfer unit (N_{te}).

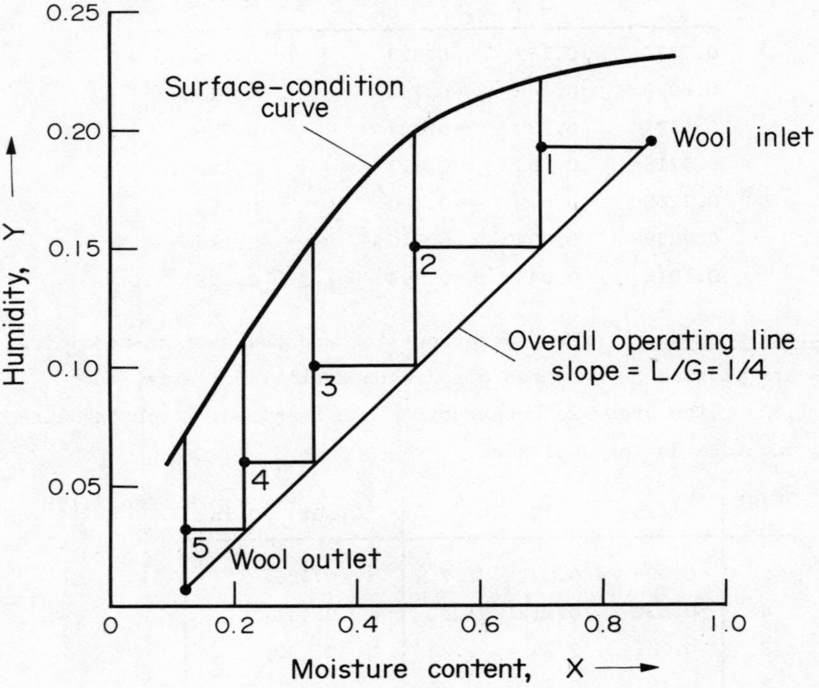

Fig. 7.35 Graphical analysis for Example 7.12

From the graph given in Fig. 7.35, we can read off the moisture content of the wool as it leaves each drum:

stage	moisture content/kg kg^{-1} inlet	outlet
1	0.86 (0.88)	0.70
2	0.70	0.495
3	0.495	0.33
4	0.33	0.215
5	0.215	0.12

Like most countercurrent mass-transfer operations, the separation is easiest in the middle of the concentration range.

The stepwise reduction in moisture contents along the dryer can also be represented by a single finite-difference equation, which is soluble when the surface-condition curve can be linearized.[(5)] In most cases, however, the graphical method is to be preferred, being simpler and capable of handling non-linearities, besides giving a visual display of the moisture concentrations throughout the dryer.

REFERENCES

1. Beran, Z. and J. Lůtcha, Optimising particle residence time in a fluidised bed dryer, Chem.Engr 303, 678-681 (1975).
2. Beyer, D. and D.C.T. Pei, The effect of different modes of operation on the drying processes, Int.J.Heat Mass Transfer, 18, 707-9 (1975).
3. Catherall, N.F., The Process Design of Continuous Dryers, B.E. Report, Univ. Canterbury, N.Z. (1970).
4. Grosvenor, W.M., Calculation for dryer design, Trans.AIChE, 1, 184-202 (1907).
5. Jenson, V.G. and G.V. Jeffreys, "Mathematical Methods in Chemical Engineering", p.322-6, Academic Press, London (1963).
6. Keey, R.B., "Drying Principles and Practice", p.247-8, Pergamon Press, Oxford (1972).
7. Keey, R.B., Process optimisation of a conveyor dryer, N.Z.Eng., 30, 53-7 (1975).
8. Keey, R. B. Process design of continuous drying equipment, AIChE Symposium Series, 73 (163), 1-11 (1977).
9. King, C. J., "Separation Processes", p. 606 McGraw Hill, New York (1971).
10. Kröll, K., "Trockner und Trocknungsverfahren", p. 129, Springer Verlag, Berlin/Göttingen/Heidelburg (1959).
11. Poersch, W., Berechnung der Verweilzeit in Gleich- oder Gegenstromtrocknern mit Hilfe von Austauscheinheiten, Verfahrenstechn., 5, 160-8, 186-192 (1971).
12. Schlünder, E.U., Fortschritte und Entwicklungstendenzen bei der Auslegung von Trocknern für vorgeformte Trocknungsgüter, Chem.-Ing.-Techn., 48, 190-8 (1976).
13. Tuiyayev T.Ya., A.L. Tsailingold' and A.B. Builov, Non-homogenity of suspended catalyst beds (in Russian), Zh.Prikl.Khim., 34, 558-564 (1961).
14. Van Meel, D.A., Adiabatic convection batch drying with recirculation of air, Chem.Eng.Sci., 9, 36-44 (1958).

PLATE 7. A twin-compartment, gas-fired batch dryer with cross-circulation of air for ceramic refractory blocks. [Casburt Ltd., with permission.]

Chapter 8

BATCH DRYING

8.1 Drying Ovens

Originally all drying was done in "batches". Goods to be dried were laid out in the sun, or exposed to the wind. The housewife still pegs out her washing on the clothes-line, and timber boards are often just stacked in the open air to season "naturally". Later, as industry developed, heated drying rooms were set aside to take the wet articles. The modern batch-drying chamber is simply a room shrunken to the size of a large cabinet which has controlled heating and ventilation.

The analysis of the way the moist material dries in such a cabinet is perforce more complex than in the case of continuous progressive drying. In batch drying, the conditions vary with both time and place.

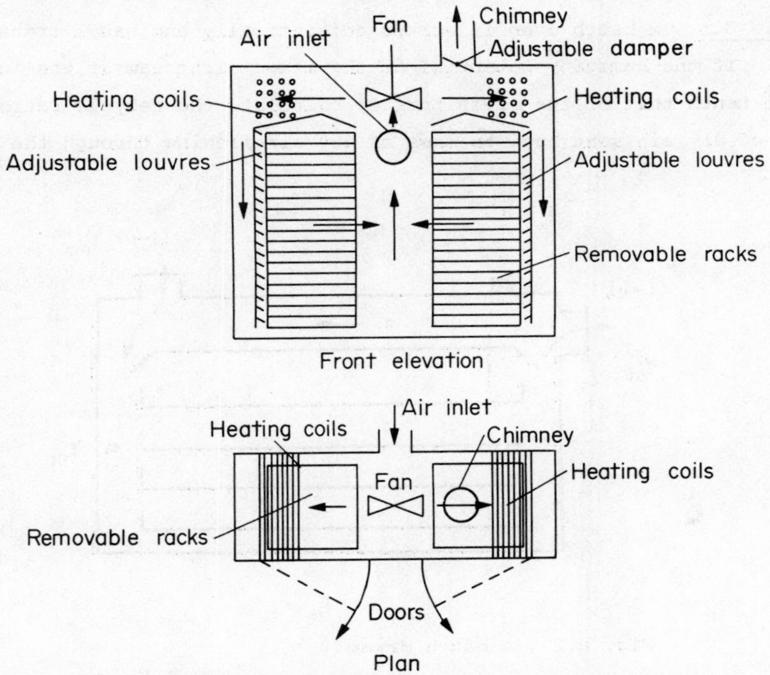

Fig. 8.1 A twin-rack drying oven with side-to-centre airflow.

Consider the drying oven shown in Fig. 8.1. As the air is drawn over the shelves holding the moist goods, it becomes damper and its drying potential falls. Thus, the stuff dries out unevenly, and if the airflow is restricted, the material towards the centre can still be quite moist when that at the sides of the cabinet is almost dry. Further, unless the fan is properly sited and the air uniformly distributed over the shelves, there will be a considerable variation in the rate at which the material dries between shelves. Variations in drying rate can also arise through temperature gradients caused by excessive heat losses through poorly insulated or ill-fitting doors. Thus, to reach a specified degree of dryness, some of the material must be overdried. When timber boards are dried in large chambers called <u>kilns</u>, the airflow direction is sometimes reversed periodically to reduce board-to-board variations in moisture content at the end of the drying schedule.

To improve the thermal efficiency, part of the air leaving the shelves is diverted to the intake and only part is expelled to the outside. The evaluation of the heat quantities for an oven has already been outlined in Section 3.3. Often the fraction of the air that is recycled is considerable, as the following worked example shows.

<u>Example 8.1</u> A batch oven is worked adiabatically and has 1 transfer unit. If the humidity potential of the mixed airstream at the inlet is one tenth that of the fresh intake, calculate the recycle ratio (mass of dry air sent back to mass of dry air flowing through the dryer).

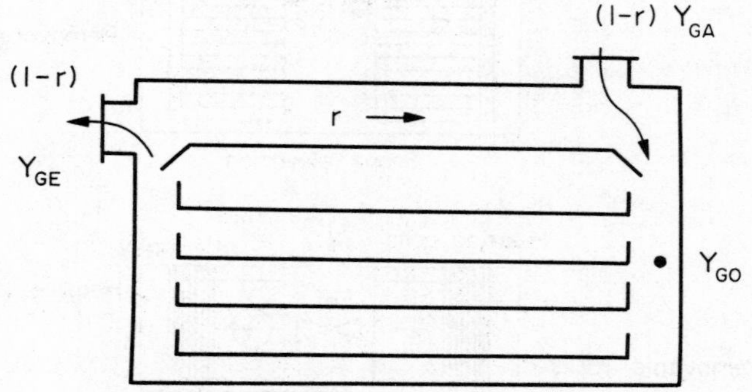

Fig. 8.2 A batch dryer.

A mass balance at the air intake yields:

$$rY_{GE} + (1-r)Y_{GA} = Y_{GO}$$

Whence

$$r = \frac{Y_{GO} - Y_{GA}}{Y_{GE} - Y_{GA}} = \frac{\Pi_A - \Pi_O}{\Pi_A - \Pi_E}$$

where $\Pi_O = (Y_W - Y_{GO})$ and so on.

But $\Pi_E/\Pi_O = N_{VE}/N_{VO} = \exp(-N_t)$ from eq. 4.72

So

$$r = \frac{\Pi_A/\Pi_O - 1}{\Pi_A/\Pi_O - \Pi_E/\Pi_O}$$

$$= \frac{\Pi_A/\Pi_O - 1}{\Pi_A/\Pi_O - \exp(-N_t)}$$

$$= \frac{(10 - 1)}{10 - \exp(-1)} = 0.934$$

Thus only 6.6 per cent of the through-air is discharged from the
dryer.   As this quantity is reduced, so the concomitant thermal
loss decreases at the penalty of lengthening the duration of drying.
Considerable labour is needed to load, unload and clean the trays.   Thus,
whenever possible, tray dryers are being superseded by other equipment such as
fluid-bed dryers.   The variation of drying conditions in the direction of
the airflow at any instant is small in these dryers unlike that in tray dryers.

8.2   Deep Beds

If the dryer is extensive, or the air-rate very small, the number of
transfer units $K_o \phi aZ/G$ becomes large.   The loading ratio for unhindered
drying,

$$\varepsilon_{max} = 1 - \exp(-N_t) \tag{6.98}$$

is thus close to unity, and the air leaves the dryer virtually saturated.
The drying now takes place essentially within a narrow zone, rather than
throughout the whole dryer, albeit at different rates.   The sequence of
events during the drying process is illustrated in Fig. 8.3.

Thus, when a deep bed of moist material such as binned grain is gently
aerated, a drying zone of limited extent progresses slowly from the
inlet to the outlet.   In effect, a desorption wave moves through the
dryer in a way that is similar to the behaviour of fixed-bed ion-exchange
columns.   However, this wave is not the sole one to pass through.   If
the aeration is slow enough, so that local equilibrium exists everywhere

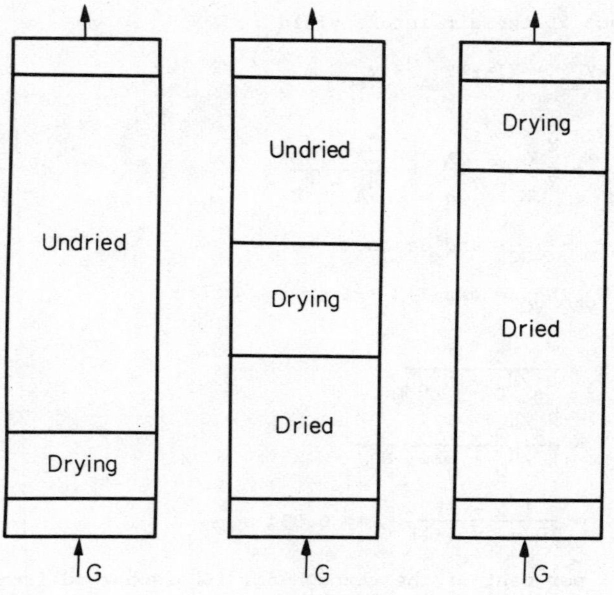

Fig. 8.3 Zones in deep-bed drying

(the number of transfer units being infinite), then mass and energy balances over a differential shell in the bed yield the expressions:[3]

$$\frac{\partial Y}{\partial \tau} + \frac{\rho_S(1-\varepsilon)}{\rho_G \varepsilon} \frac{\partial X}{\partial \tau} + \hat{u}_G \frac{\partial Y}{\partial z} = 0 \qquad (8.1)$$

$$\frac{\partial T}{\partial \tau} + \frac{\Delta H_{VS}}{C_S} \frac{\partial X}{\partial \tau} + \frac{C_{PG}}{C_S} \hat{u}_G \frac{\partial T}{\partial z} = 0 \qquad (8.2)$$

where $\hat{u}_G$ is the cup-mixing velocity of the air. This dual relationship can be resolved into one by defining[2] a new potential F, which is a linear combination of T and Y:

$$F_i = T + \alpha_i Y \qquad i = 1,2 \qquad (8.3)$$

where α_i is some positive coefficient. The single equation becomes

$$\frac{\partial F_i}{\partial \tau} + \left[\frac{\hat{u}_G}{1 + \mu \gamma_i} \right] \frac{\partial F_i}{\partial z} = 0 \qquad (8.4)$$

with $\mu = \rho_S(1-\varepsilon)/\rho_G \varepsilon$

and $\gamma_i = \left[\frac{\partial X}{\partial Y} \right]_{F_j} \qquad i,j = 1,2 \; ; \; i \neq j$

Equation 8.4 is a kinematic wave equation. Two waves F_i propagate through the bed at a velocity $[1 + \mu\gamma_i]^{-1}\ \hat{u}_G$. The F_1 wave sweeps through at constant F_2, and vice versa. Whenever the heat- and moisture-transfer processes are weakly coupled, as anticipated with slow drying, these two waves appear at quite different times. Under limiting conditions, γ_1 is very small and the first front moves through at the air velocity $\hat{u}_G$, while γ_2 is very large and the second front is much delayed. Similar behaviour occurs when the mass-transfer coefficients are finite, but the fronts are somewhat wider then.

Illustrated in Fig. 8.4 is the appearance of these fronts at the outlet of a small pad of woollen felt following a step change in the inlet-air temperature. Shortly afterwards, the temperature rises at the outlet, but the relative humidity and thus the equilibrium moisture content scarcely alter. After a considerable delay, about 250 times longer than the space-time for the bed, a more sluggish change takes place, the state path for which closely follows an adiabatic saturation path. These changes are mapped out on a psychrometric chart in Fig. 8.5.

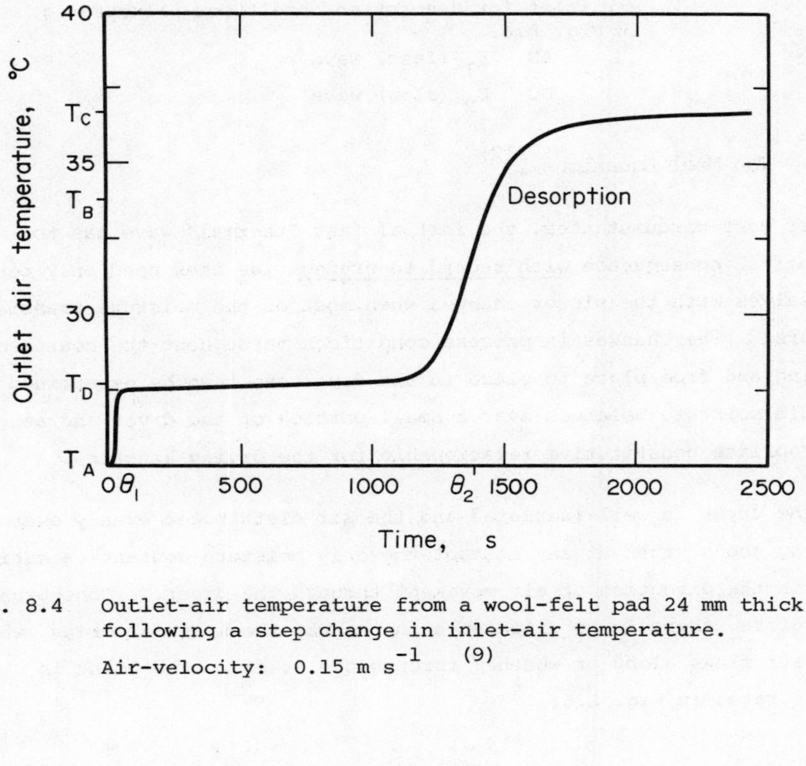

Fig. 8.4 Outlet-air temperature from a wool-felt pad 24 mm thick following a step change in inlet-air temperature. Air-velocity: 0.15 m s^{-1} (9)

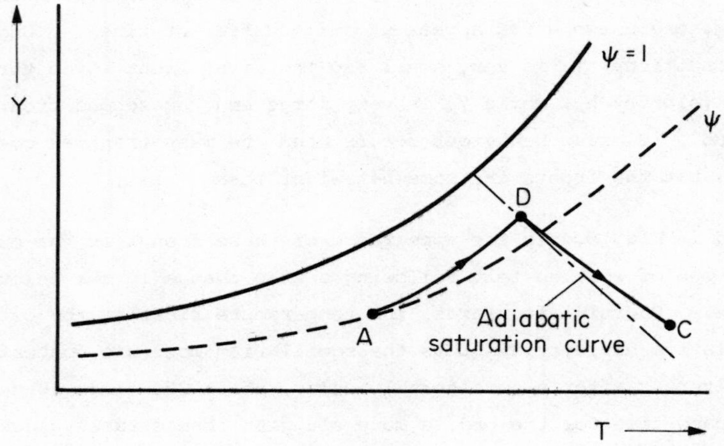

Fig. 8.5 Grosvenor chart illustrating state changes at air inlet for desorption conditions illustrated in Fig. 8.4.

AD F_1 (fast) wave

DC F_2 (slow) wave

8.3 Van Meel Equations (12)

Under most circumstances, the initial fast "thermal" wave has no practical consequence with regard to drying. We thus need only concern ourselves with the slower changes when most of the moisture transfer occurs. The changes in process conditions throughout the course of drying and from place to place in the dryer can then be determined from simple moisture balances over a small portion of the dryer and an appropriate constitutive relationship for the drying kinetics.

If the dryer is well-insulated and the air distributed evenly over the drying goods, then at any instant the only moisture-content variations lie in the direction of air movement through the dryer. Conceptually, therefore, there is no difference in the way the material dries, whether the air flows along or whether through the stuff. The point is illustrated in Fig. 8.6.

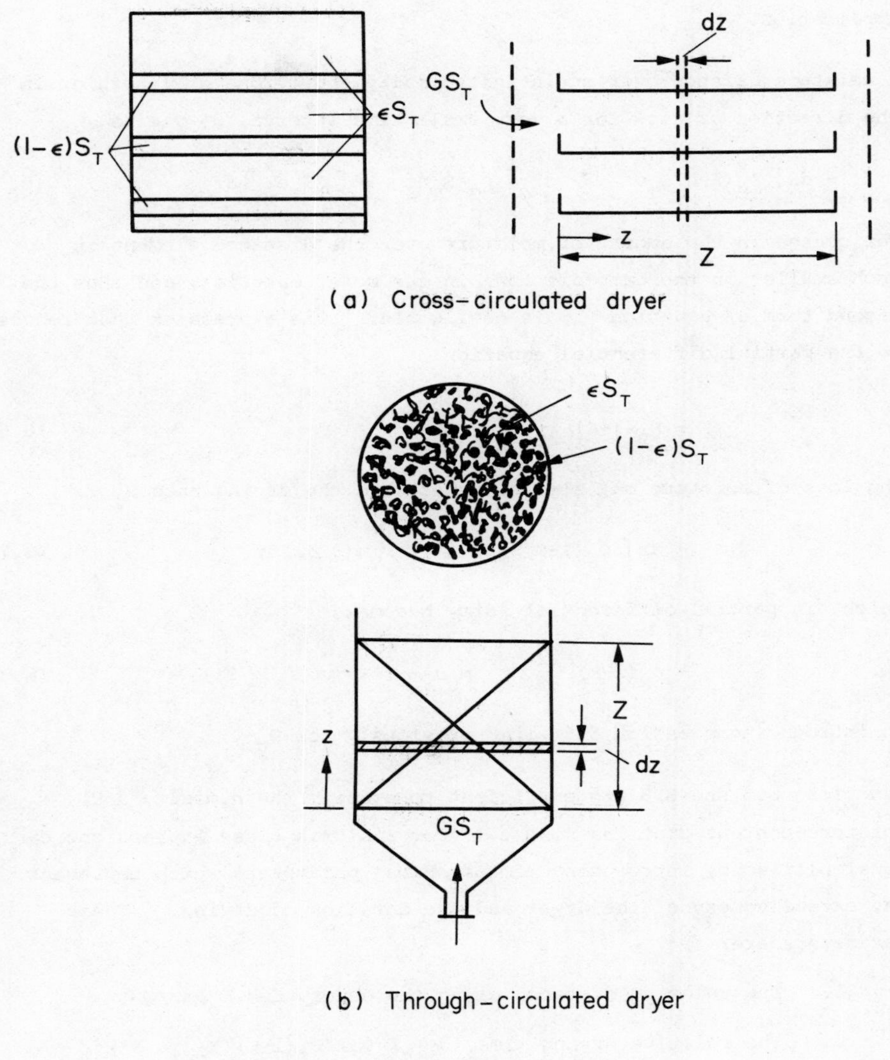

Fig. 8.6 Batch-drying modes

The analysis may be further simplified by assuming that the local mass-transfer coefficient does not vary significantly from the distance-averaged value and edge-effects, such as those caused by the peripheries of trays, for example, are slight. We thus use, for the local drying rate over a unit exposed surface, the constitutive relationship

$$N_V = f\, K_o\, \phi\, (Y_W - Y_G) \qquad (5.5)$$

which is essentially that first suggested by Van Meel[12] for this application.

A moisture balance over an infinitesimally short zone of length dz in the direction of flow for a very small time interval dτ yields

$$- dX \left[\rho_S (1-\varepsilon) S_T dz \right] - dY_G \left[\rho_G \varepsilon S_T dz \right] = \left[G S_T d\tau \right] dY_G \qquad (8.5)$$

The change in the amount of moisture over the distance element is much smaller in the damp air than in the moist material, and thus the second term of equation 8.5 is negligible. The expression then reduces to the partial differential equation

$$- \rho_S (1-\varepsilon) \frac{\partial X}{\partial \tau} = G \frac{\partial Y_G}{\partial z} \qquad (8.6)$$

The loss of moisture may also be related to the drying rate N_V :

$$- dX \left[\rho_S (1-\varepsilon) S_T dz \right] = N_V a (S_T dz) d\tau \qquad (8.7)$$

which, in partial differential form, becomes

$$- \rho_S (1-\varepsilon) \frac{\partial X}{\partial \tau} = N_V a = f K_o \phi a (Y_W - Y_G) \qquad (8.8)$$

on introducing equation 5.5 cited previously for N_V .

Equations 8.6 and 8.8 are sufficient to specify the humidity and moisture-content profiles sought. For analysis these expressions can be simplified by introducing dimensionless parameters which represent the extensiveness of the dryer and the duration of drying. These parameters are:

1.   The relative distance along the dryer, $\zeta = K_o \phi a z / G$

2.   The relative drying time, $\theta = K_o \phi a \tau / \rho_S (1-\varepsilon)(X_{cr} - X^*)$

The first parameter is the number of transfer units. The significance of the latter may be seen by lumping together related terms as follows:

$$\theta = \left[\frac{K_o \phi a Z}{G} \right] \left[\frac{G S_T}{\rho_S S_T Z (1-\varepsilon)(X_{cr} - X^*)} \right] \tau \qquad (8.9)$$

$$= \left[\text{number of transfer units} \right] \times \left[\frac{\text{air flow rate}}{\text{mass of bound moisture}} \right]$$

$$\times \text{ time}$$

Thus, θ is a measure of the duration of drying that reduces dryers of different performance and with different loads of bound moisture to a common basis.

If these parameters, ζ and θ , are introduced into equations 8.6 and 8.8, together with the humidity potential $\Pi = (Y_W - Y_G)$ and the characteristic moisture content $\Phi = (X - X^*)/(X_{cr} - X^*)$, we find

$$\frac{\partial\Phi}{\partial\theta} = \frac{\partial\Pi}{\partial\zeta} \tag{8.10a}$$

$$-\frac{\partial\Phi}{\partial\theta} = \Pi f \tag{8.10b}$$

The humidity potential can be eliminated from this pair of equations by noting

$$\frac{\partial\Phi}{\partial\theta} = \frac{\partial}{\partial\zeta}(\Pi) = \frac{\partial}{\partial\zeta}\left(-\frac{1}{f}\frac{\partial\Phi}{\partial\theta}\right) \tag{8.11}$$

or

$$\frac{\partial\Phi}{\partial\theta} = \frac{\partial}{\partial\theta}\left(-\frac{1}{f}\frac{\partial\Phi}{\partial\zeta}\right) \tag{8.12}$$

One obtains an intermediate solution to this equation by integrating both sides with respect θ :

$$\Phi + \frac{1}{f}\frac{\partial\Phi}{\partial\zeta} = P(\zeta) \tag{8.13}$$

where $P(\zeta)$ is an arbitrary function determined from the boundary conditions. Assume the dryer to be loaded with material that is uniformly moist. Then, at the start, $\partial\Phi/\partial\zeta$ is zero, so $P(\zeta)$ is equal to Φ_o , the initial moisture content.

Equation 8.12 is a second-order, partial differential equation of the hyperbolic type, for which the solution yields a semi-infinite strip (infinite in time, but finite in distance) of values of Φ. Some of the difficulties encountered in the numerical solution of hyperbolic equations in related applications are described by Rosenbrock and Storey.[10] Tetzlaff[11] finds that a strip with a width of 7 distance intervals and a length of 500 time intervals is needed for a dryer of one transfer unit. The partial differential equation (eq.8.13) is then replaced by a set of total differential equations, one for each value of ζ. These equations are put into finite-difference form, and Heun's predictor-corrector method, as illustrated in Fig. 8.7, is used to find the progress of Φ with time.

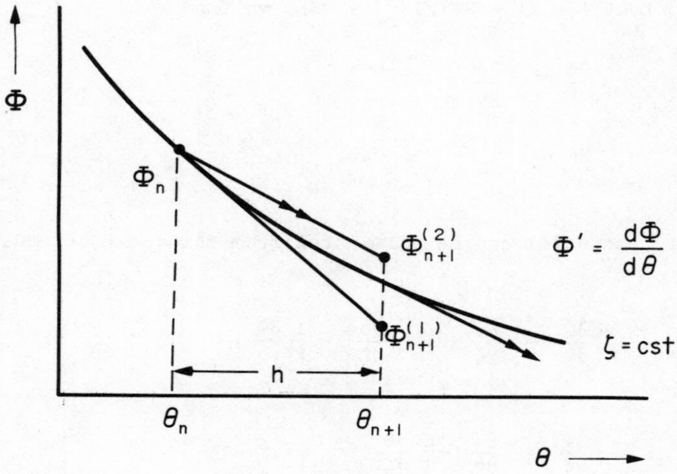

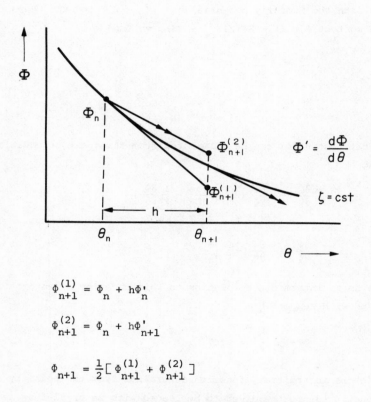

$$\Phi^{(1)}_{n+1} = \Phi_n + h\Phi'_n$$

$$\Phi^{(2)}_{n+1} = \Phi_n + h\Phi'_{n+1}$$

$$\Phi_{n+1} = \frac{1}{2}[\Phi^{(1)}_{n+1} + \Phi^{(2)}_{n+1}]$$

Fig. 8.7 Heun's predictor-corrector method.

Ashworth,[1] by adopting another finite-difference scheme, reduces the number of grid points from 3500 to 400 for the same size of problem. The procedure involves obtaining the inlet moisture content at the (n+1)th time level from the finite-difference analogue of equation 8.10(b). Equation 8.13, on being put into its finite-difference form, centred on the point $(\zeta_{i-\frac{1}{2}}, \theta_{n+1})$, can then be solved successively for the unknown variable $\Phi_{i,\,n+1}$ by a Wegstein iteration. The scheme is shown in Fig. 8.8

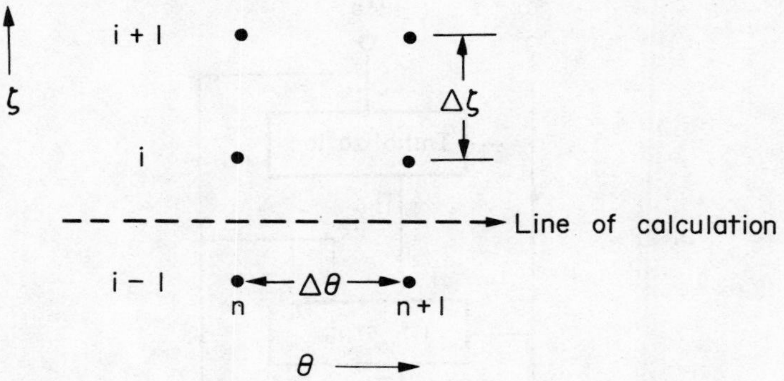

Fig. 8.8 Finite-difference grid for solving equation 8.13

The finite-difference expression to be solved is

$$\Phi_{i,n+1} = \Phi_{i-1,n+1} + \Delta\zeta\, f\left(\frac{\Phi_{i-1,n+1} + \Phi_{i,n+1}}{2}\right)\left[\Phi_o - \frac{(\Phi_{i,n+1} + \Phi_{i-1,n+1})}{2}\right] \tag{8.14}$$

To minimize truncation errors, the step-size ratio between $\Delta\zeta$ and $\Delta\theta$ is fixed by considering the relative magnitude of the gradients $\partial\Phi/\partial\tau$ and $\partial\Phi/\partial\zeta$; in this way, Ashworth suggests that these increments should be interrelated by

$$\Delta\theta = (\Phi_{o,\zeta} - \Phi_{\theta,o})\Delta\zeta \tag{8.15}$$

It is also possible to solve the set of total differential equations, derived from equation 8.10, namely

$$-\left.\frac{d\Phi}{d\theta}\right|_{\zeta} = \Pi f_{\zeta} = -\frac{d\Pi}{d\zeta} \tag{8.16}$$

by analogue computation.[1,4] The space derivative can be estimated by a three-point difference scheme

$$\frac{d\Pi}{d\zeta} = \frac{1}{2h}\left[-3\Pi_i + 4\Pi_{i-1} - \Pi_{i-2}\right] \tag{8.17}$$

where h is the distance increment. This scheme is possible because the dryer is "narrow" in space, and three steps appear suitable for a dryer of 1 transfer unit with a half-stage procedure for obtaining the initial

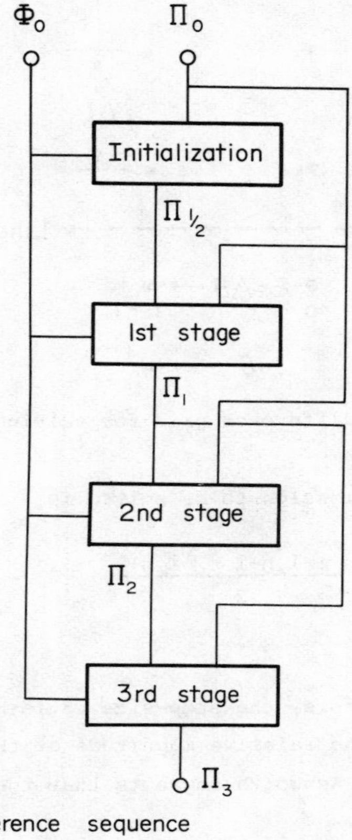

(a) Finite-difference sequence

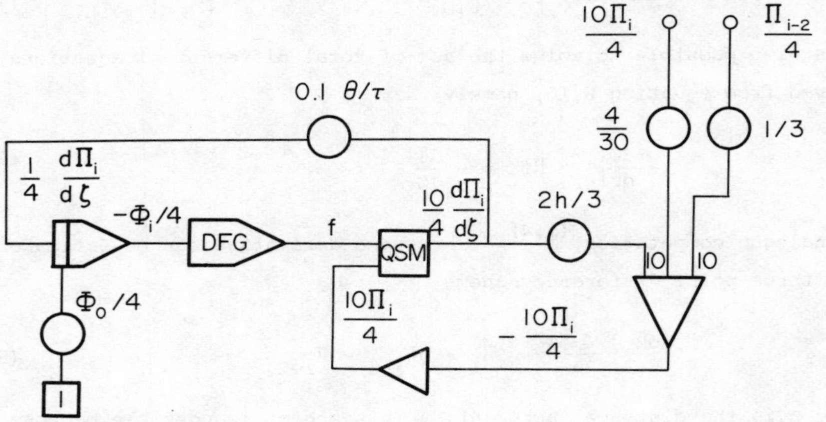

(b) Scaled analogue-circuit block for one stage

Fig. 8.9 Analogue solution of Van Meel equations

value of Π, as shown in Fig. 8.9(a). The relative drying rate f is obtained as an expression composed of ten linear segments from the value of Φ by means of a dionide function generator (DFG). The scaled analogue circuit to represent one distance step is shown in Fig. 8.9(b).

At the time of writing (1977), analogue computers are often forsaken for the lure of the increasing power of digital computing systems. However, the oscillographic output of the analogue machine provides an "instant" picture of the course of drying, while external drying conditions can be readily changed by altering the setting of the potentiometers. This facility is particularly useful when following complex batch-drying schedules, when inlet humidity potentials and airflow directions may change during the course of drying. The same kind of interactive capability is more difficult and more expensive to get with a digital computing system.

A third method of solution is by way of graphical integration. This was the method used by Van Meel himself[12] and is fully described elsewhere.[5] The method is somewhat tedious and is subject to increasing error as the drying rate falls off towards the end of the process.

Equation 8.13 can be solved directly if there is an analytical expression for the relative drying rate f. By way of example, the commonly adopted approximation of a single linear falling-rate curve will be considered in detail in the following sections. As noted before, it is always possible to match the drying curve by a few linear segments.

8.4 Drying Without Recycle

First Stage of Drying. In the first stage of drying, unhindered drying prevails, so f = 1. Equation 8.10 can be replaced by the set

$$\frac{d\Phi}{d\theta} = \frac{d\Pi}{d\zeta} = -\Pi \text{ , } \zeta = \text{cst} \tag{8.18}$$

Integration of equation 8.18 with respect to distance gives

$$-[\ln \Pi]_{\Pi_o}^{\Pi} = [\zeta]_o^{\zeta} \tag{8.19}$$

which becomes

$$\Pi = \Pi_o e^{-\zeta} \tag{8.20}$$

This exponential decay of humidity potential is characteristic of a co-current process (in this case, the velocity of the solids is zero relative to the dryer). It follows that from equation 8.18

$$\frac{d\Phi}{d\theta} = -\Pi = -\Pi_o e^{-\zeta}, \; \zeta = \text{cst} \tag{8.21}$$

Integration of this equation with respect to time at constant distance yields the moisture-content profile at any instant:

$$[\Phi]_{\Phi_o}^{\Phi} = -\Pi_o e^{-\zeta}[\theta]_o^{\theta} \tag{8.22}$$

that is

$$\Phi - \Phi_o = [-\Pi_o e^{-\zeta}]\theta \tag{8.23}$$

The end of the first period is reached when $\Phi = 1$ at the inlet ($\zeta = 0$). When this occurs, equation 8.23 reduces to the expression

$$1 - \Phi_o = -\Pi_o \theta \tag{8.24}$$

so that the first period of drying lasts for a duration θ_1 where

$$\theta_1 = [\Phi_o - 1]/\Pi_o \tag{8.25}$$

At a dry-bulb temperature of 100°C and a wet-bulb depression of 40°C, the humidity potential Π is 0.0215. Even if the initial free moisture content is only 10 per cent above the critical value, the first drying period will extend to a relative time of 0.1/0.0215 = 4.65, which is considerably greater than the relative distance (or number of transfer units) for the normal shelf dryer. This difference demonstrates the greater extensiveness in time compared with that in space, a common feature of batch drying which has already been alluded to.

Second Stage of Drying. During this stage of drying, the moisture content is less than the critical value towards the inlet, but at positions further in the dryer, unhindered drying takes place. Conceptually, we may describe this stage of drying in terms of the critical point slowly sweeping through the dryer until, at the end of this stage, the critical point reaches the outlet.

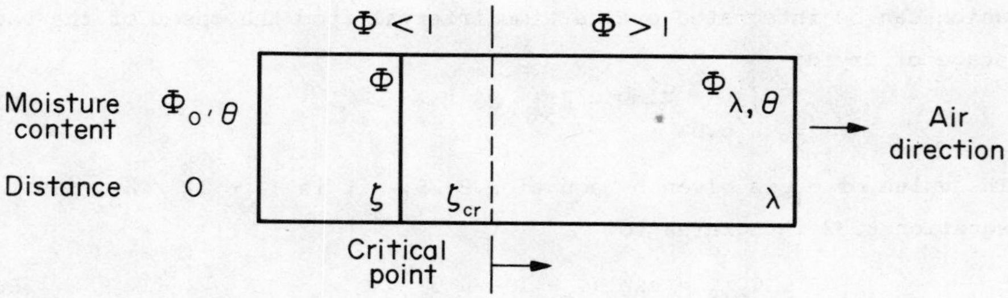

Fig. 8.10 Second stage of batch drying

Equation 8.13, when f = Φ, becomes

$$\Phi = -\frac{1}{\Phi}\frac{\partial\Phi}{\partial\zeta} + \Phi_o \quad , \quad \zeta = cst \tag{8.26}$$

For convenience, we consider the unhindered and hindered-drying zone separately in integrating equation 8.26. From the inlet to the critical-point position ζ_{cr}, the moisture content Φ at a relative length ζ from the inlet is given

$$\int_{\Phi_{o,\theta}}^{\Phi} \frac{d\Phi}{\Phi(\Phi_o - \Phi)} = \int_o^{\zeta} d\zeta = \zeta \tag{8.27}$$

By expanding the left-hand integral, we find

$$\zeta = \frac{1}{\Phi_o}\left[\ln \frac{\Phi}{(\Phi_0 - \Phi)}\right]_{\Phi_{o,\theta}}^{\Phi} \tag{8.28}$$

whence

$$\zeta = \frac{1}{\Phi_o} \ln \frac{\Phi}{\Phi_{o,\theta}} \frac{(\Phi_o - \Phi_{o,\theta})}{(\Phi_o - \Phi)} \tag{8.29}$$

Equation 8.29 can be re-arranged after some manipulation to give an explicit expression for the moisture content

$$\frac{\Phi}{\Phi_o} = \left[\frac{(\Phi_o - \Phi_{o,\theta})}{\Phi_{o,\theta}} e^{-\Phi_o \zeta} + 1\right]^{-1} \tag{8.30}$$

Thus, if we know the inlet moisture content at the given instant, $\Phi_{o,\theta}$, we can get the interior values of Φ as well.

Now, equation 8.10(b) for the inlet position, with $f = \Phi$, yields

$$\frac{d\Phi}{d\theta} = - \Pi_o \Phi \quad , \zeta = 0 \tag{8.31}$$

which can be integrated over a time interval from the onset of the second stage of drying

$$\int_{\Phi_{o,\theta}}^{1} \frac{d\Phi}{\Phi} = - \Pi_o \int_{\theta}^{\theta_1} d\theta \tag{8.32}$$

The value of θ_1 is given by equation 8.25; it is $(\Phi_o - 1) / \Pi_o$. So equation 8.32 integrates to

$$\Phi_{o,\theta} = \exp\,(\Phi_o - 1 - \Pi_o \theta) \tag{8.33}$$

Insertion of this expression into equation 8.30 yields the result:

$$\Phi = \frac{\Phi_o}{\Phi_o[\exp(\Pi_o \theta + 1 - \Phi_o) - 1]\,\exp(-\,\Phi_o \zeta) + 1} \tag{8.34}$$

which may be compared with the equivalent expression, equation 7.74, for continuous crossflow drying. In the latter case, the residence time along the band is the analogue of the time in batch drying.

The position of the critical point ($\zeta = \zeta_{cr}$) can be found by setting $\Phi = 1$ in the foregoing equation, that is

$$1 = \frac{\Phi_o}{\Phi_o[\exp(\Pi_o \theta + 1 - \Phi_o) - 1]\,\exp(-\,\Phi_o \zeta_{cr}) + 1} \tag{8.35}$$

which can be re-arranged to yield an explicit expression for ζ_{cr} :

$$\zeta_{cr} = \frac{1}{\Phi_o} \ln \left[\frac{\Phi_o \exp(\Pi_o \theta + 1 - \Phi_o) - 1}{\Phi_o - 1} \right] \tag{8.36}$$

Let us now consider the drying from this position onward in the dryer where unhindered drying still pertains. Equation 8.26 for this zone becomes

$$\int_{\Phi}^{1} \frac{d\Phi}{(\Phi_o - \Phi)} = \int_{\zeta}^{\zeta_{cr}} d\zeta \tag{8.37}$$

which has the solution

$$- \ln \frac{(\Phi_o - 1)}{(\Phi_o - \Phi)} = \zeta_{cr} - \zeta \tag{8.38}$$

or

$$\Phi = \Phi_o - (\Phi_o - 1)\, e^{(\zeta_{cr} - \zeta)} \tag{8.39}$$

The value of ζ_{cr} is given by equation 8.36; thus

$$\Phi = \Phi_o - (\Phi_o - 1)\, e^{-\zeta} \left[\frac{\Phi_o \exp(\Pi_o \theta + 1 - \Phi_o) - 1}{\Phi_o - 1}\right]^{1/\Phi_o} \tag{8.40}$$

Again, we may compare equation 8.40 to the crossflow analogue, eq. 7.72.

Third Stage of Drying. Finally, the moisture contents will everywhere fall below the critical point. Under these circumstances, equation 8.34 applies throughout the whole dryer. Mathematically, the critical point has shifted to beyond the confines of the dryer and $\zeta_{cr} > \lambda$.

Time of Drying. The expressions for the rate of drying and the duration of the first stage of drying, when unhindered drying takes place everywhere, have already been obtained (eq. 8.21 and 8.25). The speed of the drying process thereafter hinges on the rate at which the critical point sweeps through the dryer. On integrating equation 8.26 from the position ζ to the point ζ_{cr} where the critical point lies, one has

$$\int_{\Phi}^{1} \frac{d\Phi}{\Phi(\Phi_o - \Phi)} = \zeta_{cr} - \zeta \tag{8.41}$$

which, on differentiating with respect to θ, reduces to

$$\frac{d\Phi/d\theta}{\Phi(\Phi_o - \Phi)} = \frac{d\zeta_{cr}}{d\theta} \tag{8.42}$$

to give an expression for the rate at which the critical point shifts. Now $d\Phi/d\theta = -f\Pi = -\Phi\Pi$ (eq. 8.10b), so that equation 8.42 simplifies to

$$\frac{d\zeta_{cr}}{d\theta} = -\frac{\Pi}{(\Phi_o - \Phi)} \tag{8.43}$$

Equation 8.43 holds for any position, but for the inlet conditions in particular

$$\frac{d\zeta_{cr}}{d\theta} = -\frac{\Pi_o}{(\Phi_o - \Phi_{o,\theta})} \tag{8.44}$$

Equations 8.43 and 8.44 together yield the humidity-potential profile from known moisture levels:

$$\frac{\Pi}{\Pi_o} = \frac{(\Phi_o - \Phi)}{(\Phi_o - \Phi_{o,\theta})} \tag{8.45}$$

In the second stage of drying, the moisture content at the inlet has been reduced below the critical value, and equation 8.41 can be integrated to give the moisture-content profile in terms of the approach $(\zeta_{cr}-\zeta)$ to the critical point. The result is

$$\Phi = \frac{\Phi_o}{(\Phi_o-1)\exp\Phi_o(\zeta_{cr}-\zeta) + 1} \tag{8.46}$$

Therefore

$$\frac{d\Phi}{d\theta} = \frac{\Phi_o^2(\Phi_o-1)\ \exp\Phi_o(\zeta_{cr} - \zeta)}{[(\Phi_o-1)\exp\ \Phi_o(\zeta_{cr}-\zeta) + 1]^2} \cdot \left(\frac{d\zeta_{cr}}{d\theta}\right) \tag{8.47}$$

Now $d\zeta_{cr}/d\theta$ is given by equation 8.44, which requires a value for the change in moisture content at the inlet, $(\Phi_o - \Phi_{o,\theta})$. This can be obtained from equation 8.46 by substituting $\zeta = 0$. Thus

$$\frac{d\zeta_{cr}}{d\theta} = \frac{-\Pi_o}{\Phi_o-\Phi_o/[(\Phi_o-1)\exp\Phi_o\zeta_{cr} + 1]}$$

$$= \frac{-\Pi_o[\Phi_o-1 + \exp(-\Phi_o\zeta_{cr})]}{\Phi_o(\Phi_o - 1)} \tag{8.48}$$

By combining equations 8.47 and 8.48, we get the required drying rate, namely

$$\frac{d\Phi}{d\theta} = - \frac{\Pi_o\Phi_o e^{\Phi_o(\zeta_{cr}-\zeta)}[\Phi_o - 1 + \exp(-\Phi_o\zeta_{cr})]}{[(\Phi_o-1)\exp\Phi_o(\zeta_{cr}-\zeta) + 1]^2} \tag{8.49}$$

The value of $\Phi_o\zeta_{cr}$ is given by equation 8.36.

A similar analysis reveals the corresponding parameters for the rearward section where unhindered drying is still taking place. The moisture content is given by equation 8.39, namely

$$\Phi = \Phi_o - (\Phi_o-1)e^{(\zeta_{cr}-\zeta)} \quad (8.39$$

which, on integration, yields

$$\frac{d\Phi}{d\theta} = (\Phi_o-1)e^{(\zeta_{cr}-\zeta)} \cdot \left(\frac{d\zeta_{cr}}{d\theta}\right) \tag{8.50}$$

The critical speed $d\zeta_{cr}/d\theta$ has already been determined by equation 8.48 and, when substituted in equation 8.50, the required expression is found:

$$\frac{d\Phi}{d\theta} = \frac{-\Pi_o[\Phi_o - 1 + \exp(-\Phi_o\zeta_{cr})]\exp(\zeta_{cr}-\zeta)}{\Phi_o} \tag{8.51}$$

Integration of equations 8.49 and 8.51 with respect to θ gives the time of drying as a function of ζ_{cr}. When drying down to low moisture levels, the final moisture-content distribution will be almost uniform, so that the time of drying can often be calculated simply from the conditions at the air inlet. The overall drying time θ_e is thus given by

$$\theta_e = \theta_1 + \int_{\Phi_e}^{1} \frac{d\Phi}{f\Pi_o} \tag{8.52}$$

where θ_1 is the duration of the first stage of drying and Φ_e is the final moisture content. Since $f = \Phi$, equation 8.52 has the solution

$$\theta_e = \theta_1 - \frac{1}{\Pi_o} \ln \Phi_e \tag{8.53}$$

and, with equation 8.25 for θ_1, one gets

$$\theta_e = \frac{1}{\Pi_o} [\Phi_o - 1 - \ln \Phi_e] \tag{8.54}$$

The moisture content Φ_e is reached at the outlet after a further relative time (Θ) of $\Phi_o \lambda / \Pi_o$. This is the time that the drying wave takes to pass through.

The following worked example illustrates this method of estimating the drying time for a shelf dryer.

Example 8.2. Solids of bone-dry density 1600 kg m^{-3} are laid to a depth of 30 mm in a batch oven with shelves that are spaced 100 mm apart. The solids initially have a free moisture content of 1 kg kg^{-1} and are to be dried to a free moisture content of 0.05 kg kg^{-1} at the air inlet. (The free moisture content at the critical point is 0.8.) Estimate the drying time if the air conditions at the inlet are held constant at a dry-bulb temperature of 160°C and a wet-bulb temperature of 60°C.

Data: mass-transfer coefficient $K_o = 30$ g $m^{-2} s^{-1}$.

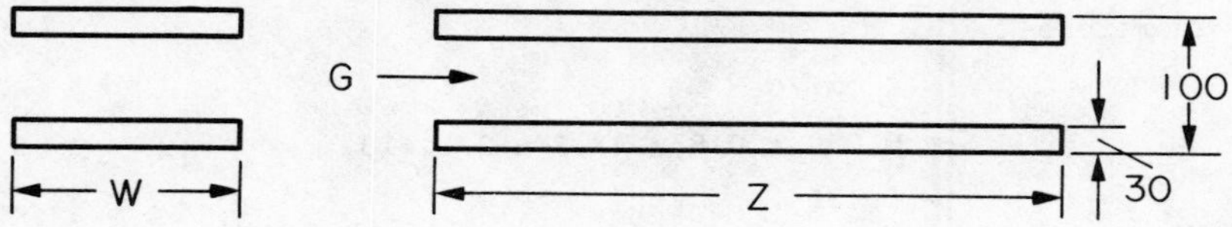

Fig. 8.11 Shelf-dryer arrangement.

Relative drying time, θ_e. Now,

$\Phi_o = 1/0.8 = 1.25$

$\Phi_e = 0.05/0.8 = 0.0625$

$\Pi_o = 0.0215$, from the humidity difference chart for $T_G = 100$ and $T_W = 60$.

Thus, from equation 8.54,

$$\theta_e = \frac{1}{\Pi o} [(\Phi_o - 1) - \ln \Phi_e]$$

$$\theta = \frac{1}{0.0215}\,[0.25 - \ln 0.0625] \quad = 140.6$$

Drying time.

$$a = \frac{WZ}{0.03WZ} = \frac{1}{0.03} = 33.3\ m^{-1}$$

$$(1-\varepsilon) = 30/100 = 0.3$$

$$K_o = 0.030\ kg\ m^{-2}s^{-1}$$

$$\phi \simeq D/(D+Y_W) = \frac{0.622}{(0.622 + 0.155)} = 0.8$$

Therefore

$$\tau = \frac{\rho_S(1-\varepsilon)(X_{cr} - X^*)}{K_o\,\phi\,a}\,.\,\theta$$

$$= [\frac{1600 \times 0.3 \times 0.8}{0.030 \times 0.8 \times 33.3}] \times 140.6$$

$$= 67\ 600\ s$$

$$\text{or } 67\ 600/3600 = 18.8\ h$$

Example 8.3. If the dryer in example 8.2 has trays 2 m long (in the airflow direction), estimate the drying-rate profile with time at the air outlet.

Data: Air velocity over shelves = $1.6\ m\ s^{-1}$

Number of transfer units, λ.

The specific gas rate is $\rho_G\hat{u}_G$.

At the air inlet conditions the air density is $0.883\ kg\ m^{-3}$.

$$\therefore\ G = 0.883 \times 1.6 = 1.413\ kg\ m^{-2}s^{-1}\ .$$

Therefore

$$\lambda = K_o\phi aZ/G$$

$$= 0.030 \times 0.8 \times 33.3 \times 2/1.413$$

$$= 1.131$$

The drying rate will be evaluated at increments $\Delta\theta$ of 20.

Stage 1 drying. This stage lasts until

$$\theta = (\Phi_o - 1)\Pi_o \quad = \quad 0.25/0.0215$$

$$= 11.6$$

i.e. after the first interval of time the drying will have entered the second stage.

The initial drying rate, if this is attained straight away, is found from equation 8.21:

$$\frac{d\Phi}{d\theta} = -\Pi_o e^{-\zeta}$$
$$= 0.0215\ e^{-1.131}$$
$$= -\ 0.00694$$

Critical position, ζ_{cr}. Use equation 8.36:

$$\zeta_{cr} = \frac{1}{1.25} \ln\left[\frac{(1.25 \exp(0.0215\theta - 0.25) - 1}{0.25} \right]$$

Θ	ζ_{cr}	Θ	ζ_{cr}
20	0.549	100	2.706
40	1.319	120	3.087
60	1.854	140	3.454
80	2.301		

Stage 2 Drying. At $\Theta = 20$, $\zeta_{cr} < \lambda$ and eq. 8.51 for unhindered-drying conditions should be used, namely:

$$\frac{d\Phi}{d\Theta} = \frac{-0.0215\,[0.25 + \exp(-1.25\,\zeta_{cr})]\ \exp(\zeta_{cr} - 1.131)}{1.25}$$

The drying is in the third stage from $\Theta = 40$ onwards as $\zeta_{cr} > \lambda$; eq. 8.49 applies then:

$$\frac{d\Phi}{d\Theta} = \frac{-\,[\,(0.0215 \times 1.25)\ \exp 1.25\ (\zeta_{cr} - 1.131)]\,(0.25 + \exp - 1.25\zeta_{cr})}{[\,0.25 \exp 1.25\ (\zeta_{cr} - 1.131) + 1]^2}$$

The values are tabulated:

Θ	ζ_{cr}	$-d\Phi/d\Theta$	Θ	ζ_{cr}	$-d\Phi/d\Theta$
20	0.549	0.00724	100	2.706	0.00702
40	1.319	0.00868	120	3.087	0.00557
60	1.854	0.00884	140	3.454	0.00418
80	2.301	0.00831			

Apart from an apparently curious upswing in rate in the beginning, the drying rates seem to decrease steadily with time at the outlet. However, a steady fall-off is an illusion wrought by the coarseness of the time intervals. A clue may be found by comparing the early rates for $\theta < 60$ with those in the first stage of drying. The following worked example brings out this point.

Example 8.4. Estimate the percentage change of the drying rate at the outlet, for the conditions of example 8.3, when the critical point has just reached there.

Equation 8.51 holds for the case where unhindered drying is just persisting. In this instance, $\zeta = \lambda$, so equation 8.51 becomes

$$\left.\frac{d\Phi}{d\theta}\right|_{\lambda,\theta} = \frac{-\Pi_o\left[(\Phi_o - 1) + \exp(-\Phi_o\zeta_{cr})\right]\exp(\zeta_{cr} - \lambda)}{\Phi_o}$$

When the critical point reaches the end of the dryer, $\zeta_{cr} = \lambda$ and thus

$$\left.\frac{d\Phi}{d\theta}\right|_{\lambda,\theta} = \frac{-\Pi_o\left[(\Phi_o - 1) + \exp(-\Phi_o\lambda)\right]}{\Phi_o}$$

Initially the drying rate is given by

$$\left.\frac{d\Phi}{d\theta}\right|_{\lambda,o} = -\Pi_o e^{-\lambda}$$

from equation 8.20. So the ratio of rates R between the time the critical point has reached the outlet and at the start is given by

$$R = \frac{(\Phi_o - 1) + \exp(-\Phi_o\lambda)}{\Phi_o\exp(-\lambda)}$$

For $\Phi_o = 1.25$, $\lambda = 1.131$, $R = 1.223$.

The drying rate is 22.3 per cent greater when the solid has been dried down to the critical point.

The foregoing phenomenon can be qualitatively explained. Initially the drying rates are reduced to a fraction $\exp(-\lambda)$ of that at the inlet. As soon as the critical point begins to progress through the dryer, the amount of moisture carried into the air falls off as the drying rates dwindle from their initial values in the hindered-drying zone. In the rear section of the dryer, the surface humidity Y_S is unchanged as unhindered drying still persists, so the humidity potential ($\Pi = Y_S - Y_G$) there begins to increase and continues to increase as long as the critical point is upstream. Clearly, the maximum effect is observed at the outlet. Further, the actual enhancement is greater the longer the dryer and wetter the feed material. The course of the drying rates at the outlet is sketched in Fig. 8.12.

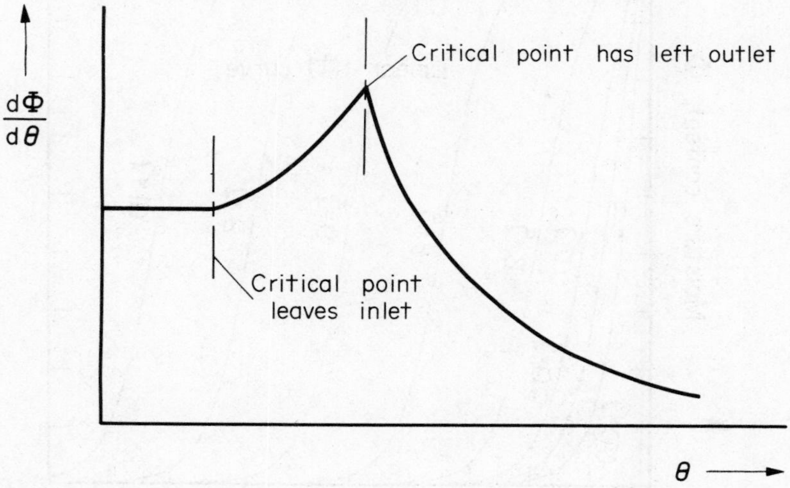

Fig. 8.12 Variation of drying rates at the outlet of a batch dryer without recycle.

The "standard, preferred" commercial tray dryer, described by Nonhebel and Moss(9) and illustrated in Fig. 8.1, has 30 mm deep trays which are stacked close together in two tiers only 800 mm long in the airflow direction with the trays 75 mm apart. The inlet air is admitted separately to each stack of trays so that λ is order 1/2. Therefore, the variations in process conditions for this arrangement will be somewhat less than those illustrated in the foregoing worked examples, but the comparison cannot be direct as there is recycle of the outlet air back to the intake in the two-tier dryer. Drying with recycle is considered in Section 8.5.

In the following diagrams, Figs. 8.13(a) to 8.13(d), the variation of process conditions in a dryer of various lengths without air recycle is reproduced from Ashworth's calculations for the case when $f = \Phi$. Should the solids be well-mixed, as in batch fluid-bed drying, then uniform external drying conditions may be assumed, and the progress of drying follows that for $\zeta = 0$. The Van Meel equations can also be solved for the case when the air flows through without recirculation in a dryer where the solids are well-mixed.

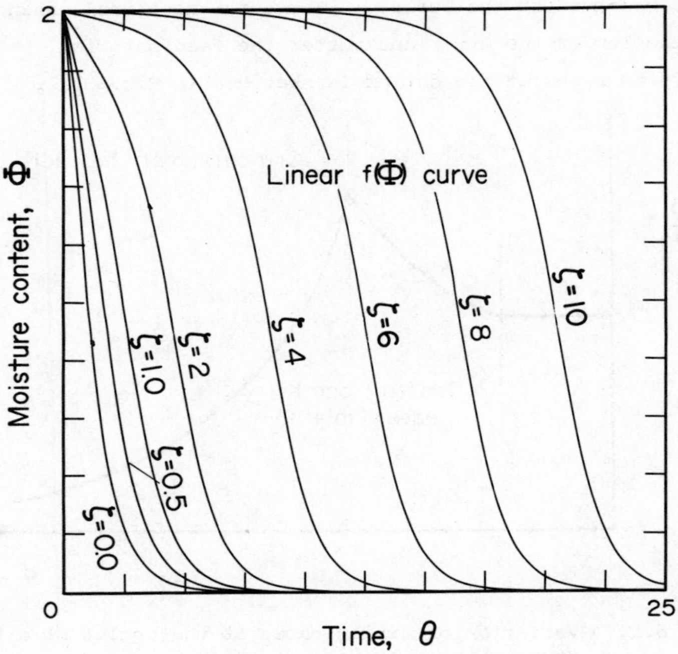
Moisture content, Φ
Linear f(Φ) curve
ζ=0.0
ζ=0.5
ζ=1.0
ζ=2
ζ=4
ζ=6
ζ=8
ζ=10
2
0
25
Time, θ

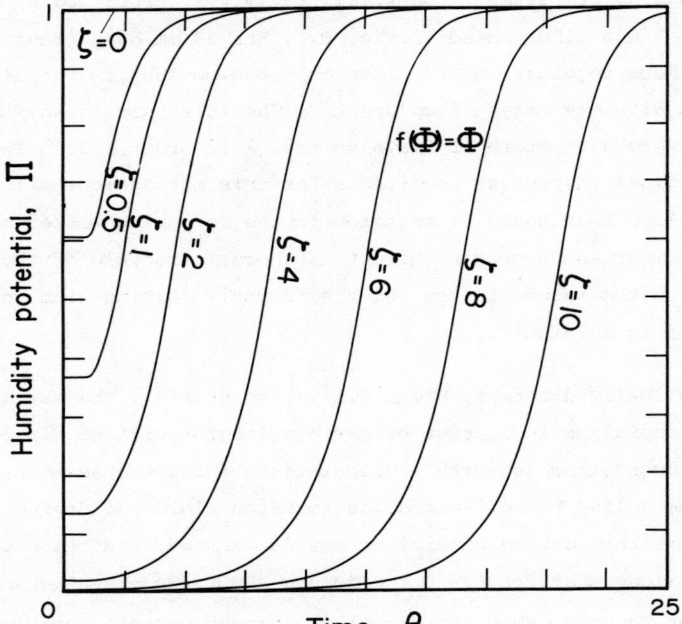
Humidity potential, Π
f(Φ)=Φ
ζ=0
ζ=0.5
ζ=1
ζ=2
ζ=4
ζ=6
ζ=8
ζ=10
1
0
25
Time, θ

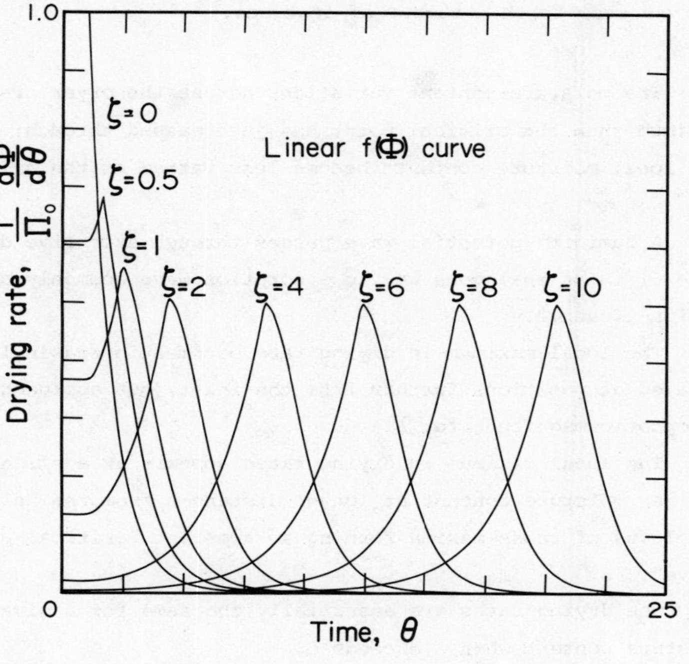
1.0
ζ=0
Linear f(Φ) curve
ζ=0.5
ζ=1
ζ=2
ζ=4
ζ=6
ζ=8
ζ=10
Drying rate, 1/Π₀ dΦ/dθ
0
25
Time, θ

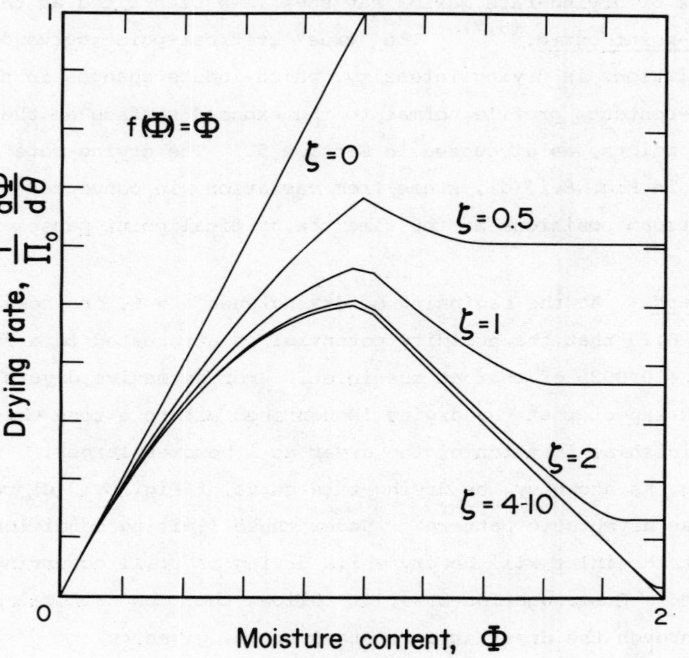
1
f(Φ)=Φ
ζ=0
ζ=0.5
ζ=1
ζ=2
ζ = 4-10
Drying rate, 1/Π₀ dΦ/dθ
0
2
Moisture content, Φ

These curves show forth a number of features:

1. The moisture-content variations across the dryer are a maximum when the critical point has just passed through; thereafter, the local moisture contents become less varied as the solid dries out.

2. A humidity-potential wave passes through extensive dryers ($\lambda > 6$) in an analogous way to a sorption wave commonly observed in ion-exchange.

3. The local maximum in drying rate becomes increasingly delayed at positions further from the inlet, but approaches an asymptotic magnitude for $\lambda > 6$.

4. The local maximum in drying rates appears at a slightly smaller moisture content at longer distances from the inlet, the locus of these maxima forming an apparent "critical-point curve".

5. The drying rates are essentially the same for a given moisture content when λ exceeds 6.

The locus of drying-rate maxima has sometimes been cited as the critical-point curve.[5,7] The "true" critical-point curve derives from variations in drying intensity, which induce changes in the moisture-content profile normal to the exposed surface at the observed critical points, as discussed in Section 5. The drying-rate locus, depicted in Fig. 8.13(d), stems from variations in convected moisture from upstream positions at the time the critical point passes by.

Long dryers. At the beginning of drying when $\lambda = 6$, one sees from equation 8.20 that the humidity potential is attenuated to a fraction (exp -6) = 0.0025 of that at the inlet. For extensive dryers when $\lambda > 6$, one would expect that the drying is confined within a zone which takes up a diminishing fraction of the dryer as λ becomes larger. The course of drying, as shown by the drying-rate curve of Fig. 8.13(d), tends towards as asymptotic pattern. Under these limiting conditions, the solids at the inlet will be dry while drying is still occurring elsewhere. From equation 8.44, it follows that the critical point is moving through the dryer at a constant speed, given by

$$\frac{d\zeta_{cr}}{d\theta} = \frac{-\Pi_o}{\Phi_o} \tag{8.55}$$

since $\Phi_{o,\theta} = 0$. From equation 8.42, it follows that

$$\frac{d\Phi}{d\theta} = \Phi(\Phi_o - \Phi)\frac{d\zeta_{cr}}{d\theta} \quad , \Phi < 1$$

$$= -\Pi_o \Phi(\Phi_o - \Phi) / \Phi_o \tag{8.56}$$

which leads to the conclusion that the drying rates depend only on moisture content. The time of drying is the sum of the time for the critical point to reach the end of the dryer and the time for the solids to dry out thereafter. This time can be found simply from equations 8.55 and 8.56

$$\theta_e = \frac{-\Phi_o}{\Pi_o}\int_\lambda^o d\zeta_{cr} - \frac{\Phi_o}{\Pi_o}\int_{\Phi_e}^{1}\frac{d\Phi}{\Phi(\Phi_o - \Phi)} \tag{8.57}$$

which, on integrating, becomes

$$\theta_e = \frac{\Phi_o}{\Pi_o}\left[\lambda - \frac{1}{\Phi_o}\ \ln\ \frac{(\Phi_o - \Phi_e)}{\Phi_o\Phi_e}\right] \tag{8.58}$$

In the case where the dryer is very extensive, the width of the drying wave is comparatively slender and the time of drying is essentially the time for the drying wave to pass through. This time is $\Phi_o\lambda/\Pi_o$.

Other effects. The analysis so far has made two implicit assumptions: the solids are charged to the batch dryer uniformly wet and the solids do not shrink on drying.

Should the solids be charged to a cold chamber, even if the material is uniformly moist, then by the time the steady-state temperatures for the first period of drying have been attained, the solids closer to the air inlet will have dried off somewhat more than those nearer the outlet. Providing the material is charged wet enough, the moisture content Φ at the air inlet at the end of this induction period will lie above the critical value. Furthermore, this moisture content will appear progressively throughout the dryer at a time θ_o, which is a function of the distance ζ. Therefore we may assume that the wet stock is uniformly moist provided that Φ_o is interpreted in this way and the calculated time of drying is extended by θ_o. Details of this procedure may be found elsewhere.[5]

The influence of shrinkage has been studied by Ashworth.[1] He shows that for a linear shrinkage relationship, typical of the behaviour of wheat and wood-chips, shrinkage effects become significant should the material shrink by more than 10 per cent. Shrinkage of the material in a direction across the airflow will have a secondary influence by producing "edge-effects".

8.5 Drying with Recycle

Most batch dryers are worked by recycling part of the humid gas from the outlet to the intake to conserve heat, as noted in Section 8.1. Under these circumstances, the inlet-air humidity potential Π_o no longer remains constant, but varies throughout the course of drying as the moisture convected into the airstream changes.

Suppose a quantity r kg dry air is recycled for every kilogram of dry air that passes through the dryer, as shown in Fig. 8.14.

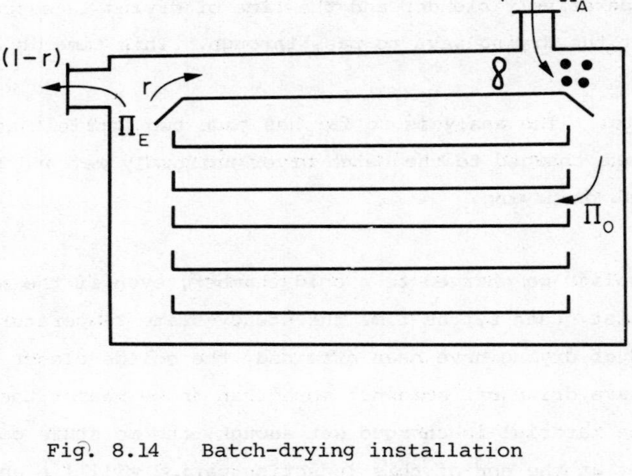

Fig. 8.14 Batch-drying installation

Further, suppose the drying is adiabatic, so that the wet-bulb temperature is constant throughout the dryer. The situation is similar to that described in Example 8.1. A moisture balance at the inlet leads to the expression:

$$r\,\Pi_E + (1-r)\Pi_A = \Pi_o \tag{8.59}$$

where Π_A , the fictitious humidity potential of the fresh air, is equal to the difference between the wet-bulb humidity and the ambient-air humidity.

It is possible to develop expressions for Π, Φ and $d\Phi/d\theta$ in the same way as that previously described for the case of drying without recycle. This is the situation that is considered by Van Meel in his original paper.(12) Only an outline of his analysis is given here.

<u>First stage of drying</u>. Unhindered drying proceeds everywhere. The humidity potential is given by

$$\Pi = \frac{(1-r)\Pi_A}{(1 - re^{-\lambda})} \cdot e^{-\zeta} \tag{8.60}$$

which reduces to equation 8.20 when $r = 0$ and $\Pi_A = \Pi_o$. The moisture-content distribution becomes for a uniformly wet feed

$$\Phi - \Phi_o = -\frac{(1-r)\Pi_A}{(1 - re^{-\lambda})} \cdot e^{-\zeta} \cdot \theta \tag{8.61}$$

which simplifies to equation 8.23 when $r = 0$ and $\Pi_A = \Pi_o$. Finally, the first drying period lasts for a time

$$\theta_1 = (\Phi_o - 1) \cdot \frac{(1 - re^{-\lambda})}{(1-r)\Pi_A} \tag{8.62}$$

which likewise simplifies to the equivalent expression for no recycle (eq. 8.25) when the air goes straight through.

Consider a shelf dryer for which $r = 0.9$ and $\lambda = 1$. With $\Pi_A = 0.0215$ (the conditions for air at 100°C with a wet-bulb depression of 40°C), θ_1 is 77.8 when $\Phi_1 = 1.25$. This drying time compares with a value of 4.65 without recycle for the same inlet humidity potential. Clearly, tolerable drying times are only possible by raising the humidity potential of the introduced air. If fresh air is being used, then $\Pi_A \sim Y_W$ and, in the case just examined, the humidity potential will be about 0.15 for the considered wet-bulb temperature of 60°C. The drying time is now 11.15.

Second stage of drying. The critical point is moving through the dryer. The humidity potential is now

$$\frac{\Pi}{\Pi_A} - (1-r) \cdot \left[\frac{(\Phi_o - \Phi)}{\Phi_o - \Phi_{o,\theta} - r(\Phi_o - \Phi_{\lambda,\theta})}\right] \tag{8.63}$$

where $\Phi_{\lambda,\theta}$ is the moisture content at the air outlet. (Compare equation 8.45 for the limit when $r = 0$). The expressions for the local moisture contents are the same as those derived for the case without recycle; namely, for the hindered-drying zone by the inlet

$$\Phi = \frac{\Phi_o}{(\Phi_o - 1)\exp \Phi_o(\zeta_{cr} - \zeta) + 1} \tag{8.46}$$

and for the unhindered-drying zone towards the outlet

$$\Phi = \Phi_o - (\Phi - 1)\exp(\zeta_{cr} - \zeta) \tag{8.39}$$

The time for drying to reach a moisture content Φ_e at the outlet is found to be

$$\theta_e = [\Phi - 1 - \ln\Phi_e + \Phi_o\lambda/(1-r)] / \Pi_A \tag{8.64}$$

Comparison with the equivalent expression for the case of no recycle (eq. 8.54) shows that the time of drying has been extended by an amount $r\Phi_o\lambda/(1-r)\Pi_A$, on noting that the moisture content Φ_e appears at the outlet at a time $\Phi_o\lambda/\Pi_o$ after appearing at the inlet.

Example 8.5. Estimate the change of drying time if, in the dryer cited in example 8.2, the fresh air enters with a dewpoint of 10^oC, but five-sixths of the outlet air is recycled.

The humidity of the fresh air is 0.0077 kg kg^{-1} (saturated at 10^oC).
The wet-bulb humidity at 60^oC is 0.1547 kg kg^{-1}

$$\therefore \Pi_A = Y_W - Y_{GA}$$

$$= 0.1547 - 0.0077 = 0.147 \text{ kg kg}^{-1}$$

The number of transfer units is calculated in Example 8.3,

i.e. $\lambda = 1.131$

From equation 8.64 :

$$\theta_e = \frac{1}{\Pi_A}[\Phi_o - 1 - \ln \Phi_E + \Phi_o \lambda/(1-r)]$$

$$= \frac{1}{0.147}[0.25 - \ln 0.0625 + 1.25 \times 1.131/\tfrac{1}{6}]$$

$$= 78.3$$

$$\therefore \tau = 18.8 \times \frac{78.3}{140.6} = 10.5 \text{ h}$$

Thus, the drying time is reduced by

$$\frac{(18.8 - 10.5)}{18.8} \times 100 = 44.2 \text{ per cent.}$$

This reduction is achieved by the greatly improved humidity potential of the air intake.

It has already been noted that the interior drying rates in batch drying without recycle rise as the critical point sweeps through the unit upstream. While the maximum drying rate reaches an asymptotic magnitude as the dryer becomes longer, the ratio of the maximum to the initial rate progressively increases. When dryers are operated with the return of the exhaust air to the inlet, the changes in local drying rates become amplified by the feed-back effect of the recycled air.[6] As soon as hindered drying begins at the inlet, the smaller quantity of convected moisture in the airstream is returned back to the air intake to enhance the humidity potentials still further in the unhindered-drying zone.

We may get a quantitative assessment of the effect by looking at the rates at the outlet when the critical point has just reached there. In drying with recycle, the humidity potential at the outlet is given by

$$\Pi = \frac{\Pi_A(1 - r)(\Phi_o - 1)}{(\Phi_o - \Phi_{o,\theta}) - r(\Phi_o - 1)} \tag{8.65}$$

from equation 8.63 for $\Phi = \Phi_{\lambda,\theta} = 1$. The value of $\Phi_{o,\theta}$ may be found from equation 8.46 for $\zeta_{cr} = \lambda$ and $\zeta = 0$; it is

$$\Phi_{o,\theta} = \frac{\Phi_o}{(\Phi_o - 1)\exp\Phi_o\lambda + 1} \tag{8.66}$$

On combining equations 8.65 and 8.66, after some manipulation, one finds

$$\left.\frac{d\Phi}{d\theta}\right|_{\lambda,\theta_{cr}} = -\Pi_{\lambda,\theta_{cr}} = \frac{\Pi_A(1-r)}{\left[\frac{\Phi_o \exp \Phi_o \lambda}{(\Phi_o - 1)\exp\Phi_o \lambda + 1} - r\right]} \tag{8.67}$$

since f = 1. At high initial moisture contents this expression approaches a limiting value of Π_A.

The initial humidity potential at the other end of the dryer is given by equation 8.60 for $\zeta = 0$. Therefore

$$\left.\frac{d\Phi}{d\theta}\right|_{o,o} = -\Pi_{o,o} = \frac{(1-r)\Pi_A}{(1 - re^{-\lambda})} \tag{8.68}$$

This expression approaches a limiting value of 0 when the recycle ratio r approaches 1. It follows then, whenever Φ_o is very high and r almost 1, the ratio of the outlet drying rate as the critical point reaches there to the initial rate at the inlet will tend towards a limiting value of $\Pi_o/0 = \infty$! Unexpectedly high drying rates may thus appear towards the outlet end of the dryer.

Higher drying rates will be found at the outlet, as the critical point sweeps by, compared with that at the inlet initially, whenever

$$1 - r\exp(-\lambda) \geq \left[\frac{\Phi_o \exp \Phi_o \lambda}{[(\Phi_o - 1)\exp \Phi_o \lambda + 1]} - r\right] \tag{8.69}$$

by setting equation 8.67 against 8.68. Equation 8.69 can be re-arranged to yield an explicit inequality for r, namely

$$r \geq \frac{1 - \exp(-\Phi_o \lambda)}{[\Phi_o - 1 + \exp(-\Phi_o \lambda)][1 - \exp(-\lambda)]} \tag{8.70}$$

$$r < 1$$

The lower bound to r may be called the critical recycle ratio. Plots of the ratio of the "critical-outlet" to inlet drying rates are drawn in Fig. 8.15 for dryers of 1 and 2 transfer units (λ) in extent as a function of the initial moisture content Φ_o and the recycle ratio r as parameter. It is seen that the material must be very wet, with a considerable quantity of unbound moisture present, before the critical-outlet rate exceeds the initial rate at the air inlet.

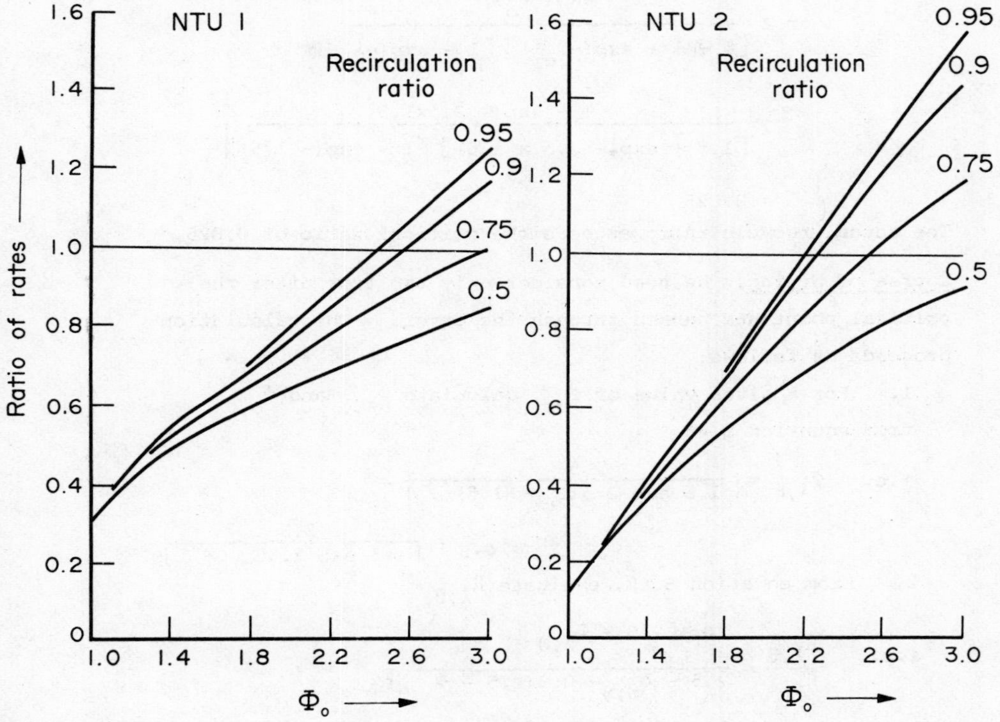

Fig. 8.15 Ratio of drying rate at outlet when $\Phi = 1$ to that at inlet initially. After Keey.[6]

A small shift of the maximum drying rate away from the critical point towards lower moisture contents has been noted in the case of drying without recycle. It is thus likely that the maximum drying rate at the outlet will not be found just as the critical point moves past, but at some time afterwards. The following worked example considers the course of drying over this period for a specific instance.

Example 8.6. Very wet material ($\Phi_o = 2.5$) is put into a batch dryer of 1.5 transfer units. Confirm that, if the recycle ratio is 0.9, the drying rate at the outlet when the critical point reaches there is greater than the initial rate at the inlet, and find the outlet moisture content at which the drying rate is a maximum.

Critical recycle ratio. From the inequality 8.70

$$r \geq \frac{1 - \exp(-\Phi_o\lambda)}{[\Phi_o - 1 + \exp(-\Phi_o\lambda)][1 - \exp(-\lambda)]}$$

$$= \frac{1 - \exp(-2.5x1.5)}{[1.5 + \exp(-2.5 \times 1.5)][1 - \exp(-1.5)]}$$

$$= 0.825$$

The actual recycle ratio exceeds the critical ratio of 0.825.

Course of drying. We need consider only the time after the critical point has passed through the dryer. The calculation proceeds as follows:

1. For a given value of ζ_{cr}, calculate $\Phi_{\lambda,\theta}$ and $\Phi_{o,\theta}$ from equation 8.46.

i.e. $\Phi_{\lambda,\theta} = \frac{2.5}{[1.5 \exp 2.5(\zeta_{cr} - 1.5) + 1]}$

and $\Phi_{o,\theta} = \frac{2.5}{[1.5 \exp 2.55_{cr} + 1]}$

2. From equation 8.63, evaluate $\Pi_{\lambda,\theta}$

i.e. $\frac{\Pi_{\lambda,\theta}}{\Pi_A} = \frac{0.1[2.5 - \Pi_{\lambda,\theta}]}{2.5 - \Phi_{o,\theta} - 0.9(2.5 - \Phi_{\lambda,\theta})}$

3. Evaluate the rate as $f\Pi_{\lambda,\theta}/\Pi_A = \Phi_{\lambda,\theta}\Pi_{\lambda,\theta}/\Pi_A$.

The calculations are done for increments of ζ_{cr} of 0.05.

ζ_{cr}	$\Phi_{\lambda,\theta}$	$\Phi_{o,\theta}$	$\Pi_{\lambda,\theta}/\Pi_A$	$f\Pi_{\lambda,\theta}/\Pi_A$
1.5	1.000	0.0386	0.1350	0.1350
1.55	0.926	0.0341	0.1500	0.1389
1.6	0.854	0.0302	0.1665	0.1422
1.65	0.786	0.0267	0.1842	0.1448
1.7	0.720	0.0235	0.2035	0.1465
1.75	0.657	0.0208	0.2246	0.1475
1.8	0.599	0.0184	0.2467	0.1478
1.85	0.544	0.0162	0.2704	0.1471
1.9	0.492	0.0143	0.2959	0.1456
1.95	0.445	0.0127	0.3222	0.1434
2.0	0.401	0.0112	0.3500	0.1404

The rate passes through a broad maximum when the moisture content is about 0.6.

In the foregoing example, the maximum rate appears when the moisture content has been reduced to 60 per cent of the free moisture content at the critical point. However, the rate is only 9.5 per cent greater.

Knowledge of the movement of the maximum drying rate through the dryer is important in determining operational policies with complex drying schedules, during which the humidity potential and the airflow direction are changed. Such changes must be carefully programmed to avoid excessive boosts to local drying rates, which may in turn induce dangerous moisture-content gradients in the material being dried. The impact of these changes on the way boards of pine sapwood dry out has been explored in some detail.[1, 4, 5].

8.6 Drying Experiments

The convenience of batch experiments in the laboratory, compared with continuous-drying installations, has provided most of the evidence for the worth of the concept of a unique drying curve to characterise hindered-drying behaviour. Much of this work is concerned with very small quantities of material, but Krischer and Jaeschke[7] describe a series of pilot-plant experiments with concrete bricks, wooden matches and wheat grains. The latter material has a highly non-linear drying curve, as the hard shells of the particles considerably restrict moisture movement. The various materials were either stacked or heaped in discrete layers, with dummy layers fore and aft to minimize end-effects, to present in each layer exposed areas that varied between 0.085 and 0.35 m^2. The layers were through-circulated without recycle. Fig. 8.16 shows a comparison of the drying-rate profile for the last layer of a six-layered stack of bricks with that estimated on the basis of a single characteristic drying curve. Although the estimated value of the maximum is only 5 per cent lower than the measured one in this instance, the calculated drying times are between 10 to 20 per cent depending upon the air velocity. In general, the mean drying rate deviated no more than 10 per cent from the calculated average for the systems studied.

Ashworth[1] uses a single characteristic drying curve to estimate the time to season sapwood boards in a commercial kiln taking a load 2.5 m wide. For an air velocity of 2.5 m s^{-1}, the estimated time to reach a moisture content of 0.12 kg kg^{-1} from green timber is about 80 h for 25 mm thick timber and 160 h for 50 mm thick material, which are close to the actual times. The drying times for thicker boards are over-estimated, since the influence of the intensity of drying, which is essentially the product of the drying rate and the thickness, has been ignored in the use of an identical characteristic curve.

Fig. 8.16 Drying-rate profile for the last layer of a six-layered stack of concrete bars.
Air velocity 0.845 m s^{-1}.
Layer surface 0.0869 m^2. Layer thickness 11.5 mm.

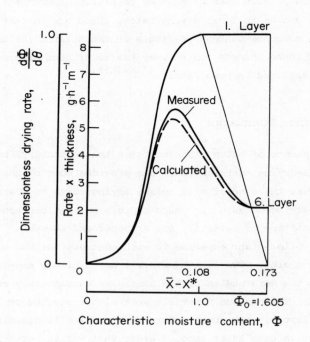

REFERENCES

1. Ashworth, J.C., The mathematical simulation of the batch-drying of softwood timber Ph. D. thesis, Univ. Canterbury(1977).
2. Banks,P.J., Coupled equilibrium heat and single adsorbate transfer in fluid flow through a porous medium - I. Characteristic potentials and capacity ratios, Chem.Eng.Sci., 27, 1143-1155 (1972).
3. Cassie, A.B.D., Propagation of Temperature Changes through Textiles in Humid Atmospheres, Part II, Theory of Propagation of Temperature Change, Trans.Farad.Soc., 36, 453-8 (1940).
4. Keey, R.B., Simulation of the kiln seasoning of timber, D5, Proc. 4 Internat.Congress, CHISA, Praha (1972).
5. Keey, R.B., "Drying Principles and Practice," Chapter 8, Pergamon Press, Oxford (1972).
6. Keey, R.B., Batch drying with air recirculation, Chem.Eng.Sci., 23, 1299-1308 (1968).
7. Krischer, O and L. Jaeschke, Trocknungsverlauf in durchströmten Haufwerken bei geordneter und ungeordneter Verteilung, Chem.-Ing.-Tech., 33, 592-8(1961).
8. Nordon, P. and Banks, P.J. in "Interacting Heat and Mass Transfer - An Australian View", 1st Australasian Heat Mass Transfer Conf., R45, Monash, Melbourne (1973).
9. Nonhebel, G. and A.A.H. Moss, "Drying of Solids in the Chemical Industry", p.71, Butterworths, London (1971).
10. Rosenbrock, H.H. and C. Storey, "Computational Techniques for Chemical Engineers", Chapter 7, Pergamon Press, Oxford (1966).
11. Tetzlaff, A.R., An investigation of drying schedules when kiln-drying radiata pine timber, B.E. Report, Univ. Canterbury (1967).
12 Van Meel, D.A., Adiabatic convection batch drying with recirculation of air, Chem.Eng.Sci., 9, 36-44 (1958).

APPENDIX

TABLE A1 HYGROTHERMAL PROPERTIES OF AIR SATURATED WITH WATER VAPOUR AT 100 kPa

Temperature (°C)	Humidity (kg/kg dry air)	Moisture vapour pressure	Moisture concentration (kg/m³)	Enthalpy of vaporization (kJ/kg)
0	0·003 821	0·610 8	0·004 846	2 500·8
2	0·004 418	0·705 4	0·005 557	2 495·9
4	0·005 100	0·812 9	0·006 358	2 491·3
6	0·005 868	0·934 6	0·007 257	2 486·6
8	0·006 749	1·072 1	0·008 267	2·481·9
10	0·007 733	1·227 1	0·009 396	2 477·2
12	0·008 849	1·401 5	0·010 66	2 472·5
14	0·010 105	1·597 4	0·012 06	2 467·8
16	0·011 513	1·816 8	0·013 63	2 463·1
18	0·013 108	2·062	0·015 36	2 458·4
20	0·014 895	2·337	0·017 29	2·453·1
22	0·016 892	2·642	0·019 42	2 449·0
24	0·019 131	2·982	0·021 77	2 442·0
26	0·021 635	3·360	0·024 37	2 439·5
28	0·024 435	3·778	0·027 23	2 434·8
30	0·027 558	4·241	0·030 36	2 430·0
32	0·031 050	4·753	0·033 80	2 425·3
34	0·034 950	5·318	0·037 58	2 420·5
36	0·039 289	5·940	0·041 71	2 415·8
38	0·044 136	6·624	0·046 22	2 411·0
40	0·049 532	7·375	0·051 14	2 406·2
42	0·055 560	8·198	0·056 50	2 401·4
44	0·062 278	9·010	0·062 33	2 396·6
46	0·069 778	10·085	0·068 67	2 391·8
48	0·078 146	11·161	0·075 53	2 387·0
50	0·087 516	12·335	0·082 98	2 382·1
52	0·098 018	13·613	0·091 03	2 377·3
54	0·109 76	15·002	0·099 74	2 372·4
56	0·122 97	16·509	0·109 1	2 367·6
58	0·137 90	18·146	0·119 3	2 362·7
60	0·154 72	19·92	0·130 2	2 357·9
62	0·173 80	21·84	0·141 9	2 353·0
64	0·195 41	23·91	0·154 5	2 348·1
66	0·220 21	26·14	0·168 0	2 343·1
68	0·248 66	28·55	0·182 6	2 338·2
70	0·281 54	31·16	0·198 1	2 333·3
72	0·319 66	33·96	0·214 6	2 328·3
74	0·364 68	36·96	0·232 4	2 323·3
76	0·417 90	40·19	0·251 4	2 318·3
78	0·480 48	43·65	0·271 7	2 313·3
80	0·559 31	47·36	0·293 3	2 308·3
82	0·655 73	51·33	0·316 2	2 303·2
84	0·777 81	55·57	0·340 6	2 298·1
86	0·937 68	60·50	0·366 6	2 293·0
88	1·152 44	64·95	0·394 2	2 287·9
90	1·458 73	70·11	0·423 5	2 282·8
92	1·927 18	75·61	0·454 5	2 277·6
94	2·731 70	81·46	0·487 3	2 272·4
96	4·426 70	87·69	0·522 1	2 267·1
98	10·303 06	94·30	0·558 8	2 261·9
100	∞	101·325	0·597 7	2 256·7

The table above has been compiled from the following sources:

1. LANDHOLT-BÖRNSTEIN, *Zahlenwerte und Funktionen*, 6th edn., Vol. 2, Springer–Verlag, Berlin/Göttingen/Heidelberg (1955).
2. MAYHEW, Y. R. and ROGERS, G. F. C., *Thermodynamic and Transport Properties of Fluids*, S.I. units, 2nd ed., Blackwell, Oxford, 1968.
3. *VDI-Wasserdampftafeln*, 4th ed., Springer-Verlag, Berlin/Göttingen/Heidelberg (1956).

TABLE A2 TRANSPORT PROPERTIES OF DRY AIR AT 100 kPa

Temperature T/K	Density ρ_G/kg m^{-3}	Kinematic Viscosity $10^6 \nu$/m^{-2} s^{-1}	Heat Capacity C_p/kJ kg^{-1} K^{-1}	Thermal Conductivity $10^6 \lambda$/kW m^{-1} K^{-1}	Pr
250	1.412	11.32	1.0031	22.27	0.720
275	1.284	13.43	1.0038	24.28	0.713
300	1.177	15.68	1.0049	26.24	0.707
325	1.086	18.07	1.0063	28.16	0.701
350	1.009	20.56	1.0082	30.03	0.697
375	0.9413	23.17	1.0106	31.86	0.692
400	0.8824	25.91	1.0135	33.65	0.688
450	0.7844	31.68	1.0206	37.10	0.684
500	0.7060	37.82	1.0295	40.41	0.680
550	0.6418	44.39	1.0398	43.57	0.680
600	0.5883	51.28	1.0511	46.61	0.680
650	0.5430	58.53	1.0629	49.54	0.682
700	0.5043	66.07	1.0750	52.36	0.684
750	0.4706	73.99	1.0870	55.09	0.687

Data have been abstracted from the following source :

Mayhew, Y.R. and G.F.C. Rogers, "Thermodynamics and Transport Properties of Fluids, S.I. units, 2/e, Blackwell,Oxford (1968)

TABLE A3 TRANSPORT PROPERTIES OF WATER SUBSTANCE

	Vapour			Heat Capacity		Thermal Conductivity	
Temperature $T/^{o}C$	Saturation Pressure P_W^o/k Pa	Density $\rho_V/kg\,m^{-3}$	Kinematic Viscosity $10^6\nu/m^{-2}s^{-1}$	Liquid $C_L/kW\,kg^{-1}K^{-1}$	Vapour $C_{PV}/kW\,kg^{-1}K^{-1}$	Liquid $\lambda_L/Wm^{-1}K^{-1}$	Vapour $10^6\lambda_V/kW\,m^{-1}K^{-1}$
0.01	0.6112	0.822	10.3	4.210	1.86	0.569	16.3
20	2.337	0.758	12.1	4.183	1.87	0.603	17.9
40	7.375	0.708	14.0	4.179	1.89	0.632	19.5
60	19.92	0.663	16.0	4.185	1.91	0.652	21.2
80	47.36	0.624	17.8	4.198	1.95	0.670	22.9
100	101.325	0.589	20.4	4.219	2.01	0.681	24.8
120	198.5	0.558	22.9	4.248	2.09	0.687	26.8
140	361.4	1.530	25.4	4.29	2.21	0.688	28.8
160	618.1	0.504	28.2	4.35	2.38	0.684	31.3
180	1003	0.482	31.1	4.42	2.62	0.676	34.1
200	1555	0.460	34.1	4.51	2.91	0.665	37.5
220	2320	0.441	36.9	4.63	3.25	0.648	41.5
240	3348	0.424	40.3	4.78	3.68	0.628	46.5
260	4694	0.408	43.9	4.98	4.22	0.603	52.8
280	6419	0.393	47.8	5.24	4.98	0.574	61.0
300	8592	0.379	52.2	5.65	6.18	0.541	72.0

Data abstracted and calculated from Mayhew, Y.R. and G.F.C. Rodgers, loc.cit.

Vapour densities are estimated from compressibility factors given in Table 2.3 for a total pressure of 100 kPa and are approximate below 380K (107°C).

The values for saturated steam can be used with only moderate accuracy below saturation pressures at temperature greater than 200°C.

TABLE A4 LUIKOV MOISTURE-SORPTION PARAMETERS

$$\frac{1}{X^*} = \frac{1}{X^*_{max}} - a \ln \psi$$

X^* = equilibrium moisture content

X^*_{max} = maximum hygroscopic moisture content

a = coefficient proportional to temperature

ψ = relative humidity

Correlation is claimed to be valid over the range $0.1 < \psi < 1$, but in some cases the range may be narrower. Values of a and X^*_{max} are abstracted from Luikov's more extensive compilation in "Teoriya Sushki" 2/e, p 60-1, Energiya, Moskva (1968).

Substance	Temperature/°C	X* max	a
brick, red	40	0.0100	1090
	20	0.0111	770
	0	0.0125	600
cellulose	-	0.111	9
clay	-	0.147	9
concrete	40	0.0290	108
	20	0.0294	88
	0	0.0307	70
cotton sheet	24	0.1175	15
flax fibre	-	0.139	7
jute	24	0.25	8
kieselguhr	24	0.04	71
newsprint	24	0.13	16
paper, Kraft	24	0.172	98
pine	40	0.1315	23
	20	0.1428	20
	0	0.1428	17
rubber tyre	-	0.0125	163
soap	-	0.400	12
viscose	-	0.178	7
wood	-	0.238	9
wood, predried	-	0.222	15
wool	-	0.312	67

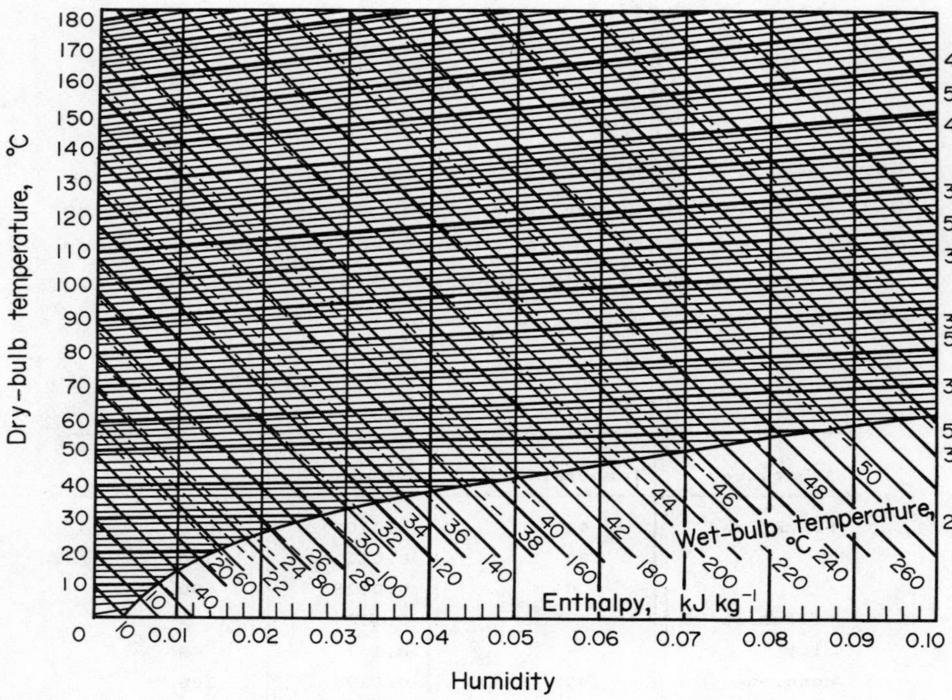

A1. Enthalpy-humidity chart for air-water vapour at 100 kPa. (SIDI 1012 : 1977, NZIE Chemical Engineering Group, by permission).

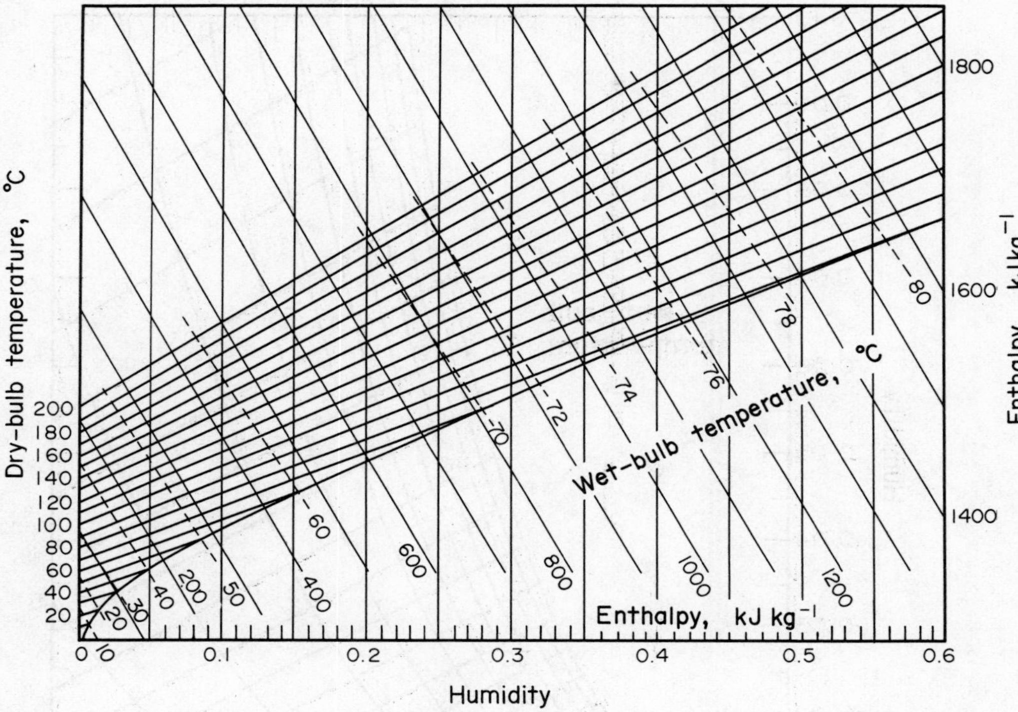

A2. Enthalpy-humidity chart for air-water vapour at high humidities at 100 kPa.
(SIDI 1012 : 1977, NZIE Chemical Engineering Group, by permission).

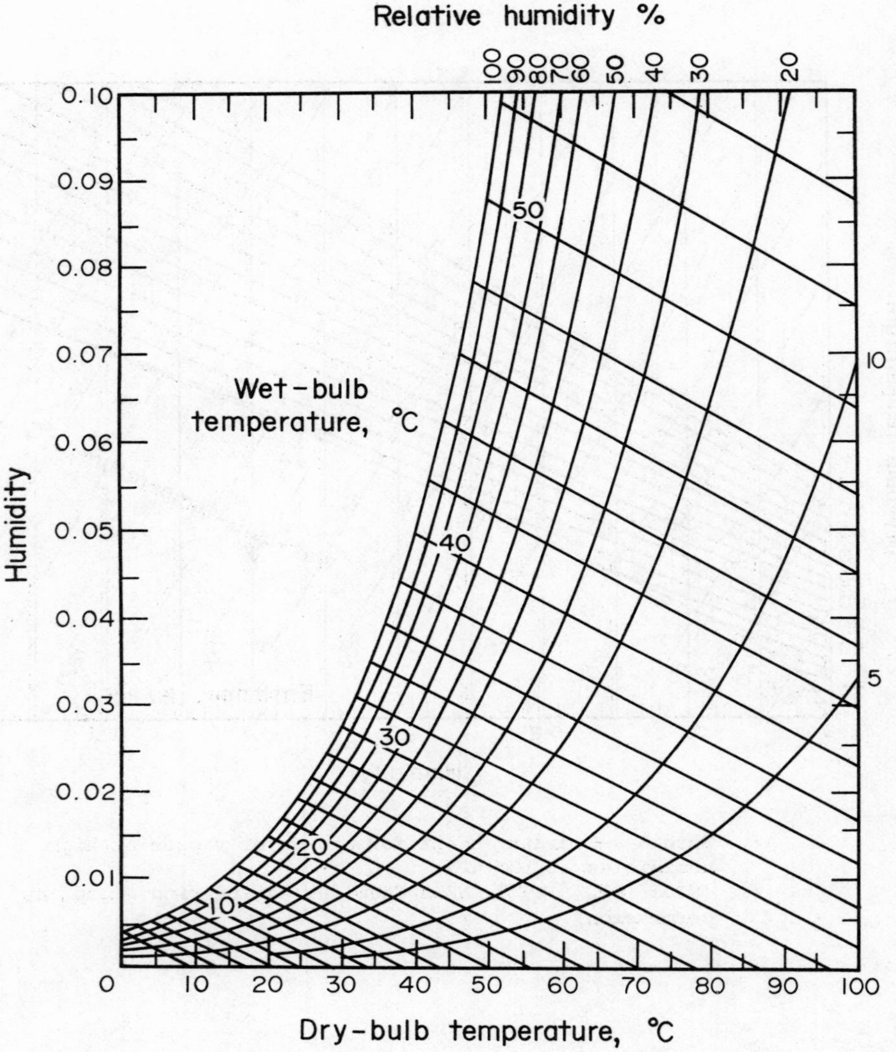

A3. Humidity-temperature chart for air-water vapour at at 100 kPa.
(SIDI 1012 : 1977, NZIE Chemical Engineering Group, by permission).

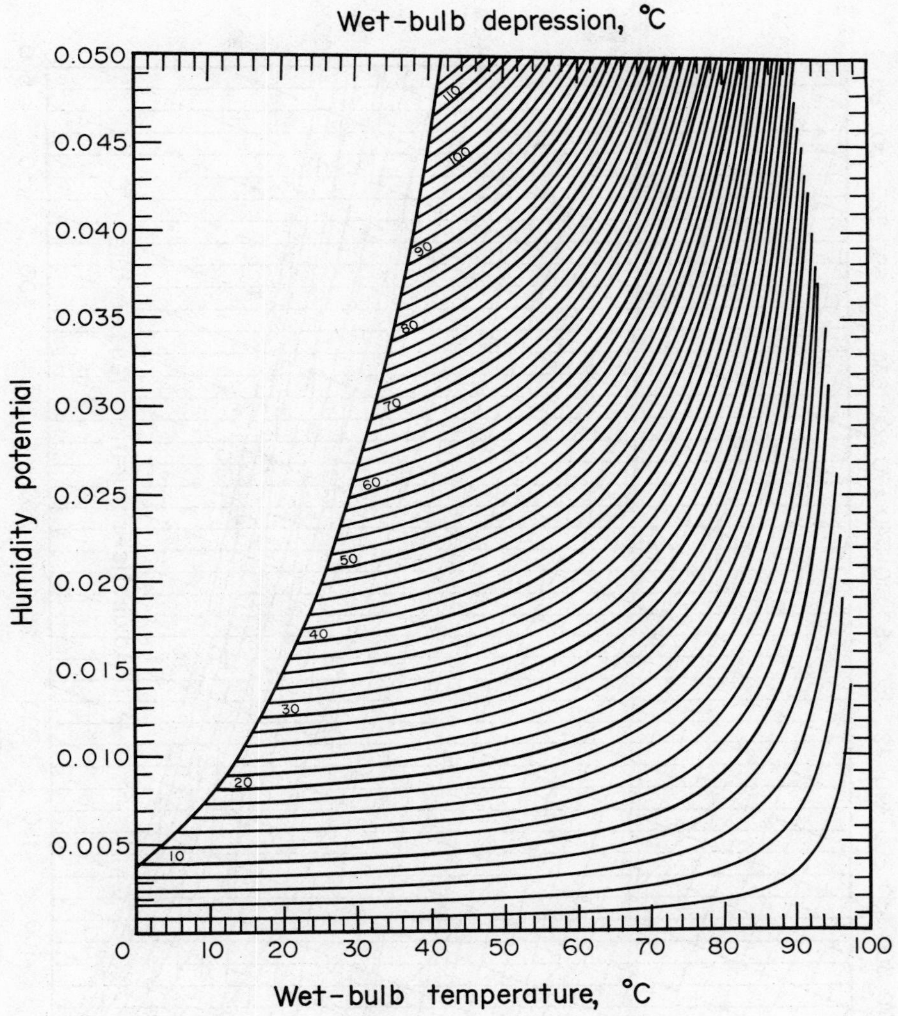

A4. Humidity-difference chart for air-water vapour at 100 kPa. (SIDI 1012 : 1977, NZIE Chemical Engineering Group, by permission).

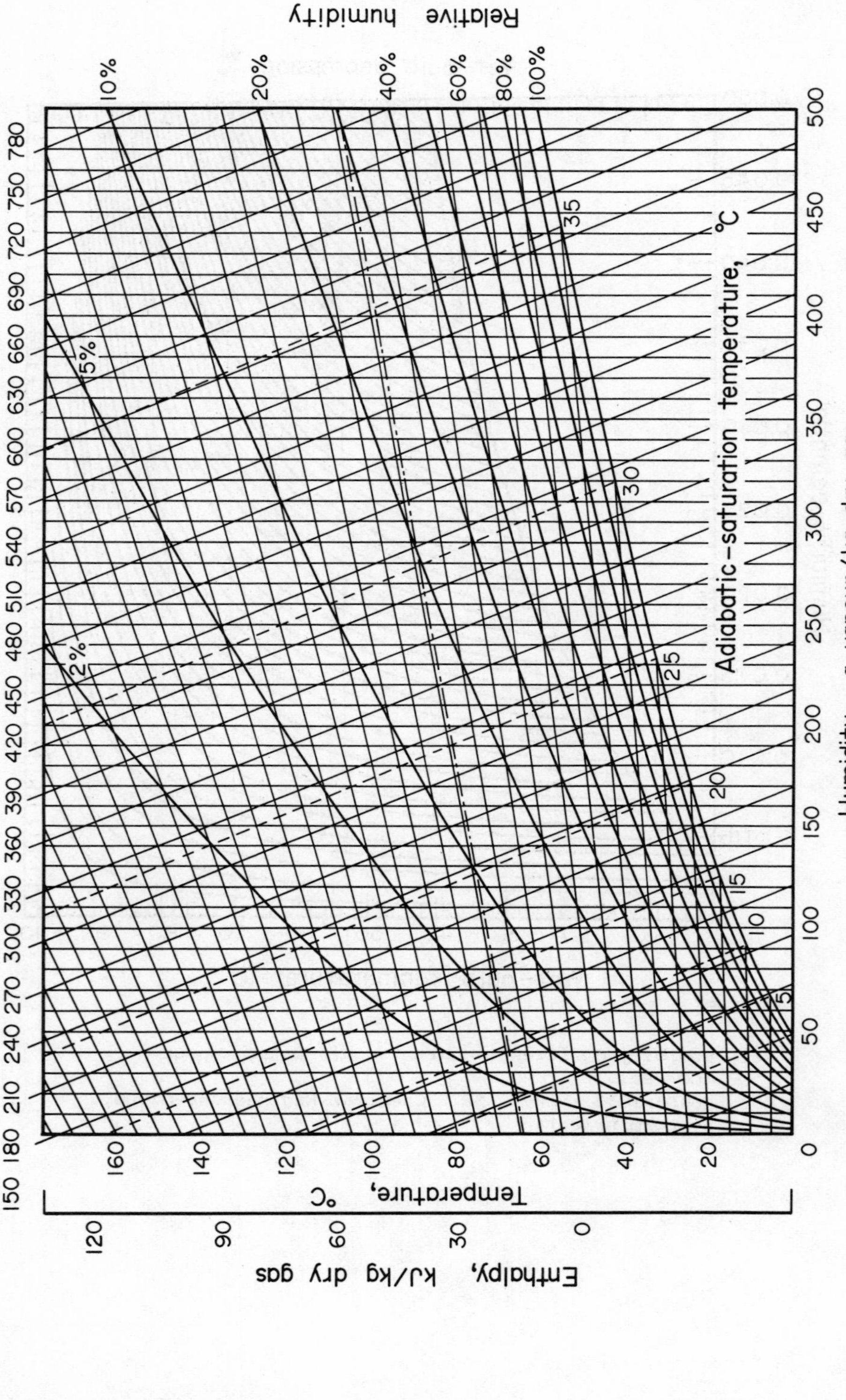

A5. Enthalpy-humidity chart for nitrogen-methanol vapour at 100 kPa. (By permission Separation Processes Service, AERE Harwell).

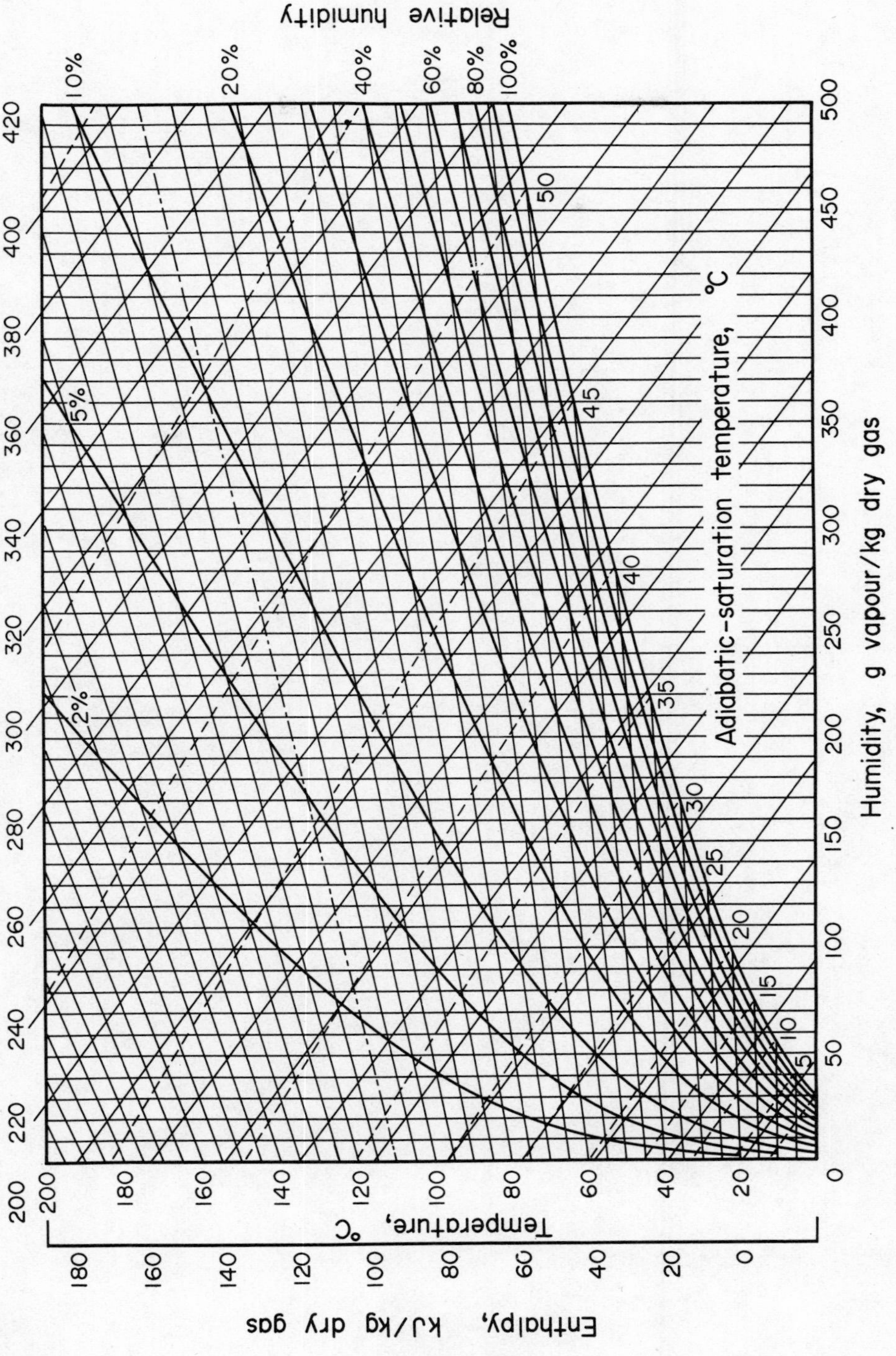

A6. Enthalpy-humidity chart for nitrogen-toluene vapour at 100 kPa. (By permission Separation Processes Service, AERE Harwell).

INDEX